Thermodynamic Properties of Nonelectrolyte Solutions

Thermodynamic Properties of Nonelectrolyte Solutions

WILLIAM E. ACREE, JR.
Department of Chemistry
Kent State University
Kent, Ohio

1984

ACADEMIC PRESS

(Harcourt Brace Jovanovich, Publishers)
Orlando   San Diego   San Francisco   New York   London
Toronto   Montreal   Sydney   Tokyo   São Paulo

COPYRIGHT © 1984, BY ACADEMIC PRESS, INC.
ALL RIGHTS RESERVED.
NO PART OF THIS PUBLICATION MAY BE REPRODUCED OR
TRANSMITTED IN ANY FORM OR BY ANY MEANS, ELECTRONIC
OR MECHANICAL, INCLUDING PHOTOCOPY, RECORDING, OR ANY
INFORMATION STORAGE AND RETRIEVAL SYSTEM, WITHOUT
PERMISSION IN WRITING FROM THE PUBLISHER.

ACADEMIC PRESS, INC.
Orlando, Florida 32887

United Kingdom Edition published by
ACADEMIC PRESS, INC. (LONDON) LTD.
24/28 Oval Road, London NW1 7DX

Library of Congress Cataloging in Publication Data

Acree, William Eugene.
　Thermodynamic properties of nonelectrolyte solutions.

　　Includes bibliographical references and index.
　　1. Nonaqueous solvents--Thermal properties. 2. Solution (Chemistry)--Thermal properties. I. Title.
QD544.5.A26 1983　　　541.3'416　　　83-9998
ISBN 0-12-043020-7

PRINTED IN THE UNITED STATES OF AMERICA

84 85 86 87　　9 8 7 6 5 4 3 2 1

Contents

Preface   ix

1   Mathematical Relationships

    A.   Concentration Variables   1
    B.   Partial Derivatives   1
    C.   Total Derivative   2
    D.   Implicit Differentiation   2
    E.   Chain Rule   3
    F.   Cycle Rule   4
    G.   Homogeneous Functions   4
    H.   Euler's Theorem   5
    I.   Exact Differentials   7
    J.   Method of Least Squares   8
    Problems   9

2   Partial Molar Quantities

    A.   Partial Molar Volumes   12
    B.   Other Partial Molar Quantities   16
    C.   The Gibbs–Duhem Equation   19
    D.   Apparent Partial Molar Quantities   19
    E.   Determination of Partial Molar Quantities   21
    F.   Interpretive Descriptions   25
    Problems   27

3   Ideal and Nonideal Solutions

    A.   Fugacity   30
    B.   Activity   33
    C.   The Ideal Solution   37
    D.   Vapor–Liquid Equilibrium in an Ideal Solution   39
    E.   Behavior of Ideal Dilute Solutions   42
    F.   Thermodynamic Excess Functions   45
    G.   Vapor–Liquid Equilibrium in Nonideal Solutions   48
    Problems   57

4 Empirical Expressions for Estimating Multicomponent Properties from Binary Data
Text 62
Problems 70

5 Binary and Ternary Mixtures Containing Only Nonspecific Interactions
A. The Theory of Van Laar 77
B. The Scatchard–Hildebrand Model 81
C. The Flory–Huggins Model 88
D. The Wilson Model 90
E. The Nonrandom Two-Liquids (NRTL) Model 97
F. The UNIQUAC and Effective UNIQUAC Models 101
G. Summary 108
Problems 111

6 Prediction of Thermodynamic Excess Properties of Liquid Mixtures Based on Group Contribution Methods
A. Analytical Solution of Groups Model (ASOG) 117
B. The Analytical Group Solution Model (AGSM) 124
C. The UNIQUAC Functional Group Activity Coefficients Model (UNIFAC) 125
D. Summary 131
Problems 132

7 Simple Associated Solutions
A. Thermodynamic Properties in Associated Solutions 135
B. The Ideal Associated Solution Model: Systems Having a Single AB-type Complex 137
C. The Ideal Associated Solution Model: Systems Having Both AB-type and AB_2-type Complexes 140
D. Summary 144
Problems 146

8 Estimation of Thermodynamic Excess Properties of Ternary-Alcohol Hydrocarbon Systems from Binary Data
A. Kretschmer–Wiebe Association Model 151
B. Mecke–Kempter Association Model 161
C. Attenuated Equilibrium Constant (AEC) Model 165
D. Two-Constant Kretschmer–Wiebe Association Model 171
E. Summary 173
Problems 173

Contents

9 Vapor–Liquid Equilibria and Azeotropic Systems

A. Isothermal Vapor–Liquid Equilibrium in Binary Systems 176
B. Isothermal Vapor–Liquid Equilibrium in Ternary Systems 179
C. Isobaric Vapor–Liquid Equilibrium in Binary Systems 181
D. Isobaric Vapor–Liquid Equilibrium in Ternary Systems 184
E. Azeotropes in Binary Mixtures Under Isobaric Conditions 185
F. Prediction of Ternary Azeotropes from Binary Data 189
Problems 193

10 Solubility Behavior of Nonelectrolytes

A. Solid-Liquid Equilibrium in an Ideal Solution 199
B. The Scatchard-Hildebrand Solubility Parameter Model 205
C. Stoichiometric Complexation Model of Higuchi 210
D. The Nearly Ideal Binary Solvent Theory (NIBS), Solubility in Binary Solvents of Nonspecific Interactions 216
E. The Nearly Ideal Binary Solvent Model (NIBS), Monomeric and Dimeric Treatment of Carboxylic Acids 227
F. Extension of the Nearly Ideal Binary Solvent Model to Systems Having Solute-Solvent Complexation 229
G. Solubility Predictions Using the UNIFAC Group Contribution Method 233
Problems 236

11 Liquid–Liquid Equilibrium: Distribution of a Solute between Two Immiscible Liquid Phases

A. Liquid–Liquid Equilibrium in Binary Systems 246
B. Distribution of a Solute between Two Immiscible Solvents 247
C. Partitioning of a Solute between a Binary Organic Phase and Water 252
Problems 254

12 Physio-Chemical Applications of Gas–Liquid Chromatography to Nonelectrolyte Solutions

A. The Nearly Ideal Binary Solvent Theory: Gas–Liquid Partition Coefficients in Noncomplexing Systems 258
B. The Kretschmer-Wiebe Association Model: Gas–Liquid Partition Coefficients of Alcohol Solutes on Binary Solvent Mixtures of Inert Hydrocarbons 261
C. The Nearly Ideal Binary Solvent Theory: Gas–Liquid Partition Coefficients in Systems Containing Solute-Solvent Complexation 262
D. Gas–Liquid Chromatographic Partition Coefficients of Inert Solutes on Self-Associating Binary Solvent Systems 264
E. Summary 269
Problems 269

Appendix A Solubility Parameters and Molar Volumes at 25°C   271

Appendix B UNIQUAC Structural Parameters   274

Appendix C Constants and Conversion Factors   275

Appendix D Answers to Selected Problems   276

References   287

Index   295

Preface

For many years the chemical industry has recognized the importance of thermodynamic and physical properties in design calculations involving chemical separations, fluid flow, and heat transfer. The development of flow calorimeters, continuous dilution dilatometers, and vibrating-tube densimeters has enabled the experimental determination of excess enthalpies, heat capacities, and volumes of liquid mixtures with convenience and accuracy. The utilization of continuous dilution techniques has reduced the experimental time needed for the determination of Gibbs free energy through conventional vapor–pressure measurements. Recent advances in gas–liquid chromatography have enabled infinite dilution activity coefficients to be measured with accuracies once believed unattainable. But even with modern instrumentation it is not possible to measure the thermodynamic properties of all conceivable multicomponent mixtures.

To overcome this problem, researchers have turned to predictive methods as a way to generate desired quantities. Although much progress has been made in recent years, we are still far from a "perfect method," as is demonstrated by the large number of new methods appearing in the literature each year. For mixtures of nonpolar molecules, the Scatchard–Hildebrand solubility parameter theory provides a good first approximation of vapor–liquid equilibria. Better predictions can often be made if one uses a more sophisticated solution model based on nonrandom mixing or local compositions. Group contribution methods are available for those wanting to predict multicomponent properties from molecular structure. Associated solution theories provide reasonable approximations whenever there is independent evidence that strong chemical forces operate in the liquid mixture. To a large extent the selection of a predictive method depends on the type of solution and on the information already available.

My purpose in writing this book is to assemble several of the more classical solution theories into a single reference book. Basic thermodynamic principles are reviewed in the first three chapters, with the remaining nine chapters being devoted to predictive methods and molecular thermodynamics. Molecular thermodynamics constitutes a means for going beyond the limitations of classical thermodynamics in solving chemical problems. Despite the many accomplishments of classical thermodynamics, it can be applied usefully only to the relationships between properties. It does not offer any source other than experiment for the initial properties needed in a given procedure. On the other hand, molecular thermodynamics surpasses the classical treatment by relating macroscopic thermodynamics of chemical physics. By using a microscopic approach it is frequently possible to start with independently known physicochemical properties and from these to estimate $P-V-T$ relationships, enthalpies, entropies, free energies, and especially fugacities. The reason for establishing such relationships is to facilitate meaningful correlations for calculating from a minimum number of experimental determinations such phase-equilibrium data as are needed for either engineering design or laboratory applications.

I have elected to use most of the literature data in its original form. Because predictive methods often require a priori knowledge of pure component properties and sometimes binary mixing data, it is my firm belief that engineers will need to have familiarity not only with the "new SI units," but also with many of the "older units." In many respects this will serve as a valuable lesson to those learning predictive thermodynamics for the first time.

This book is intended primarily for professional chemists and researchers; others may find it useful. For example, the book also should be of value to university seniors or first-year graduate students in chemistry or chemical engineering who have completed a standard one-year course in physical chemistry and who have had some previous experience in classical thermodynamics. Problems at the end of each chapter are intended for those wishing to use the book as a self-study guide.

1
Mathematical Relationships

A. Concentration Variables

The composition of a liquid solution can be expressed in a variety of ways, as (1) concentration in grams or moles per unit volume, (2) the ratio of the number of moles of one component in a binary solution to the number of moles of a second component, n_1/n_2, (3) *mole fraction*

$$X_1 = n_1/(n_1 + n_2 + \cdots) \tag{1.1}$$

or (4) *volume fraction*

$$\phi_1 = V_1/(V_1 + V_2 + \cdots) = n_1\bar{V}_1/(n_1\bar{V}_1 + n_2\bar{V}_2 + \cdots) \tag{1.2}$$

Strictly speaking, the *true* volume of a *real* solution is not equal to the sum of the volumes of its individual components but is the fractional sum of partial molar volumes, which for a binary solution is

$$V = n_1\bar{V}_1 + n_2\bar{V}_2 \quad \text{and} \quad \bar{V} = X_1\bar{V}_1 + X_2\bar{V}_2 \tag{1.3}$$

For the purposes of this book, we shall use volume fractions defined in terms of the molar volumes of the pure liquid components, $\bar{V}_i^{\bullet}$ (molecular weight divided by density)

$$\phi_1 = n_1\bar{V}_1^{\bullet}/(n_1\bar{V}_1^{\bullet} + n_2\bar{V}_2^{\bullet} + \cdots) \tag{1.4}$$

B. Partial Derivatives

If a function has more than one variable, then partial derivatives must be used to express mathematically how this function varies with a particular variable. For example, the van der Waals equation of state for one mole is

$$P = RT/(V - b) - (a/V^2) \tag{1.5}$$

The pressure is a function of both temperature and volume,

$$P = P(T, V)$$

To describe how the pressure varies with temperature, it is written[†]

$$(\partial P/\partial T)_V = R/(V-b) \tag{1.6}$$

The symbol ∂ reminds us that it is a partial derivative, that is, the differentiation is performed at constant volume. Consider functions having more than two variables

$$y = y(x_1, x_2, x_3, \ldots)$$

in which case the partial derivative of y with respect to x_1 must be written as

$$(\partial y/\partial x_1)_{x_2, x_3, \ldots}$$

C. Total Derivative

Next consider the case where the independent variables of the van der Waals equation are allowed to vary simultaneously. A theorem in basic calculus relates the total differential dP to the partial differentials dV and dP as follows:

$$dP = (\partial P/\partial V)_T \, dV + (\partial P/\partial T)_V \, dT \tag{1.7}$$

Because

$$(\partial P/\partial T)_V = R/(V-b) \tag{1.8}$$

and

$$(\partial P/\partial V)_T = -RT/(V-b)^2 + 2a/V^3 \tag{1.9}$$

we obtain

$$dP = [-RT/(V-b)^2 + 2a/V^3] \, dV + R/(V-b) \, dT \tag{1.10}$$

Partial and total derivatives find considerable application in thermodynamics.

D. Implicit Differentiation

In examples involving the van der Waals equation of state, the pressure is explicitly expressed in terms of temperature and volume. Partial differentiation of pressure with respect to temperature and/or volume is relatively

---

[†] The terms a and b are considered to be numerical values in this example.

E. Chain Rule

straightforward. Equation (1.5) is cubic in volume and it is not feasible to explicitly express V as a function of T and P. The partial derivative of V with respect to P, $(\partial V/\partial P)_T$ is found via term-by-term differentiation:

$$\left(\frac{\partial P}{\partial P}\right)_T = 1 = \frac{-RT}{(V-b)^2}\left(\frac{\partial V}{\partial P}\right)_T + \frac{2a}{V^3}\left(\frac{\partial V}{\partial P}\right)_T \tag{1.11}$$

Algebraic manipulation of Eq. (1.11) gives

$$\left(\frac{\partial V}{\partial P}\right)_T = \frac{V^3(V-b)^2}{2a(V-b)^2 - RTV^3} \tag{1.12}$$

In summary, implicit differentiation involves differentiating the equation and solving the resulting equation for the desired derivative. Alternatively, $(\partial V/\partial P)_T$ could have been obtained as the reciprocal of Eq. (1.9). In general terms, this important relationship is

$$(\partial Z/\partial Y)_X = 1/(\partial Y/\partial Z)_X \tag{1.13}$$

E. Chain Rule

Suppose a function Z exists

$$Z = Z(x, y) = 3x^2 + 6xy + y^2 \tag{1.14}$$

$$x = x(t) = t^3 + 1 \tag{1.15}$$

that is described in terms of two variables, x and y, and that x was a function of a third variable t. The partial derivative $(\partial Z/\partial t)_y$ can be found by substituting Eq. (1.15) into Eq. (1.14), followed by differentiation

$$Z = 3(t^3 + 1)^2 + 6(t^3 + 1)y + y^2$$

Therefore

$$(\partial Z/\partial t)_y = 18(t^3 + 1)t^2 + 18t^2 y$$

The chain rule enables one to perform this differentiation without first expressing Z in terms of t, simply as

$$\left(\frac{\partial Z}{\partial t}\right)_y = \left(\frac{\partial Z}{\partial x}\right)_y \frac{dx}{dt} \tag{1.16}$$

and

$$(\partial Z/\partial t)_y = (6x + 6y)(3t^2) = 18(t^3 + 1)t^2 + 18t^2 y$$

The partial derivative is the same regardless of the method used. A more complex form of the basic chain rule is: If $Z = Z(x, y)$ and $x = x(u, v)$, $y = y(u, v)$, then

$$\left(\frac{\partial Z}{\partial u}\right)_v = \left(\frac{\partial Z}{\partial x}\right)_y \left(\frac{\partial x}{\partial u}\right)_v + \left(\frac{\partial Z}{\partial y}\right)_x \left(\frac{\partial y}{\partial u}\right)_v \tag{1.17}$$

and

$$\left(\frac{\partial Z}{\partial v}\right)_u = \left(\frac{\partial Z}{\partial x}\right)_y \left(\frac{\partial x}{\partial v}\right)_u + \left(\frac{\partial Z}{\partial y}\right)_x \left(\frac{\partial y}{\partial v}\right)_u \tag{1.18}$$

F. Cycle Rule

Consider the pressure as a function of volume and temperature, while imposing the additional constraint that $dP = 0$, then

$$dP = (\partial P/\partial V)_T \, dV + (\partial P/\partial T)_V \, dT \tag{1.19}$$

Equation (1.19) can then be rearranged as

$$\left(\frac{\partial P}{\partial T}\right)_V = -\left(\frac{\partial P}{\partial V}\right)_T \left(\frac{\partial V}{\partial T}\right)_P \tag{1.20}$$

to give a specific example of the cycle rule.

G. Homogeneous Functions

In connection with the development of the thermodynamic concept of partial molar quantities, it is desirable to be familiar with a mathematical relationship known as *Euler's theorem*. Because this theorem is stated with reference to *homogeneous functions*, the nature of these functions needs to be considered. Consider again the function

$$Z(x, y) = 3x^2 + 6xy + y^2 \tag{1.14}$$

If the variables x and y are replaced by kx and ky, in which k is a parameter,

H. Euler's Theorem

z can be written

$$\begin{aligned} Z^* = Z(kx, ky) &= 3(kx)^2 + 6(kx)(ky) + (ky)^2 \\ &= k^2 3x^2 + k^2 6xy + k^2 y^2 \\ &= k^2(3x^2 + 6xy + y^2) \\ &= k^2 Z(x, y) \end{aligned} \quad (1.21)$$

Because the net result of multiplying each independent variable by the parameter k has been the same as to multiply the entire function by k^2, the function is called homogeneous of the second degree.

H. Euler's Theorem

Consider a function $f(a, b, x, y)$ that is homogeneous to the degree h in x and y. By definition, if the variables x and y are each multiplied by a factor k, the value of $f(a, b, kx, ky)$ will be increased by a factor of k^h. Thus, for any value of k

$$f(a, b, x^*, y^*) = k^h f(a, b, x, y) \quad (1.22)$$

where

$$x^* = kx \quad \text{and} \quad y^* = ky$$

Equating the total differentials of Eq. (1.22) and treating k as a variable because Eq. (1.22) is valid for all values of k, one obtains

$$\frac{\partial}{\partial a}[f(a, b, x^*, y^*)]_{b,x^*,y^*} da + \frac{\partial}{\partial b}[f(a, b, x^*, y^*)]_{a,x^*,y^*} db$$

$$+ \frac{\partial}{\partial x^*}[f(a, b, x^*, y^*)]_{a,b,y^*} dx^* + \frac{\partial}{\partial y^*}[f(a, b, x^*, y^*)]_{a,b,x^*} dy^*$$

$$= k^h \frac{\partial}{\partial a}[f(a, b, x, y)]_{b,x,y} da + k^h \frac{\partial}{\partial b}[f(a, b, x, y)]_{a,x,y} db$$

$$+ k^h \frac{\partial}{\partial x}[f(a, b, x, y)]_{a,b,y} dx + k^h \frac{\partial}{\partial y}[f(a, b, x, y)]_{a,b,x} dy$$

$$+ h(k)^{h-1}[f(a, b, x, y)] dk, \quad (1.23)$$

with

$$dx^* = k\, dx + x\, dk \quad \text{and} \quad dy^* = k\, dy + y\, dk \quad (1.24)$$

Substituting Eq. (1.24) into Eq. (1.23), and collecting like terms

$$\left\{\frac{\partial}{\partial a}[f(a, b, x^*, y^*)] - k^h \frac{\partial}{\partial a}[f(a, b, x, y)]\right\} da$$

$$+ \left\{\frac{\partial}{\partial b}[f(a, b, x^*, y^*)] - k^h \frac{\partial}{\partial b}[f(a, b, x, y)]\right\} db$$

$$+ \left\{(k)\frac{\partial}{\partial x^*}[f(a, b, x^*, y^*)] - k^h \frac{\partial}{\partial x}[f(a, b, x, y)]\right\} dx$$

$$+ \left\{(k)\frac{\partial}{\partial y^*}[f(a, b, x^*, y^*)] - k^h \frac{\partial}{\partial y}[f(a, b, x, y)]\right\} dy$$

$$+ \left\{(x)\frac{\partial}{\partial x^*}[f(a, b, x^*, y^*)] + (y)\frac{\partial}{\partial y^*}[f(a, b, x^*, y^*)]\right.$$

$$\left. - h(k)^{h-1}[f(a, b, x, y)]\right\} dk = 0. \qquad (1.25)$$

Because a, b, x, y, and k are independent variables, Eq. (1.25) is valid only if the coefficients of da, db, dx, dy, and dk are each equal to zero. Thus,

$$\frac{\partial}{\partial a}[f(a, b, x^*, y^*)] = k^h \frac{\partial}{\partial a}[f(a, b, x, y)] \qquad (1.26a)$$

$$\frac{\partial}{\partial b}[f(a, b, x^*, y^*)] = k^h \frac{\partial}{\partial b}[f(a, b, x, y)] \qquad (1.26b)$$

$$\frac{\partial}{\partial x^*}[f(a, b, x^*, y^*)] = k^{h-1} \frac{\partial}{\partial x}[f(a, b, x, y)] \qquad (1.26c)$$

$$\frac{\partial}{\partial y^*}[f(a, b, x^*, y^*)] = k^{h-1} \frac{\partial}{\partial y}[f(a, b, x, y)] \qquad (1.26d)$$

$$x\frac{\partial}{\partial x^*}[f(a, b, x^*, y^*)] + y\frac{\partial}{\partial y^*}[f(a, b, x^*, y^*)] = h(k)^{h-1}[f(a, b, x, y)]. \qquad (1.26e)$$

Combination of Eqs. (1.26a–e) gives

$$x\frac{\partial}{\partial x}[f(a, b, x, y)] + y\frac{\partial}{\partial y}[f(a, b, x, y)] = h[f(a, b, x, y)] \qquad (1.27)$$

Equation (1.27) is the general form of Euler's theorem. Notice it contains only terms in those variables for which f is homogeneous to degree h. The thermodynamic functions of interest are special cases of homogeneous func-

I. Exact Differentials

tions. In particular, all thermodynamic functions are either homogeneous to the first degree in mass (extensive) or homogeneous to the zeroth degree in mass (intensive).

I. Exact Differentials

Consider expressions of the general type

$$\delta Z = M(x, y)\, dx + N(x, y)\, dy \qquad (1.28)$$

where δZ represents an infinitesimal quantity and $M(x, y)$ and $N(x, y)$ are functions of the independent variables x and y. This type of expression may or may not be the total differential of a function such as Eq. (1.7). If there is such a function $Z(x, y)$, then

$$M(x, y) = (\partial Z/\partial x)_y \quad \text{and} \quad N(x, y) = (\partial Z/\partial y)_x$$

and

$$\frac{\partial M}{\partial y} = \frac{\partial^2 Z}{\partial x\, \partial y} = \frac{\partial N}{\partial x} \qquad (1.29)$$

The quantity $\partial M/\partial y = \partial N/\partial x$ is a necessary and sufficient condition that an expression of the type of Eq. (1.28) is an exact differential of a function Z. If Eq. (1.29) holds, the differential may be integrated directly between two states

$$\Delta Z = Z(x_2, y_2) - Z(x_1, y_1) \qquad (1.30)$$

without ever specifying the path from x_1, y_1 to x_2, y_2. Conversely, if Eq. (1.29) does not hold, it is impossible to integrate Eq. (1.28) unless the path is specified. The numerical result of this later integration is strongly dependant on the path chosen.

For three independent variables,

$$dZ = M(x, y, u)\, dx + N(x, y, u)\, dy + Q(x, y, u)\, du \qquad (1.31)$$

is an exact differential provided all of the following equations are satisfied:

$$(\partial M/\partial y)_{x,u} = (\partial N/\partial x)_{y,u} \qquad (1.32a)$$

$$(\partial M/\partial u)_{x,y} = (\partial Q/\partial x)_{y,u} \qquad (1.32b)$$

and

$$(\partial N/\partial u)_{x,y} = (\partial Q/\partial y)_{x,u} \qquad (1.32c)$$

In Eq. (1.10) notice that dP is an exact differential for the van der Waals equation of state because

$$\frac{\partial}{\partial T}\left(\frac{-RT}{(V-b)^2} + \frac{a}{V^3}\right) = \frac{\partial}{\partial V}\left(\frac{R}{(V-b)}\right) \quad (1.33)$$

This is true of equations of state in general. The reader will have the opportunity to test other equations of state in the problems at the end of this chapter.

J. Method of Least Squares

Whenever an experiment is performed, an attempt to obtain a limited number of experimental values that is representative of a much larger universe of values is desired. The experimenter then must try to estimate the characteristics of the larger universe from this experimental sample. The problem is to obtain the most probable values of the characteristic parameters of that universe with respect to the experiments and assumptions concerning the nature of the universe of measurements.

A common method of estimating the most probable values is the method of least squares, which can be stated as: The most probable value of a quantity is that value for which the sum of the squares of the deviations of the observed values from the most probable value is a minimum.

If the assumption is made that the series of observations represents random variations in the experimental value of a single quantity, then the quantity to be minimized is

$$\sum_{i=1}^{n}(w_i - W)^2 \quad (1.34)$$

where w_i is an observed value, W the most probable value sought, and n the number of observations performed. The w_i are constant and the summation is to be minimized with respect to the variation in the parameter W. Thus,

$$\frac{d}{dW}\sum_{i=1}^{n}(w_i - W)^2 = \frac{d}{dW}\left[\sum_{i=1}^{n}w_i^2 - 2\sum_{i=1}^{n}w_iW + \sum_{i=1}^{n}W^2\right]$$

$$= -2\sum_{i=1}^{n}w_i + 2nW = 0 \quad (1.35)$$

Solving for W

$$W = \sum_{i=1}^{n}w_i\bigg/n \quad (1.36)$$

the most probable value for a series of observations *assumed* to represent a single quantity is the arithmetic mean.

If the assumption is made that two variables, y_i and x_i, are connected by the linear relationship

$$y = mx + b \tag{1.37}$$

then the quantity to be minimized is

$$R = \sum_{i=1}^{n} [y_i - (mx_i + b)]^2 \tag{1.38}$$

The y_i and x_i are known constants obtained from experiment, and the quantity is minimized with respect to variation in m and b. The conditions for the minimum are

$$\left(\frac{\partial R}{\partial m}\right)_b = -2 \sum_{i=1}^{n} x_i y_i + 2 \sum_{i=1}^{n} (mx_i + b)x_i = 0 \tag{1.39}$$

$$\left(\frac{\partial R}{\partial b}\right)_m = -2 \sum_{i=1}^{n} y_i + 2 \sum_{i=1}^{n} (mx_i + b) = 0 \tag{1.40}$$

The solutions of the two simultaneous equations are

$$m = \left[\sum_{i=1}^{n} (x_i y_i) - \left(\sum_{i=1}^{n} x_i\right)\left(\sum_{i=1}^{n} y_i\right)\right] \bigg/ \left[n \sum_{i=1}^{n} (x_i)^2 - \left(\sum_{i=1}^{n} x_i\right)^2\right] \tag{1.41}$$

and

$$b = \left[\left(\sum_{i=1}^{n} x_i^2\right)\left(\sum_{i=1}^{n} y_i\right) - \left(\sum_{i=1}^{n} x_i\right)\left(\sum_{i=1}^{n} x_i y_i\right)\right] \bigg/ \left[n \sum_{i=1}^{n} x_i^2 - \left(\sum_{i=1}^{n} x_i\right)^2\right] \tag{1.42}$$

It must be emphasized that a least squares treatment gives only the most probable values of the parameters of an *assumed* relationship; it does *not* show that the assumed relationship is valid.

Problems

1.1 Write the complete differential dP for each of the three equations of state listed below and verify that dP is an exact differential. Consider volume and temperature as the only independent variables.

$$P = \begin{cases} RT/V & \text{(Ideal gas law)} \\ RT/(V-b) - a/V(V+b) & \text{(Modified Redlich–Kwong equation of state)} \\ RT \exp(-a/RTV)/(V-b) & \text{(Dieterici equation of state)} \end{cases}$$

1.2 The volume and total surface area of a rectangular parallelepiped are

$$V = abc$$
$$A = 2(ab + ac + bc)$$

where a, b, and c are the lengths of the three sides. Write the complete differentials dV and dA. Verify that both are exact differentials.

1.3 Show that Eq. (2.16) holds for one mole of a van der Waal's fluid. (Hint: Two of the partial derivatives are given in this chapter.)

1.4 Examine the following functions for homogeneity and degree of homogeneity:

$$u = 3x^2y + xy^2 + xyz, \qquad u = x^2 - y^2/x - y$$
$$u = (x^2 + y)^2, \qquad u = 3x^2 + 6x + 9$$
$$u = \ln(x/y), \qquad u = (x^2 + y^2)^{1/2}$$

1.5 Decide if the following are exact or inexact differentials:

$$du = \frac{2x + y}{x^2 + xy + y^2} dx + \frac{2y + x}{x^2 + xy + y^2} dy$$

$$du = [-y(x^2 + y^2)\exp(-y) + 2x\exp(-y)] dx$$
$$\quad + [-x(x^2 + y^2)\exp(-y) + 2y\exp(-y)] dy$$

$$du = (y/x) dx + (\ln x) dy$$

$$du = [3x^2 + 2xy^2 + y] dx + [y^2 + x + 2y(x^2 + y)] dy$$

$$du = \frac{-y^2}{(x-y)^2} dx + \frac{-x^2}{(x-y)^2} dy$$

1.6 The coefficient of thermal expansion, α, and the coefficient of compression, K_T, are defined by the following two equations:

$$\alpha = \frac{1}{V}\left(\frac{\partial V}{\partial T}\right)_P \quad \text{and} \quad K_T = -\frac{1}{V}\left(\frac{\partial V}{\partial P}\right)_T$$

Show that:

$$(\partial P/\partial T)_V = \alpha/K_T$$

and

$$(\partial \alpha/\partial P)_T + (\partial K_T/\partial T)_P = 0$$

1.7 Thermodynamic properties of binary mixtures are generally reported in the chemical literature in tabular form with an accompanying mathematical expression describing how the property varies with liquid

TABLE I
Excess Molar Volumes of 1,2-Dichloroethane (1) + n-Nonane (2) Mixtures[a]

X_1	$\Delta \bar{V}^{ex}$ (cm^3/mol)
0.1502	0.605
0.2793	0.911
0.4385	1.090
0.5431	1.097
0.6746	0.974
0.8154	0.698
0.8732	0.532
0.9263	0.337

[a] Experimental values were reproduced with permission from Krishnaiah and Naidu (1980). Copyright 1980 American Chemical Society.

phase composition. Assuming one wished to parametrize the excess volume, $\Delta \bar{V}^{ex}$ of 1,2-dichloroethane + n-nonane mixtures by

$$\Delta \bar{V}^{ex} = X_1(1 - X_1)[A + B(2X_1 - 1)^2]$$

describe how to use a linear least squares procedure to determine the two parameters A and B. What values of A and B *best* describe the experimental data given in Table I?

2
Partial Molar Quantities

A. Partial Molar Volumes

If a solution is formed by mixing at constant temperature and pressure, $n_1, n_2, \ldots, n_r$ moles of components $1, 2, \ldots, r$, the total volume of the *unmixed* components is the sum of the individual volumes

$$V^{um} = n_1 \bar{V}_1^\bullet + n_2 \bar{V}_2^\bullet + \cdots + n_r \bar{V}_r^\bullet = \sum_{i=1}^{r} n_i \bar{V}_i^\bullet \tag{2.1}$$

where the solid circles ($\bullet$) denote the property of pure substances and the superscript um indicates unmixed. The molar volume of pure component i at the specified temperature and pressure is $\bar{V}_i^\bullet$. After mixing it is found that the volume of the solution is generally not equal to V^{um}; that is $V \neq V^{um}$. The same situation holds for the other extensive thermodynamic properties such as U, H, S, A, and G.

The development of expressions for the volume of multicomponent solutions and other extensive properties is desired. Each property is a function of the solution's state, which can be described by the independent variables $T, P, n_1, n_2, \ldots, n_r$. That is

$$\begin{aligned} V &= V(T, P, n_1, n_2, \ldots, n_r) \\ H &= H(T, P, n_1, n_2, \ldots, n_r) \\ G &= G(T, P, n_1, n_2, \ldots, n_r) \end{aligned} \tag{2.2}$$

with analogous equations for A, S, etc. The total differential dV is given by

$$\begin{aligned} dV = &\left(\frac{\partial V}{\partial T}\right)_{P, n_i} dT + \left(\frac{\partial V}{\partial P}\right)_{T, n_i} dT + \left(\frac{\partial V}{\partial n_1}\right)_{T, P, n_i \neq n_1} dn_1 \\ &+ \left(\frac{\partial V}{\partial n_2}\right)_{T, P, n_i \neq n_2} dn_2 + \cdots + \left(\frac{\partial V}{\partial n_r}\right)_{T, P, n_i \neq n_r} dn_r \end{aligned} \tag{2.3}$$

A. Partial Molar Volumes

The subscript n_i in the first two partial derivatives indicates that the number of moles of all the components is held constant, and the subscript $n_i \neq n_1$ indicates all mole numbers are held constant except for n_1. The *partial molar volume* $\bar{V}_j$ in the multicomponent solution is defined as

$$\bar{V}_j = (\partial V / \partial n_j)_{T, P, n_i \neq n_j} \tag{2.4}$$

Using the partial molar volumes defined by Eq. (2.4), dV becomes

$$dV = \left(\frac{\partial V}{\partial P}\right)_{T, n_i} dP + \left(\frac{\partial V}{\partial T}\right)_{P, n_i} dT + \sum_{i=1}^{r} \bar{V}_i \, dn_i \tag{2.5}$$

As seen from Eq. (2.4), partial molar volumes are partial derivatives of V and consequently depend on the same variables as V:

$$\bar{V}_j = \bar{V}_j(T, P, n_1, n_2, \ldots, n_r)$$

The partial molar volume of a pure substance is identical to its molar volume $\bar{V}^\bullet$. The partial molar volume of component i in a multicomponent solution, however, is not necessarily equal to the molar volume of pure i. In order to distinguish between these two quantities, the partial molar volume of component i in solution will be denoted by $\bar{V}_i$ and the molar volume of the pure component by $\bar{V}_i^\bullet$.

Equation (2.5) describes how the change in the volume of the solution is related to the partial molar volumes of the individual components. But more useful are the expressions for the thermodynamic properties of the solution rather than only the changes in these properties. At a fixed temperature and pressure, it is known that increasing the amounts of all components by some factor k produces an identical increase in the solution's volume. (Recall that V is a homogeneous function of the first degree. Refer to Section 1.6 of Chapter 1.) Thus, if the final amounts of the individual components are $n_1^*, n_2^*, \ldots, n_r^*$, related to the initial amounts $n_1, n_2, \ldots, n_r$ by

$$n_1^* = kn_1$$
$$n_2^* = kn_2$$
$$\vdots$$
$$n_r^* = kn_r$$

then the final volume $V(n_1^*, n_2^*, \ldots, n_r^*)$ is related to the initial volume $V(n_1, n_2, \ldots, n_r)$ by

$$V(n_1^*, n_2^*, \ldots, n_r^*) = kV(n_1, n_2, \ldots, n_r) \tag{2.6}$$

Differentiation of the left-hand side of Eq. (2.6) with respect to k gives

$$\frac{\partial V(n_1^*, n_2^*, \ldots, n_r^*)}{\partial k} = \frac{\partial V(n_1^*, n_2^*, \ldots, n_r^*)}{\partial n_1^*}\left(\frac{\partial n_1^*}{\partial k}\right)$$
$$+ \frac{\partial V(n_1^*, n_2^*, \ldots, n_r^*)}{\partial n_2^*}\left(\frac{\partial n_2^*}{\partial k}\right) + \cdots$$
$$+ \frac{\partial V(n_1^*, n_2^*, \ldots, n_r^*)}{\partial n_r^*}\left(\frac{\partial n_r^*}{\partial k}\right) \quad (2.7)$$

with

$$(\partial n_i^*/\partial k) = n_i$$

The derivative of the right-hand side of Eq. (2.6) is

$$\frac{\partial}{\partial k}[kV(n_1, n_2, \ldots, n_r)] = V(n_1, n_2, \ldots, n_r) \quad (2.8)$$

because $V(n_1, n_2, \ldots, n_r)$ is independent of k. By equating Eq. (2.7) and (2.8), one obtains

$$[\partial V(n_1^*, n_2^*, \ldots, n_r^*)/\partial n_1^*]n_1 + [\partial V(n_1^*, n_2^*, \ldots, n_r^*)/\partial n_2^*]n_2$$
$$+ \cdots + [\partial V(n_1^*, n_2^*, \ldots, n_r^*)/\partial n_r^*]n_r = V(n_1, n_2, \ldots, n_r) \quad (2.9)$$

Equation (2.9) is valid for all values of k, and a useful expression is developed when k is set equal to 1. Then $n_1^* = n_1$, $n_2^* = n_2, \ldots, n_r^* = n_r$ and

$$n_1[\partial V(n_1, n_2, \ldots, n_r)/\partial n_1] + n_2[\partial V(n_1, n_2, \ldots, n_r)/\partial n_2]$$
$$+ \cdots + n_r[\partial V(n_1, n_2, \ldots, n_r)/\partial n_r] = V(n_1, n_2, \ldots, n_r) \quad (2.10)$$

which is written as

$$V = n_1(\partial V/\partial n_1) + n_2(\partial V/\partial n_2) + \cdots + n_r(\partial V/\partial n_r) \quad (2.11)$$

Because all of the partial derivatives taken in Eq. (2.11) were assuming constant T and P, they correspond to partial molar quantities. This enables the volume of the solution to be expressed as

$$V = n_1\bar{V}_1 + n_2\bar{V}_2 + \cdots + n_r\bar{V}_r \quad (2.12)$$

Dividing both sides of Eq. (2.12) by the total number of moles enables the expression of the solution volume (an extensive thermodynamic property) in terms of the *average molar volume* of the solution $\bar{V}$ (an intensive thermodynamic property):

$$\bar{V} = V/n_{\text{tot}} = \sum_{i=1}^{r} X_i \bar{V}_i \quad (2.13)$$

A. Partial Molar Volumes

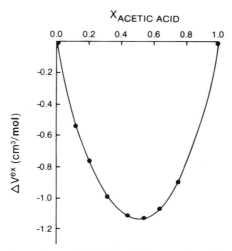

Fig. 2.1 Excess molar volumes ($\Delta \bar{V}^{ex}$) for binary acetic acid + water mixtures. The experimental data were determined by Casanova et al. (1981).

As seen from Eq. (2.13), the average molar volume is the mole fraction average of the individual components partial molar volumes. Generally, extensive properties are described as $Z(T, P, n_1, n_2, \ldots, n_r)$ and intensive properties (such as partial molar properties) are described as $\bar{Z}_i(T, P, X_1, X_2, \ldots, X_r)$ or with other intensive composition variables such as molarities, molalities, volume fractions, etc.

Generally, volumetric data for multicomponent systems is reported in the literature as the solution's actual volume minus the total volume of the unmixed components

$$\Delta V = V - \sum_{i=1}^{r} n_i \bar{V}_i^{\bullet} = \sum_{i=1}^{r} n_i (\bar{V}_i - \bar{V}_i^{\bullet}) \tag{2.14}$$

$$\Delta \bar{V} = \bar{V} - \sum_{i=1}^{r} X_i \bar{V}_i^{\bullet} = \sum_{i=1}^{r} X_i (\bar{V}_i - \bar{V}_i^{\bullet}) \tag{2.15}$$

In subsequent chapters it will be shown that the volume change upon mixing $\Delta \bar{V}$ is the excess molar volume, formally denoted as $\Delta \bar{V}^{ex}$. The $\Delta \bar{V}$s for binary mixtures containing acetic acid and water are graphically depicted in Fig. 2.1. Because the numerical values of $\Delta \bar{V}$s are negative, the actual volumes of acetic acid and water mixtures are less than the total volumes of the two unmixed components, in other words, the solution contracts upon mixing. For most systems commonly encountered, the magnitude of $\Delta \bar{V}$ rarely exceeds ± 2 cm^3/mol.

B. Other Partial Molar Quantities

The ideas developed for the volume V apply to any extensive property of the solution. For example, the solution's Gibbs free energy G is a function of $T, P, n_1, n_2, \ldots, n_r$. By analogy with Eq. (2.4), the partial molar Gibbs free energy of component j in the solution is defined by

$$\bar{G}_j \equiv (\partial G/\partial n_j)_{T,P,\,n_i \neq n_j} \equiv \mu_j \tag{2.16}$$

The partial molar Gibbs function is of special interest because it is identical to the chemical potential.

The total differential dG is given by

$$dG = \left(\frac{\partial G}{\partial T}\right)_{P,n_i} dT + \left(\frac{\partial G}{\partial P}\right)_{T,n_i} dP + \sum_{j=1}^{r} \bar{G}_j \, dn_j \tag{2.17}$$

Replacing the symbol V by G in Eq. (2.6–2.12), it can be shown that the Gibbs free energy of the multicomponent solution is

$$G = \sum_{i=1}^{r} n_i \bar{G}_i \tag{2.18}$$

The change in the Gibbs free energy (ΔG) for the process of forming the multicomponent solution from its pure components at constant T and P is

$$\Delta G = G - G^{\text{um}} = \sum_{i=1}^{r} n_i \bar{G}_i - \sum_{i=1}^{r} n_i \bar{G}_i^{\bullet}$$

$$= \sum_{i=1}^{r} n_i (\bar{G}_i - \bar{G}_i^{\bullet}) = \sum_{i=1}^{r} n_i (\mu_i - \mu_i^{\bullet}) \tag{2.19}$$

where $\bar{G}_i^{\bullet}$ (or $\mu_i^{\bullet}$) is the molar Gibbs free energy for pure component i.

In addition, the partial molar internal energies, $\bar{U}_i$; partial molar Helmholtz free energies, $\bar{A}_i$; partial molar enthalpies, $\bar{H}_i$; partial molar entropies, $\bar{S}_i$; and partial molar heat capacities, $\bar{C}_{p_i}$ and $\bar{C}_{v_i}$[†] are defined

---

[†] Conventional partial molar properties result from the choice of temperature and pressure for independent intensive variables in the defining equation. Reis (1982) presented a discussion of partial molar properties from a generalized thermodynamic framework. The author introduced partial molar properties at contant temperature and molar volume, $\bar{Z}_i = (\partial Z/\partial n_i)_{T,\bar{V},\,n_j \neq n_i}$; at constant molar entropy and pressure, $\bar{Z}_i (\partial Z/\partial n_i)_{\bar{S},P,\,n_j \neq n_i}$; and at constant molar entropy and molar volume, $\bar{Z}_i = (\partial Z/\partial n_i)_{\bar{S},\bar{V},\,n_i \neq n_j}$. Readers are encouraged to review the article because several interesting ideas were presented.

B. Other Partial Molar Quantities

$$\bar{U}_i \equiv (\partial U/\partial n_i)_{T,P,\,n_j \neq n_i} \tag{2.20}$$

$$\bar{A}_i \equiv (\partial A/\partial n_i)_{T,P,\,n_j \neq n_i} \tag{2.21}$$

$$\bar{H}_i \equiv (\partial H/\partial n_i)_{T,P,\,n_j \neq n_i} \tag{2.22}$$

$$\bar{S}_i \equiv (\partial S/\partial n_i)_{T,P,\,n_j \neq n_i} \tag{2.23}$$

$$\bar{C}_{p_i} \equiv (\partial C_p/\partial n_i)_{T,P,\,n_j \neq n_i} \tag{2.24}$$

$$\bar{C}_{v_i} \equiv (\partial C_v/\partial n_i)_{T,P,\,n_j \neq n_i} \tag{2.25}$$

Many of the thermodynamic relationships between extensive properties of a homogeneous system have corresponding relationships involving partial molar properties. For example

$$G = H - TS$$

The partial differentiation with respect to n_i gives

$$(\partial G/\partial n_i)_{T,P,\,n_j \neq n_i} = (\partial H/\partial n_i)_{T,P,\,n_j \neq n_i} - T(\partial S/\partial n_i)_{T,P,\,n_j \neq n_i}$$

$$\bar{G}_i = \bar{H}_i - T\bar{S}_i \tag{2.26}$$

A second example relating two partial molar properties can be developed from the pressure dependence of the Gibbs free energy

$$(\partial G/\partial P)_{T,n_i} = V \tag{2.27}$$

Differentiation of Eq. (2.27) with respect to n_j gives

$$[(\partial/\partial n_j)(\partial G/\partial P)_{T,n_i}]_{T,P,\,n_i \neq n_j} = (\partial V/\partial n_j)_{T,P,\,n_i \neq n_j} \tag{2.28}$$

Because the order of differentiation is unimportant, Eq. (2.28) can be rearranged to

$$[(\partial/\partial n_j)(\partial G/\partial P)_{T,n_i}]_{T,P,\,n_i \neq n_j} = [(\partial/\partial P)(\partial G/\partial n_j)_{T,P,\,n_i \neq n_j}]_{T,n_i}$$

$$(\partial V/\partial n_j)_{T,P,\,n_i \neq n_j} = (\partial \bar{G}_j/\partial P)_{T,n_i} = \bar{V}_j \tag{2.29}$$

If Eq. (2.19) is differentiated with respect to pressure, holding temperature and mole numbers constant, Eq. (2.14) is rederived for ΔV,

$$(\partial \Delta G/\partial P)_{T,n_i} = \sum_{i=1}^{r} n_i [(\partial \bar{G}_i/\partial P)_{T,n_i} - (\partial \bar{G}_i^\bullet/\partial P)_{T,n_i}]$$

$$\Delta V = \sum_{i=1}^{r} n_i (\bar{V}_i - \bar{V}_i^\bullet)$$

The solution's remaining thermodynamic properties are also describable by an analogous set of equations. Generalizing to any extensive property Z, the following is obtained:

$$Z = \sum_{i=1}^{r} n_i \bar{Z}_i \tag{2.30}$$

$$\bar{Z} = Z \bigg/ \left(\sum_{i=1}^{r} n_i\right) = \sum_{i=1}^{r} X_i \bar{Z}_i \tag{2.31}$$

$$\Delta Z = \sum_{i=1}^{r} n_i (\bar{Z}_i - \bar{Z}_i^{\bullet}) \tag{2.32}$$

$$\Delta \bar{Z} = \sum_{i=1}^{r} X_i (\bar{Z}_i - \bar{Z}_i^{\bullet}) \tag{2.33}$$

Absolute values of partial molar enthalpies cannot be determined, similarly as with absolute values of enthalpy. In addition, it is not possible to measure absolute values of partial molar entropies and partial molar free energies. Thus, it is necessary to select some state as a reference and to express the partial molar enthalpy relative to this reference state.

Consider isothermally mixing several pure substances of mole numbers $n_1, n_2, \ldots, n_r$ to form a homogeneous solution. This process of mixing is accompanied by an enthalpy change ΔH, which will be a function of composition. If $\bar{H}_i^{\bullet}$ denotes the molar enthalpies of the pure materials, the total enthalpy of the multicomponent mixture is

$$H = \sum_{i=1}^{r} n_i \bar{H}_i^{\bullet} + \Delta H \tag{2.34}$$

Partial differentiation of Eq. (2.34) with respect to the number of moles of component j, while holding T, P and $n_i \neq n_j$ constant, gives the partial molar enthalpy of component j in the solution

$$\bar{H}_j = (\partial H / \partial n_j)_{T, P, n_i \neq n_j} = \bar{H}_j^{\bullet} + (\partial \Delta H / \partial n_j)_{T, P, n_i \neq n_j} \tag{2.35}$$

The change in the partial molar enthalpy upon dissolving a pure substance in the solution is given by the expression

$$\Delta \bar{H}_j = \bar{H}_j - \bar{H}_j^{\bullet} = (\partial \Delta H / \partial n_j)_{T, P, n_i \neq n_j} \tag{2.36}$$

It follows from Eq. (2.36) that a partial molar analysis of enthalpies of mixing (ΔH) will yield differences between the partial molar enthalpies in solution and the molar enthalpies of the pure substances ($\bar{H}_j^{\bullet}$). To simplify the notation $\Delta \bar{H}_j$ is defined to be equal to $\bar{H}_j - \bar{H}_j^{\bullet}$.

C. The Gibbs–Duhem Equation

Taking the total differential of Eq. (2.18), the change in the Gibbs free energy of the solution is expressed as

$$dG = d\sum_{i=1}^{r} n_i \bar{G}_i = \sum_{i=1}^{r} d(n_i \bar{G}_i) = \sum_{i=1}^{r} n_i d\bar{G}_i + \sum_{i=1}^{r} \bar{G}_i dn_i \qquad (2.37)$$

Substitution of Eq. (2.17) for dG gives

$$\left(\frac{\partial G}{\partial P}\right)_{T,n_i} dP + \left(\frac{\partial G}{\partial T}\right)_{P,n_i} dT + \sum_{i=1}^{r} \bar{G}_i dn_i = \sum_{i=1}^{r} n_i d\bar{G}_i + \sum_{i=1}^{r} \bar{G}_i dn_i$$

$$\left(\frac{\partial G}{\partial P}\right)_{T,n_i} dP + \left(\frac{\partial G}{\partial T}\right)_{P,n_i} dT - \sum_{i=1}^{r} n_i d\bar{G}_i = 0 \qquad (2.38)$$

Equation (2.38) is known as the Gibbs–Duhem equation and applies to any infinitesimal process. The most common application of Eq. (2.38) in thermodynamics is for a process occurring at constant temperature and pressure ($dP = 0$ and $dT = 0$):

$$\sum_{i=1}^{r} n_i d\bar{G}_i = 0 \qquad (2.39)$$

Dividing both sides of Eq. (2.39) by the total number of moles in the solution gives an alternate form involving mole fraction compositions

$$\sum_{i=1}^{r} X_i d\bar{G}_i = 0 \qquad (2.40)$$

Equations (2.39) and (2.40) are generalized to any partial molar quantity as

$$\sum_{i=1}^{r} n_i d\bar{Z}_i = 0 \qquad (2.41)$$

$$\sum_{i=1}^{r} X_i d\bar{Z}_i = 0 \qquad (2.42)$$

Inspection of the general Gibbs–Duhem equation reveals the $\bar{Z}_i$s are not all independent. Knowledge of the values of $r - 1$ of the $\bar{Z}_i$s as a function of composition permits the calculation of the remaining $\bar{Z}_i$ through the integration of Eq. (2.41) or (2.42).

D. Apparent Partial Molar Quantities

Apparent partial molar quantities are used to simplify the reporting of experimental data. The extensive property Z is considered to be composed of a contribution because one component (j) and the properties of all other components in their pure states. That is,

$$Z = n_j \bar{Z}_j^{\text{app}} + \sum_{i=1, i \neq j}^{r} n_i \bar{Z}_i^{\bullet} \tag{2.43}$$

where $\bar{Z}_j^{\text{app}}$ is the *apparent partial molar property* of component j (Many texts denote apparent partial molar properties as ϕ_{Z_j}, but this notation is not used here because the ϕs are used to represent volume fractions.) By differentiating Eq. (2.43) with respect to n_j, the following relationship between partial molar quantities $\bar{Z}_j$ and apparent partial molar quantities $\bar{Z}_j^{\text{app}}$ is obtained.

$$(\partial Z/\partial n_j)_{T,P, n_i \neq n_j} = \bar{Z}_j = \bar{Z}_j^{\text{app}} + n_j(\partial \bar{Z}_j^{\text{app}}/\partial n_j)_{T,P, n_i \neq n_j} \tag{2.44}$$

The calculation of partial molar quantities from apparent partial molar quantities generally is used when $\bar{Z}_j^{\text{app}}$ is expressible as an analytical function, such as

$$\bar{Z}_j^{\text{app}} = a + bm_j^{1/2} + cm_j,$$

in which m_j is the molality of component j, and a, b, and c are constants. Consider a series of solutions containing 1000 grams of solvent so that

$$n_j = m_j \quad \text{and} \quad dn_j = dm_j$$

It follows directly from Eq. (2.44) that

$$\begin{aligned}\bar{Z}_j &= m_j(\partial \bar{Z}_j^{\text{app}}/\partial m_j)_{T,P,n_i} + \bar{Z}_j^{\text{app}} \\ &= m_j(0.5 m_j^{-1/2} + c) + a + bm_j^{1/2} + cm_j \\ &= a + (0.5 + b)m_j^{1/2} + 2cm_j\end{aligned}$$

It is easily shown that at infinite dilution ($n_j \approx 0$) $\bar{Z}_j^{\text{app}}$ is equal to $\bar{Z}_j$. By substituting Eq. (2.30) into Eq. (2.43), one obtains

$$\bar{Z}_j^{\text{app}} = \left(Z - \sum_{i=1, i \neq j}^{r} n_i \bar{Z}_i^{\bullet}\right)\bigg/ n_j = \left[n_j \bar{Z}_j + \sum_{i=1, i \neq j}^{r} n_i(\bar{Z}_i^{\bullet} - \bar{Z}_i)\right]\bigg/ n_j \tag{2.45}$$

Because $\bar{Z}_i$ ($i \neq j$) at infinite dilution is equal to $\bar{Z}_i^{\bullet}$ ($i \neq j$)[†], the limit of $\bar{Z}_j^{\text{app}}$ at $n_j = 0$ is an indeterminate form (0/0). The limit can be evaluated by applying l'Hôpital's rule and differentiating the numerator and denominator

---

[†] Remember that by using apparent partial molar properties all deviations from additivity are attributed to component j. In other words, if component j were not present, then Z of the solution would be given by

$$Z(n_j = 0) = \sum_{i=1, i \neq j}^{r} n_i \bar{Z}_i = \sum_{i=1, i \neq j}^{r} n_i \bar{Z}_i^{\bullet}$$

E. Determination of Partial Molar Quantities

of Eq. (2.45) with respect to n_j at constant n_i, that is,

$$\lim_{n_j \to 0} \bar{Z}_j^{\text{app}} = \lim_{n_j \to 0} \bar{Z}_j/1 = \bar{Z}_j \tag{2.46}$$

Apparent partial molar properties $\bar{Z}_j^{\text{app}}$ can be calculated from ΔZ

$$\Delta Z = Z - Z^{\text{um}}$$

$$= n_j \bar{Z}_j^{\text{app}} + \sum_{i=1, i \neq j}^{r} n_i \bar{Z}_i - \sum_{i=1}^{r} n_i \bar{Z}_i^{\bullet}$$

$$= n_j (\bar{Z}_j^{\text{app}} - \bar{Z}_j^{\bullet}) \tag{2.47}$$

provided $\bar{Z}_j^{\bullet}$ is a measureable quantity. In the case of enthalpies, entropies, and free energies, the absolute values of $\bar{Z}_i^{\bullet}$'s cannot be determined, therefore it is more convenient to define a *relative apparent partial molar property* $\Delta \bar{Z}_j^{\text{app}}$ as

$$\Delta \bar{Z}_j^{\text{app}} = \bar{Z}_j^{\text{app}} - \bar{Z}_j^{\bullet} \tag{2.48}$$

Expressed in terms of $\Delta \bar{Z}_j^{\text{app}}$, Eq. (2.47) now becomes

$$\Delta Z = n_j \Delta \bar{Z}_j^{\text{app}} \tag{2.49}$$

E. Determination of Partial Molar Quantities

There are several methods for calculating partial molar properties. The selected method depends to a large extent on how the experimental data are reported. As seen in the previous section, partial molar quantities can be calculated from apparent partial molar quantities provided the experimental data are expressed in terms of $\bar{Z}_j^{\text{app}}$.

One of the most popular ways to report experimental data for binary systems is in terms of average molar properties $\bar{Z}$:

$$\bar{Z} = Z/(n_1 + n_2) \tag{2.50}$$

Differentiation of Eq. (2.50) with respect to n_1 while holding T, P, and n_2 constant gives

$$\left(\frac{\partial \bar{Z}}{\partial n_1}\right)_{T,P,n_2} = \frac{(n_1 + n_2)(\partial Z/\partial n_1)_{T,P,n_2} - Z}{(n_1 + n_2)^2}$$

$$= \frac{\bar{Z}_1}{n_1 + n_2} - \frac{\bar{Z}}{n_1 + n_2} \tag{2.51}$$

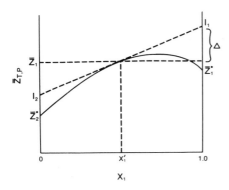

Fig. 2.2 Graphical determination of partial molar quantities from average molar quantities.

Using the chain rule, Eq. (2.51) can be rewritten as

$$\left(\frac{\partial \bar{Z}}{\partial X_1}\right)_{T,P} \left(\frac{\partial X_1}{\partial n_1}\right)_{T,P,n_2} = \frac{\bar{Z}_1}{n_1 + n_2} - \frac{\bar{Z}}{n_1 + n_2} = \left(\frac{\partial \bar{Z}}{\partial X_1}\right)_{T,P} \frac{X_2}{n_1 + n_2}$$

or expressed explicitly in terms of $\bar{Z}_1$ as

$$\bar{Z}_1 = \bar{Z} + X_2 (\partial \bar{Z}/\partial X_1)_{T,P} \tag{2.52}$$

Similarly, it can be shown that

$$\bar{Z}_2 = \bar{Z} + X_1 (\partial \bar{Z}/\partial X_2)_{T,P} \tag{2.53}$$

If $\bar{Z}$ is expressed graphically, as shown in Fig. 2.2, it is possible to evaluate $\bar{Z}_1$ and $\bar{Z}_2$ by the method of intercepts as: The tangent to the curve at X_1' is the partial derivative of $\bar{Z}$ with respect to X_1 (evaluated at X_1'), and the slope at this point is equal to the difference $(\bar{Z}_1 - \bar{Z}_2)$ at X_1'. The intercept of this tangent at $X_1 = 1.0$ can be equated to the increment $\Delta = (1 - X_1')$ $(\partial \bar{Z}/\partial X_1)$ plus the value of $\bar{Z}$ at X_1', therefore

$$I_1 = \bar{Z}' + (1 - X_1')(\partial \bar{Z}/\partial X_1')_{T,P,(\text{at } X_1')} = \bar{Z}_1' = \bar{Z}_1(\text{at } X_1') \tag{2.54}$$

Similarly

$$I_2 = \bar{Z}' - X_1'(\partial \bar{Z}/\partial X_1')_{T,P,(\text{at } X_1')} = \bar{Z}_2' = \bar{Z}_2(\text{at } X_1') \tag{2.55}$$

The straight line between $\bar{Z}_2^\bullet$ and $\bar{Z}_1^\bullet$ is described by $X_1 \bar{Z}_1^\bullet + X_2 \bar{Z}_2^\bullet$. The differences between the curve ($\bar{Z}$) and this straight line at any value of X_1 is

$$\Delta \bar{Z} = X_1(\bar{Z}_1 - \bar{Z}_1^\bullet) + X_2(\bar{Z}_2 - \bar{Z}_2^\bullet) \tag{2.56}$$

This quantity represents the change in $\bar{Z}$ for the process

X_1 component 1 (pure, T, P) + X_2 component 2 (pure T, P) $\longrightarrow$ solution (X_1, T, P)

E. Determination of Partial Molar Quantities

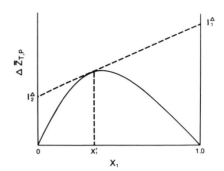

Fig. 2.3 Graphical determination of $\bar{Z}_1$ and $\bar{Z}_2$ from plots of $\Delta \bar{Z}$ versus X_1.

There are, however, two disadvantages associated with this method based on Eqs. (2.54) and (2.55). First, when Z is a quantity such as enthalpy, Gibbs free energy, or entropy, the method is not applicable because the quantity itself cannot be measured. Only differences in the quantity, that is, ΔZ can be measured. Secondly, even when Z is a quantity that can be measured (such as volume or heat capacity at constant pressure), the method is not accurate whenever the deviations from linearity of a plot of Z versus X_1 are small compared with the differences $(\bar{Z}_1 - \bar{Z}_2)$ between the end values.

Figure 2.3 shows a typical plot of $\Delta \bar{Z}$ versus X_1 for experimental binary data. The slope of a tangent drawn to the curve at some arbitrary point is $(\partial \Delta \bar{Z}/\partial X_1)_{T,P,(\text{at } X'_1)}$, and the tangent is shown to intersect the ordinate at $X_1 = 0.0$ and $X_1 = 1.0$ at points labeled I_2^A and I_1^A, respectively. From inspection of Eq. (2.56), it follows that

$$I_2^A = (\bar{Z}_1 - \bar{Z}_1^\bullet) = \Delta \bar{Z}' + (1 - X'_1)(\partial \Delta \bar{Z}/\partial X_1)_{T,P,(\text{at } X'_1)} \quad (2.57)$$

$$I_2^A = (\bar{Z}_2 - \bar{Z}_2^\bullet) = \Delta \bar{Z}' + X'_1(\partial \Delta \bar{Z}/\partial X_1)_{T,P,(\text{at } X'_1)} \quad (2.58)$$

A practical disadvantage of this tangental method based on Eqs. (2.56–2.58) is that it is difficult to draw tangents with great accuracy. Van Ness and Mrazek (1959) and Van Ness (1959) suggested an alternative method based on plotting $\Delta \bar{Z}/X_1 X_2$ versus X_1, as shown in Fig. 2.4. Now

$$\Delta \bar{Z}_1 = \bar{Z}_1 - \bar{Z}_1^\bullet = X'^2_2[(2 \Delta \bar{Z}'/X'_1 X'_2) - J_2] \quad (2.59)$$

$$\Delta \bar{Z}_2 = \bar{Z}_2 - \bar{Z}_2^\bullet = X'^2_1[(2 \Delta \bar{Z}'/X'_1 X'_2) - J_1] \quad (2.60)$$

The numerical values of $\Delta \bar{Z}_1$ and $\Delta \bar{Z}_2$ calculated *via* Eqs. (2.59) and (2.60) are much less sensitive to errors in the tangental construction than are values obtained by graphical methods based on either $\bar{Z}$ versus X_1 or $\Delta \bar{Z}$ versus X_1 plots.

Graphical evaluation of partial molar properties is unnecessary if a computer is accessible because it is possible to fit the data for $\bar{Z}$ (or $\Delta \bar{Z}$) with an

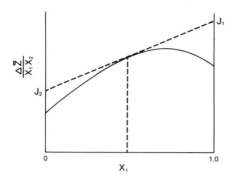

Fig. 2.4 Graphical determination of partial molar quantities from plots of $\Delta \bar{Z}/X_1 X_2$ versus X_1.

appropriate power series in X_1 or X_2 and differentiate this mathematical representation directly. As an example, consider experimental data for butyl-p-hydroxybenzoate + n-propanol mixtures. Cave et al. (1980) reported the average molar volumes in the form

$$\bar{V} = 76.451 + 99.007 X_2 + 9.282 X_2^2$$

where X_2 is the mole fraction composition of butyl-p-hydroxybenzoate. The corresponding partial molar volumes of the two components can be calculated by the two formulas

$$\bar{V}_2 = 76.451 + 99.007 + 9.282 X_2 (1 + X_1)$$
$$\bar{V}_1 = 76.451 - 9.282 X_2^2$$

Acree et al. (1981) noted these two equations gave values of $\bar{V}_1 = 75.2$ cm^3/mol and $\bar{V}_2 = 180.9$ cm^3/mol when the experimental solubility ($X_2 = 0.361$) for butyl-p-hydroxybenzoate in n-propanol was used.

Calculating partial molar quantities from analytical representations is relatively straightforward for binary mixtures. Because many of the solutions of practical importance contain more than two components, consider how to calculate partial molar properties for higher-order multicomponent systems. For a ternary solution, $\bar{Z}$ can be expressed as

$$\bar{Z} = X_1 \bar{Z}_1 + X_2 \bar{Z}_2 + X_3 \bar{Z}_3 \tag{2.61}$$

a mole fraction average of the partial molar $\bar{Z}_i$s of the individual components. Through differentiation of $\bar{Z}$ we obtain

$$d\bar{Z} = \bar{Z}_1 dX_1 + \bar{Z}_2 dX_2 + \bar{Z}_3 dX_3 + X_1 d\bar{Z}_1 + X_2 d\bar{Z}_2 + X_3 d\bar{Z}_3 \tag{2.62}$$

The Gibbs–Duhem equation $\sum X_i d\bar{Z}_i = 0$ enables Eq. (2.62) to be simplified to

$$d\bar{Z} = \bar{Z}_1 dX_1 + \bar{Z}_2 dX_2 + \bar{Z}_3 dX_3$$

The derivatives in this expression can be taken with respect to temperature, pressure, or mole fraction composition of any one of the three components. Two specific cases are

$$(\partial \bar{Z}/\partial X_3)_{T,P, X_2 = \text{constant}} = -\bar{Z}_1 + \bar{Z}_3 \qquad (2.63)$$

$$(\partial \bar{Z}/\partial X_3)_{T,P, X_1 = \text{constant}} = -\bar{Z}_2 + \bar{Z}_3 \qquad (2.64)$$

Combining Eqs. (2.61), (2.63), and (2.64) in the manner

$$\bar{Z} + X_1(\partial \bar{Z}/\partial X_3)_{T,P, X_2 = \text{constant}} + X_2(\partial \bar{Z}/\partial X_3)_{T,P, X_1 = \text{constant}}$$
$$= X_1\bar{Z}_1 + X_2\bar{Z}_2 + X_3\bar{Z}_3 - X_1\bar{Z}_1 + X_1\bar{Z}_3 - X_2\bar{Z}_2 + X_2\bar{Z}_3$$
$$= (X_1 + X_2 + X_3)\bar{Z}_3$$
$$= \bar{Z}_3 \qquad (2.65)$$

gives a relatively simple expression for $\bar{Z}_3$ in terms of the average molar properties of the ternary solution. The partial molar properties $\bar{Z}_1$ and $\bar{Z}_2$ are obtained from analogous expressions

$$\bar{Z}_1 = \bar{Z} + X_2(\partial \bar{Z}/\partial X_1)_{T,P, X_3 = \text{constant}} + X_3(\partial \bar{Z}/\partial X_1)_{T,P, X_2 = \text{constant}} \qquad (2.66)$$

$$\bar{Z}_2 = \bar{Z} + X_1(\partial \bar{Z}/\partial X_2)_{T,P, X_3 = \text{constant}} + X_3(\partial \bar{Z}/\partial X_2)_{T,P, X_1 = \text{constant}} \qquad (2.67)$$

The general ideas used in developing Eqs. (2.65–2.67) can be extended to systems containing more than three components. For the purposes of the book, only binary and ternary mixtures will be considered. The partial derivatives of the various mathematical representations become quite complex for higher-order multicomponent system.

F. Interpretive Descriptions

It is important to note partial molar quantities and apparent partial molar quantities are just two of the ways that may be chosen to describe a solution. There are many interpretive schemes that can be used for describing a system, but in using an interpretive description one should always be sure to specify the interpretive scheme in enough detail to allow conversion of the tabulated results to other descriptive schemes in the case that interpretation is later found to be faulty or inadequate. The following example illustrates how a single process can be described in terms of partial molar properties, apparent partial molar properties, and an interpretive scheme.

Consider the mixing of N_H moles of liquid H_2O with N_D moles of liquid D_2O at a specified temperature and pressure.

Case 1 The process is treated as simple mixing

$$N_H H_2O(\ell, T, P) + N_D D_2O(\ell, T, P) \longrightarrow (N_H H_2O, N_D D_2O)(\ell, T, P)$$

Applying Eqs. (2.30) and (2.32),

$$Z = N_H \bar{Z}_{H_2O} + N_D \bar{Z}_{D_2O} \tag{2.68}$$

$$Z^{um} = N_H \bar{Z}^{\bullet}_{H_2O} + N_D \bar{Z}^{\bullet}_{D_2O} \tag{2.69}$$

$$\Delta Z = Z - Z^{um}$$
$$= N_H(\bar{Z}_{H_2O} - \bar{Z}^{\bullet}_{H_2O}) + N_D(\bar{Z}_{D_2O} - \bar{Z}^{\bullet}_{D_2O}) \tag{2.70}$$

where $\bar{Z}_{H_2O}$ and $\bar{Z}_{D_2O}$ refer to partial molar quantities in a *binary* solution.

Case 2 The mixing process is treated in terms of apparent partial properties.

$$Z = N_H \bar{Z}^{\bullet}_{H_2O} + N_D \bar{Z}^{app}_{D_2O} \tag{2.71}$$

$$Z^{um} = N_H \bar{Z}^{\bullet}_{H_2O} + N_D \bar{Z}^{\bullet}_{D_2O} \tag{2.72}$$

$$\Delta Z = N_D(\bar{Z}^{app}_{D_2O} - \bar{Z}^{\bullet}_{D_2O}) \tag{2.73}$$

or in terms of the apparent partial molar properties of H_2O

$$Z = N_H \bar{Z}^{app}_{H_2O} + N_D \bar{Z}^{\bullet}_{D_2O} \tag{2.74}$$

$$\Delta Z = N_H(\bar{Z}^{app}_{H_2O} - \bar{Z}^{\bullet}_{H_2O}) \tag{2.75}$$

Case 3 Considering the same process in a different manner,

$$N_H H_2O(\ell, T, P) + N_D D_2O(\ell, T, P) \longrightarrow \begin{Bmatrix} \hat{N}_{HD} HDO \\ \hat{N}_H H_2O \\ \hat{N}_D D_2O \end{Bmatrix} (\ell, T, P, N_D/N_H)$$

with the stoichiometric restrictions

$$2N_D = 2\hat{N}_D + \hat{N}_{HD} \quad \text{and} \quad 2N_H = 2\hat{N}_H + \hat{N}_{HD}$$

Z of the *ternary* solution is now given by

$$Z = \hat{N}_{HD} \hat{Z}_{HDO} + \hat{N}_D \hat{Z}_{D_2O} + \hat{N}_H \hat{Z}_{H_2O} \tag{2.76}$$

where the $\hat{Z}$ terms refer to the partial molar quantities in a ternary system. Using this interpretive description, ΔZ is determined by

$$\Delta Z = \hat{N}_{HD}(\hat{Z}_{HDO} - \hat{Z}_{H_2O}/2 - \hat{Z}_{D_2O}/2) + N_H(\hat{Z}_{H_2O} - \bar{Z}^{\bullet}_{H_2O})$$
$$+ N_D(\hat{Z}_{D_2O} - \bar{Z}^{\bullet}_{D_2O}) \tag{2.77}$$

$$Z = \hat{N}_{HD} \hat{Z}^{app}_{HDO} + \hat{N}_H \bar{Z}^{\bullet}_{H_2O} + N_D \bar{Z}^{\bullet}_{D_2O} \quad \text{and} \tag{2.78}$$

$$\Delta Z = \hat{N}_{HD}[\hat{Z}^{app}_{HDO} - (\bar{Z}^{\bullet}_{H_2O} - \bar{Z}^{\bullet}_{D_2O})/2] \tag{2.79}$$

Equations (2.70), (2.73), (2.75), (2.77), and (2.79) are equally valid expressions describing a single experimental process.

Problems

2.1 Show that for a binary system at constant temperature and pressure
(a) $\bar{Z}_2 - \bar{Z}_1 = (\partial \bar{Z}/\partial X_2)$, (b) $\bar{Z}_1 - \bar{Z}_2 = (\partial \bar{Z}/\partial X_1)$, and (c) $\partial(\bar{Z}/X_2)/\partial X_2 = -\bar{Z}_1/X_2^2$.

2.2 Derive $\bar{V}_1$, $\bar{V}_2$, and $\bar{V}_3$ for a ternary system obeying $\bar{V} = X_1 \bar{V}_1^\bullet + X_2 \bar{V}_2^\bullet + X_3 \bar{V}_3^\bullet + X_1 X_2 \beta_{12} + X_1 X_3 \beta_{13} + X_2 X_3 \beta_{23}$. The β_{ij}s refer to numerical constants and are independent of mole fraction.

2.3 Determine if the following analytical representation of partial molar heat capacities $\bar{C}_{p_1}$ and $\bar{C}_{p_2}$ in a binary mixture $\bar{C}_{p_1} = \bar{C}_{p_1}^\bullet + \alpha X_2^2 + \beta X_2^2(3X_1 - X_2)$, and $\bar{C}_{p_2} = \bar{C}_{p_2}^\bullet + \alpha X_1^2 + \beta X_1^2(3X_2 - X_1)$ are consistent with the Gibbs–Duhem equation.

2.4 Morris et al. (1975) parametrized $\Delta \bar{H}$ for binary mixtures of acetone + chloroform, acetone + methanol and methanol + chloroform according to the formula

$$\Delta \bar{H}/RTX_1 X_2 = A_{ji} X_i + A_{ij} X_j - X_i X_j (X_i \lambda_{ji} + X_j \lambda_{ij}) + X_i^2 X_j^2 (X_i \psi_{ji} + X_j \psi_{ij})$$

where A_{ij}, A_{ji}, λ_{ij}, λ_{ji}, ψ_{ij} and ψ_{ji} are numerical constants. Derive expressions for $\Delta \bar{H}_1$ and $\Delta \bar{H}_2$.

2.5 Suppose the density of a binary solution is known as a function of mole fraction composition. (a) Show that the partial molar volume of component 1 will be given by the expression

$$\bar{V}_1 = [M_1 - X_2(X_1 M_1 + X_2 M_2)(\partial \rho/\partial X_1)_{T,P}]/\rho^2$$

where M_1 and M_2 denote molecular weights. (b) Mikhail and Kimel (1961) expressed the density of methanol (1) + water (2) mixtures at 25°C via

$$\rho(\text{g/ml}) = 0.9971 - 0.18527 X_1 + 0.22013 X_1^2 - 0.60418 X_1^3 + 0.53912 X_1^4 - 0.18012 X_1^5$$

Using the expression derived in part a, calculate the partial molar volume of methanol, $\bar{V}_1$, at $X_1 = 0.5$.

2.6 For a binary mixture, show that the relative partial molar enthalpy $\Delta \bar{H}_2$ is related to the relative apparent partial molar enthalpy $\Delta \bar{H}_2^{\text{app}}$ by

$$(\partial \Delta H/\partial n_2)_{T,P,n_1} = \Delta \bar{H}_2 = \Delta \bar{H}_2^{\text{app}} + X_1 X_2 (\partial \Delta \bar{H}_2^{\text{app}}/\partial X_2)_{T,P}$$

2.7 C. de Visser et al. (1977) determined the density of water + N-methylformamide mixtures at 25°C. Using their data given in Table I, calculate the apparent partial molar volumes of water and N-methylformamide at all 12 binary compositions.

TABLE I
Densities for Binary Mixtures of Water (1) + N-Methylformamide (2) at 25°C.[a]

X_1	ρ (grams/cm^3)
0.0000	0.998244
0.0511	0.999210
0.1284	1.000991
0.2108	1.003069
0.3130	1.006106
0.4073	1.008891
0.4994	1.011834
0.6398	1.015314
0.6979	1.015999
0.8248	1.013838
0.9016	1.008794
0.9506	1.003533
0.9797	0.999821
1.0000	0.997047

[a] Reproduced with permission from C. de Visser et al. (1977).

2.8 For a binary solution, show that the apparent partial molar volume of component 2 can be expressed in terms of $\Delta \bar{V}^{ex}$ by

$$\bar{V}_2^{app} = \bar{V}_2 + \Delta \bar{V}^{ex}/X_2$$

TABLE II
Densities of Sulfolane (1) + Water (2) Mixtures at 303.15 K[a]

X_2	ρ (grams/cm^3)
0.00000	1.262350
0.09998	1.256150
0.19839	1.249324
0.30332	1.241165
0.40086	1.232060
0.50410	1.219807
0.60708	1.203390
0.70744	1.181052
0.80763	1.147923
0.90380	1.095701
1.00000	0.995651

[a] Reproduced with permission from Castagnolo et al. (1981).

Using the density data of Castagnolo *et al.* (1981) given in Table II calculate $\Delta \bar{V}^{ex}$ and $\bar{V}_2^{app}$ at the nine binary compositions.

2.9 In a mixture of water and ethanol in which the mole fraction of water is 0.4, the partial molar volumes of water and ethanol are 16.3 and 57.5 cm³/mol, respectively. (a) Use an appropriate sketch and show that the partial molar volumes of the components in a binary mixture may both be positive as for this particular system or may be of opposite sign, but cannot be both negative at the same composition. (b) Calculate the density of the ethanol + water mixture.

2.10 The densities of mixtures of liquids A (MW = 100 g/mol) and B (MW = 200 g/mol) at 25°C are given by the following equation, which is valid up to a molality $m_B = 2.0$:

$$\rho(\text{g/cm}^3) = 0.900 + 0.0700\ m_B - 0.0200\ m_B^2$$

Calculate: (a) The apparent partial molar volume of B in a solution of $m_B = 1.000$. (b) The partial molar volume of A in a solution of $m_B = 1.000$. (c) The limiting partial molar volume of B at infinite dilution in A.

2.11 The enthalpy of mixing of two liquids C and D at 25°C obeys the equation

$$\Delta \bar{H}^{mix}(\text{J/mol}) = 1250\ X_C X_D/(1 + 1.50\ X_D)$$

Calculate ΔH for the following processes: (a) One mole of liquid C is mixed with one mole of liquid D at 25°C and 1 atm. (b) One mole of liquid C is mixed with an infinite amount of liquid D at 25°C and 1 atm. (c) One mole of liquid C is mixed with an infinite amount of an equimolar mixture of C and D at 25°C and 1 atm.

3
Ideal and Nonideal Solutions

The chemical potential is an important thermodynamic quantity because it enables us to specify the conditions of equilibrium in a multicomponent phase, and provides for the thermodynamic description of multicomponent solutions. The use of chemical potentials in the thermodynamic analysis of solutions is inconvenient primarily because the chemical potential becomes negative without bound as the concentration or gas pressure approaches zero. Furthermore, absolute values of the chemical potential are unknown and cannot be computed. For these reasons it has been found more convenient to introduce two new auxiliary thermodynamic functions (fugacity and activity) that facilitate the analysis of many problems encountered in chemical thermodynamics.

A. Fugacity

Consider one mole of a pure gaseous substance at a pressure sufficiently low enough to ensure the gas obeys the ideal gas law $P\bar{V} = RT$. The change in the chemical potential of this gas with pressure at constant temperature is given by

$$dG = d\mu_{\text{id g}} = \bar{V}_{\text{id g}} dP = RT \, d\ln P \tag{3.1}$$

where id g stands for ideal gas.

At higher pressures where the ideal gas law fails it is necessary to make substitutions for $\bar{V}$ from a more complicated equation of state. As a result, the relationship between the chemical potential and pressure becomes more complicated. To make the simple form of Eq. (3.1) applicable to nonideal gases as well as ideal gases, G. N. Lewis introduced the fugacity, f as

$$d\mu = RT \, d\ln f \tag{3.2}$$

or for a constituent of a multicomponent mixture

$$d\mu_i = RT \, d\ln f_i \tag{3.3}$$

A. Fugacity

Comparison of Eq. (3.1) reveals that the fugacity of a perfect gas must be proportional to its pressure, and for convenience the two are set equal to each other. The definition of fugacity of a real gas is then completed by setting the condition

$$\lim_{P \to 0} f/P = 1 \tag{3.4}$$

and for a component in a mixture

$$\lim_{P \to 0} f_i/P_i = \lim_{P \to 0} f_i/y_i P = 1 \tag{3.5}$$

where y_i is the mole fraction of component i in the vapor phase, $P_i = y_i P$, and P is the total pressure. The fugacity of a real gas approaches its pressure (equal to that of an ideal gas) in the limit of zero pressure.

As is evident, the fugacity is related to the chemical potential of a real gas in the same manner that the pressure is related to the chemical potential of an ideal gas. Because direct experimental measurement of fugacity and chemical potentials is impossible, it becomes necessary to relate these quantities to other directly measureable properties of the system.

At constant temperature the chemical potential of a pure real gas is related to its pressure by the equation

$$d\mu = \bar{V} \, dP$$

and for the pure ideal gas

$$d\mu_{\text{id g}} = \bar{V}_{\text{id g}} \, dP$$

Subtracting the last two equations from each other,

$$d\mu - d\mu_{\text{id g}} = (\bar{V} - \bar{V}_{\text{id g}}) \, dP \tag{3.6}$$

However, $d\mu$ and $d\mu_{\text{id g}}$ are also given by Eqs. (3.1) and (3.2), respectively, and thus,

$$d \ln f - d \ln P = (RT)^{-1} (\bar{V} - \bar{V}_{\text{id g}}) \, dP \tag{3.7}$$

Upon integrating between the limits of pressure P and a pressure approaching zero ($P^+ \to 0$), and taking into consideration that at this low pressure the fugacity of any real gas approaches its ideal gas pressure, $f^+ = P^+$, we obtain

$$\ln f/P = \ln v = 1/RT \int_{P^+ \to 0}^{P} (\bar{V} - \bar{V}_{\text{id g}}) \, dP \tag{3.8}$$

The ratio f/P is called the fugacity coefficient v and gives a direct measure of the extent that real gases deviate from ideality ($v = 1$) at a given pressure and temperature.

Equation (3.8) provides a basic formula to calculate fugacities. Knowing $\bar{V}$ as a function of pressure, the quantity $(\bar{V} - \bar{V}_{\text{id g}})/RT$ may be calculated and graphically plotted against pressure. The area under the curve from $P = 0$ to P is the numerical value of the integral. Alternatively, this value can be calculated analytically be expressing $\bar{V}$ as a function of P from any equation of state. For example, the virial equation $\bar{V} = (RT/P) + B$ gives

$$\ln f/P = \ln v = BP/RT \tag{3.9}$$

Fugacity appears to be even more useful in dealing with gaseous mixtures, liquids and solids, pure or mixed. From the direct relationship between the chemical potential and the fugacity, Eq. (3.2) and (3.3), it follows that the fugacity of any constituent is the same in all phases of the system

$$f_i^\alpha = f_i^\beta = f_i^\gamma = \cdots$$

provided the system is at equilibrium.

The concentration dependence of fugacity may be obtained by substituting Eq. (3.3) into the Gibbs–Duhem equation

$$\sum_{i=1}^{N} X_i \left(\frac{\partial \ln f_i}{\partial X_1} \right)_{T,P} = 0 \tag{3.10}$$

To solve this differential equation the partial derivatives must first be evaluated from the experimental data. In a two-component system, $dX_1 = -dX_2$, and thus,

$$X_1 (\partial \ln f_1/\partial X_1)_{T,P} = -X_2 (\partial \ln f_2/\partial X_1)_{T,P} = X_2 (\partial \ln f_2/\partial X_2)_{T,P} \tag{3.11}$$

For a binary mixture it is sufficient to know the variation of the fugacity with mole fraction for one of the components. The variation for the other component may then be found *via* Eq. (3.11).

Fugacity also varies with temperature. To derive an expression relating fugacity and temperature proceed as such: At constant temperature permit one mole of substance i to be transferred from a state of pressure P to another state of pressure P^+. The change in the Gibbs free energy for this process is given by

$$\Delta \bar{G} = \mu_i^+ - \mu_i = RT \ln(f_i^+/f_i) \tag{3.12}$$

or rearranging terms

$$\mu_i^+/T - \mu_i/T = R \ln f_i^+ - R \ln f_i \tag{3.13}$$

Partial differentiation with respect to T yields

$$(\partial(\mu_i^+/T)/\partial T)_{P^+} - (\partial(\mu_i/T)/\partial T)_P = R(\partial \ln f_i^+/\partial T)_{P^+} - R(\partial \ln f_i/\partial T)_P \tag{3.14}$$

B. Activity

The significance of the terms on the left-hand side of this expression may be seen from

$$(\partial(\mu_i/T)/\partial T)_P = (1/T^2)[-\mu_i + T(\partial \mu_i/\partial T)_P] = -\bar{H}_i/T^2 \qquad (3.15)$$

Similarly,

$$(\partial(\mu_i^+/T)/\partial T)_P = -\bar{H}_i^+/T^2 \qquad (3.16)$$

Substitution of these two expressions into Eq. (3.14) gives

$$(-\bar{H}_i^+ + \bar{H}_i)/T^2 = R(\partial \ln f_i^+/\partial T)_{P^+} - R(\partial \ln f_i/\partial T)_P \qquad (3.17)$$

If the pressure P^+ is sufficiently low, then $f_i^+ = P_i^+$ and $(\partial \ln f_i^+/\partial T)_{P^+} = 0$. As a consequence of this

$$(\partial \ln f_i/\partial T)_P = (\bar{H}_i^+ - \bar{H}_i)/RT^2 \qquad (3.18)$$

where $\bar{H}_i^+$ is the partial molar enthalpy of component i at a low pressure (at which the gaseous mixture behaves ideally), and thus, $\bar{H}_i^+ = H_i$. The difference, $H_i = \bar{H}_i$, represents the enthalpy change for the expansion of the component from pressure P to pressure P^+.

B. Activity

When describing liquid or solid solutions it is more convenient to use activity and activity coefficients instead of fugacity and fugacity coefficients. Integration of Eq. (3.3) between two states and at constant temperature gives

$$(\mu_i)_{\text{st }2} - (\mu_i)_{\text{st }1} = RT \ln[(f_i)_{\text{st }2}/(f_i)_{\text{st }1}] \qquad (3.19)$$

where st 1 and st 2 represent state 1 and state 2.

If state 1 is fixed and completely specified, then the difference in the chemical potentials will depend only on state 2. Denote the chemical potential and fugacity of this arbitrarily chosen fixed state by μ_i° and f_i°, respectively, so that

$$\mu_i - \mu_i^\circ = RT \ln(f_i/f_i^\circ) \qquad (3.20)$$

Substituting the quantity

$$f_i/f_i^\circ = a_i/a_i^\circ \qquad (3.21)$$

into Eq. (3.20) yields

$$\mu_i - \mu_i^\circ = RT \ln(a_i/a_i^\circ) \qquad (3.22)$$

where a_i and a_i° denote the activity in a given state and in the fixed state, respectively. In this fixed state, called the standard state, the activity a_i° is always assigned a numerical value of unity (in concentration units); consequently,

$$\mu_i - \mu_i^\circ = RT \ln[a_i/1(\text{concentration})] = RT \ln a_i \qquad (3.23)$$

From Eqs. (3.1) and (3.8) it follows that fugacity has the dimension of pressure, whereas the fugacity coefficient is dimensionless. Activity may be defined to be with or without dimensions. In many thermodynamic textbooks, activity is treated as a dimensionless function. Because fugacity and activity are so closely related and, as will be shown, that by using the proper standard state they might be identical, this author prefers to assign dimensions to activity.

Just as fugacity is related to the pressure through the fugacity coefficient, activity is related to the concentration via the activity coefficient

$$a_i = \gamma_i X_i \qquad (3.24)$$

$$a_i = (\gamma_i)_m M_i \qquad (3.25)$$

$$a_i = (\gamma_i)_c C_i \qquad (3.26)$$

where γ_i, $(\gamma_i)_m$, and $(\gamma_i)_c$ are the activity coefficients referred to as the mole fraction, molality, and molarity concentration scale, respectively. Assigning dimensions to the activity requires the activity coefficients to be dimensionless.

The numerical value of the activity depends on the selection of the standard state, as well as on the temperature, pressure, and composition. If f_i° is changed, then a_i and γ_i must also change. The activity is meaningless unless it is accompanied by a detailed description of the standard state that it refers to. Within certain limitations, the standard state for a component may be arbitrarily selected. The primary limitation is that the temperature of the standard state must always be the same as the temperature of the system.

The standard state of a gas at any given temperature is taken as the state in which the fugacity and pressure are equal to 1 atm, namely, $f^\circ = p^\circ = 1$. Inherent in this definition is the requirement that the gas behave ideally at a pressure of 1 atm. Because no real gases meet this requirement, this standard state is not a real one, but rather a hypothetical state. Using this standard state the activity of any gas becomes equal to its fugacity (see Eq. (3.21), remembering that $a_i^\circ = 1$)

$$\mu_i = \mu_i^\circ(T, 1\text{ atm}) + RT \ln[f_i/1(\text{atm})] = \mu_i^\circ(T, 1\text{ atm}) + RT \ln f_i \qquad (3.27)$$

where $\mu_i^\circ(T, 1\text{ atm})$ is the chemical potential in the standard state and depends only on temperature.

B. Activity

For liquid and solid solutions there are two commonly used standard states:

(a) **Symmetrical Reference State** assumes as a standard state the pure liquid or solid component at the temperature and pressure of the system; sometimes, also at a pressure of 1 atm or at the vapor pressure of the pure component. This definition is based on the equations

$$\mu_i^{\bullet}(T, P) = \lim_{X_i \to 1} (\mu_i - RT \ln X_i) \qquad i = 1, 2, \ldots, N \qquad (3.28)$$

and

$$\lim_{X_i \to 1} (a_i/X_i) = \lim_{X_i \to 1} \gamma_i = 1 \qquad (3.29)$$

(b) **Unsymmetrical Reference System** distinguishes between the solvent and solute molecules. The standard state for the solvent, the component at highest concentration, is the same as in the symmetrical reference system. The standard state of the solutes, though, is defined according to

$$(\mu_i^{\blacktriangle})_g(T, P) = \lim_{X_1 \to 1} (\mu_i - RT \ln g_i) \qquad i = 2, 3, \ldots, N \qquad (3.30)$$

Here, g_i refers to the composition variable of the solute (i.e., molarity, mole fraction, etc.), and the superscript ▲ denotes a standard state of component i at infinite dilution. As a result,

$$\lim_{g_i \to 0} (a_i/g_i) = \lim_{g_i \to 0} (\gamma_i^{\blacktriangle})_g = 1 \qquad (3.31)$$

The solute's standard state in the unsymmetrical reference system is a hypothetical one (see Fig. 3.1) and is commonly referred to as the infinite dilution state.

On the molality concentration scale, the standard state of the solvent is given by Eq. (3.28), and for solutes

$$(\mu_i^{\blacktriangle})_m(T, P) = \lim_{X_1 \to 1} (\mu_i - RT \ln M_i) \qquad \text{where} \quad i = 2, 3, \ldots, N \qquad (3.32)$$

$$\lim_{M_i \to 0} (a_i/M_i) = \lim_{M_i \to 0} (\gamma_i^{\blacktriangle})_m = 1 \qquad (3.33)$$

This standard state is also hypothetical and represents a state in which the solute would exist at unity molality but would still have the molecular environment typical of an extremely dilute solution. Similar expressions may be given for the molarity

$$(\mu_i^{\blacktriangle})_c(T, P) = \lim_{X_1 \to 1} (\mu_i - RT \ln C_i) \qquad \text{where} \quad i = 2, 3, \ldots, N \qquad (3.34)$$

$$\lim_{C_i \to 0} (a_i/C_i) = \lim_{C_i \to 0} (\gamma_i^{\blacktriangle})_c = 1 \qquad (3.35)$$

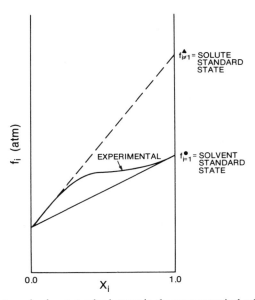

Fig. 3.1 Solute and solvent standard states in the unsymmetrical reference system.

and the mole fraction concentration scale

$$(\mu_i^\blacktriangle)_x(T, P) = \lim_{X_1 \to 1} (\mu - RT \ln X_i) \quad \text{where} \quad i = 2, 3, \ldots, N \quad (3.36)$$

$$\lim_{X_i \to 0} (a_i/X_i) = \lim_{X_i \to 0} (\gamma_i^\blacktriangle)_x = 1 \quad (3.37)$$

Problem 3.2 at the end of this chapter requires one to convert from one standard state $(\mu_2^\blacktriangle)_m$ to a second standard state $(\mu_2^\blacktriangle)_x$.

The pressure dependence of activity is given by the equation

$$(\partial \ln a_i/\partial P)_{T, X_i} = \frac{1}{RT} [(\partial \mu_i/\partial P)_{T, X_i} - (\partial \mu_i^\circ/\partial P)_T]$$

$$= (RT)^{-1}(\bar{V}_i - \bar{V}_i^\circ) \quad (3.38)$$

Pressure dependence may be ignored when dealing with liquid or solid solutions as the difference in the molar volumes $\bar{V}_i - \bar{V}_i^\circ$ is negligibly small. For gases, the change is significant and must be included.

The change of the activity with temperature is represented by the expression

$$(\partial \ln a_i/\partial T)_{P, X_i} = -(\bar{H}_i - \bar{H}_i^\circ)/RT^2 \quad (3.39)$$

The activity decreases with temperature whenever the difference in the partial molar enthalpy and molar enthalpy of the pure component i $(\bar{H}_i - \bar{H}_i^\circ)$ is positive, and decreases with temperature whenever $(\bar{H}_i - \bar{H}_i^\circ)$ is negative.

Expressions for the dependence of the activity and activity coefficient on composition (at constant T and P) are based on the Gibbs–Duhem equation. Combination of Eqs. (2.42) and (3.23) gives

$$\sum_{i=1}^{N} X_i (\partial \ln a_i / \partial X_1)_{T,P} = 0 \qquad (3.40)$$

Using the relationship $a_i = X_i \gamma_i$, we can write the variation of γ_i with mixture composition as

$$\sum_{i=1}^{N} X_i (\partial \ln \gamma_i / \partial X_1)_{T,P} = 0 \qquad (3.41)$$

The partial derivatives in Eq. (3.40) and (3.41) can be taken with respect to the remaining $N - 1$ components, and are not restricted to component 1.

As was the case for partial molar quantities, the determination of the activity and the activity coefficient dependence on composition requires experimental data. The primary value of Eqs. (3.40) and (3.41) is that they reduce the number of direct experimental observations and reveal inconsistencies and errors in the measured values.

C. The Ideal Solution

Just as the theory of gases is simplified by considering an idealized type of gas as a first approximation to the behavior of real gases, an idealized type of liquid mixture is also valuable in developing a coherent theory of solutions. A perfect (ideal) mixture may be defined in a number of ways. A simple way is to state that a solution is ideal if the chemical potential of every constituent is a linear function of the logarithm of its mole fraction $\tilde{X}_i$, according to the relation

$$\mu_i = \mu_i^\bullet (T, P) + RT \ln \tilde{X}_i \qquad (3.42)$$

where $\tilde{X}_i$ denotes a mole fraction in a gaseous, solid, or liquid state. It can be easily proved that this definition for an ideal solution also requires the change, on mixing of the pure components, to be zero for the thermodynamic quantities U, H, V, C_p, and C_v. This will be proven for the volume and enthalpy. Interested readers are encouraged to prove $\Delta Z^{\text{mix}} = 0$ for the remaining three thermodynamic properties listed above.

Taking the derivative of Eq. (3.42) with respect to T at constant P and X_i gives

$$\left(\frac{\partial(\mu_i/T)}{\partial T}\right)_{P,X_i} = \left(\frac{\partial(\mu_i^\bullet/T)}{\partial T}\right)_P \tag{3.43}$$

According to Eq. (3.15)

$$\left(\frac{\partial(\mu_i/T)}{\partial T}\right)_{P,X_i} = -\frac{\bar{H}_i}{T^2} \quad \text{and} \quad \left(\frac{\partial(\mu_i^\bullet/T)}{\partial T}\right)_P = -\frac{\bar{H}_i^\bullet}{T^2}$$

Combination with Eq. (3.43) yields

$$\bar{H}_i = \bar{H}_i^\bullet \tag{3.44}$$

Consequently, the formation of an ideal solution from the pure components is

$$\Delta H^{\text{mix}} = \sum_{i=1}^{N} n_i(\bar{H}_i - \bar{H}_i^\bullet) = 0 \tag{3.45}$$

neither an exothermic nor endothermic process.

Differentiating Eq. (3.42) with respect to pressure while maintaining constant temperature and composition gives

$$(\partial \mu_i/\partial P)_{T,X_i} = (\partial \mu_i^\bullet/\partial P)_T \tag{3.46}$$

Because

$$(\partial \mu_i/\partial P)_{T,X_i} = \bar{V}_i \quad \text{and} \quad (\partial \mu_i^\bullet/\partial P)_T = \bar{V}_i^\bullet$$

it follows that

$$\bar{V}_i = \bar{V}_i^\bullet \tag{3.47}$$

As a result, the formation of an ideal solution is not accompanied by a volumetric contraction or expansion, i.e.,

$$\Delta V^{\text{mix}} = \sum_{i=1}^{N} n_i(\bar{V}_i - \bar{V}_i^\bullet) = 0 \tag{3.48}$$

Mixing is a spontaneous process and therefore, the entropy of mixing (and of all thermodynamic functions related to entropy) will be nonzero, even for an ideal solution. For example, the Gibbs free energy of mixing and the entropy of mixing are given by Eqs. (3.49) and (3.50)

$$\Delta G^{\text{mix}} = \Delta H^{\text{mix}} - T \Delta S^{\text{mix}} = \sum_{i=1}^{N} n_i(\mu_i - \mu_i^\bullet) = RT \sum_{i=1}^{N} n_i \ln X_i \tag{3.49}$$

$$\Delta S^{\text{mix}} = -R \sum_{i=1}^{N} n_i \ln X_i \tag{3.50}$$

D. Vapor–Liquid Equilibrium in an Ideal Solution

Consider an isothermal system composed of two components and two phases; both phases, liquid and vapor, behave ideally. The chemical potential of any constituent in any phase will depend, at constant T, on the pressure and composition only. That is,

$$d\mu_i = (\partial\mu_i/\partial P)_{T,\tilde{X}_i} dP + (\partial\mu_i/\partial \tilde{X}_i)_{T,P} d\tilde{X}_i \tag{3.51}$$

where $\tilde{X}_i$ refers to the mole fraction in both phases. The first partial derivative on the right-hand side of Eq. (3.51) represents the molar volume $\bar{V}_i$. The second derivative can be shown by simple differentiation to be

$$(\partial\mu_i/\partial \tilde{X}_i)_{T,P} = RT/\tilde{X}_i \tag{3.52}$$

and combination with Eq. (3.51) gives

$$d\mu_i = \bar{V}_i dP + (RT/\tilde{X}_i) d\tilde{X}_i \tag{3.53}$$

Integration between the limits of $\tilde{X}_i = 1$ and $\tilde{X}_i$ yields

$$\mu_i(P, T) = \mu_i^\bullet(T, P) + \int_{P_i^\bullet}^{P} \bar{V}_i dP + RT \ln \tilde{X}_i \tag{3.54}$$

where $P_i^\bullet$ is the vapor pressure of pure component i, and P is the pressure above the solution of composition X_i. Comparison of Eqs. (3.42) and (3.54) shows the standard state chemical potential at the pressure of the system $\mu_i^\bullet(T, P)$ is related to the standard state chemical potential at the vapor pressure of the pure component $\mu_i^\bullet(T, P_i^\bullet)$ via

$$\mu_i^\bullet(T, P) = \mu_i^\bullet(T, P_i^\bullet) + \int_{P_i^\bullet}^{P} \bar{V}_i dP \tag{3.55}$$

both chemical potentials being at the temperature of the system.

After combining the equilibrium conditions

$$\mu_i(\ell, T, P) = \mu_i(g, T, P) \tag{3.56}$$

$$\mu_i^\bullet(\ell, T, P_i^\bullet) = \mu_i^\bullet(g, T, P_i^\bullet) \tag{3.57}$$

and with Eq. (3.53),

$$\int_{P_i^\bullet}^{P} \bar{V}_i^{\text{liq}} dP + RT \ln X_i = \int_{P_i^\bullet}^{P} \bar{V}_i^{\text{vap}} dP + RT \ln y_i \tag{3.58}$$

where X_i and y_i are the mole fractions in the liquid, and in the vapor phases, respectively.

The molar volume of the component in the liquid phase $\bar{V}_i^{\text{liq}}$ can be considered constant in the given pressure interval, and the molar volume in the vapor phase $\bar{V}_i^{\text{vap}}$ may be expressed by the virial gas equation

$$\bar{V}_i^{\text{vap}} = (RT/P) + B_{ii}$$

After evaluating the integrals, Eq. (3.58) takes the form

$$\bar{V}_i^{\text{liq}}(P - P_i^\bullet) + RT \ln X_i = B_{ii}(P - P_i^\bullet) + RT \ln (P/P_i^\bullet) + RT \ln y_i \quad (3.59)$$

When Eq. (3.59) is written in an exponential form for a two-component system, the pressure can be expressed in terms of component 1

$$P = (X_1 P_1^\bullet/y_1) \exp[(\bar{V}_1^{\text{liq}} - B_{11})(P - P_1^\bullet)/RT] \quad (3.60)$$

or in terms of component 2

$$P = (X_2 P_2^\bullet/y_2) \exp[(\bar{V}_2^{\text{liq}} - B_{22})(P - P_2^\bullet)/RT] \quad (3.61)$$

With the relationship $y_1 + y_2 = 1$, the last two equations can be combined to give

$$P = y_1 P + y_2 P = X_1 P_1^\bullet \exp[(\bar{V}_1^{\text{liq}} - B_{11})(P - P_1)/RT] \\ + X_2 P_2^\bullet \exp[(\bar{V}_2^{\text{liq}} - B_{22})(P - P_2^\bullet)/RT] \quad (3.62)$$

On comparing Eq. (3.62) with Dalton's law, $P = P_1 + P_2$, the partial pressure is found to be

$$P_i = X_i P_i^\bullet \exp[(\bar{V}_i^{\text{liq}} - B_{ii})(P - P_i^\bullet)/RT] \quad (3.63)$$

The value of the exponent is, in most cases, close to zero, and has a first approximation

$$P_i \approx X_i P_i^\bullet \quad (3.64)$$

Equation (3.64) was found empirically by Raoult in 1886. It states that the partial pressure of component i over an ideal solution P_i is equal to the product of the vapor pressure of the pure component i at the temperature of the solution P_i and the mole fraction of component i in the liquid phase X_i. Raoult's law does not specifically define an ideal solution. It follows from the exact definition, Eq. (3.63), only after making the additional assumption that the exponential term equals 1.

Using Eq. (3.64), Dalton's law may be written in the form

$$P = X_1 P_1^\bullet + X_2 P_2^\bullet = X_1 P_1^\bullet + (1 - X_1) P_2^\bullet = X_1 (P_1^\bullet - P_2^\bullet) + P_2^\bullet \quad (3.65)$$

Because $P_1^\bullet$ and $P_2^\bullet$ are constant for any system at a specified temperature, the total pressure dependence on the liquid phase composition may be represented by a straight line (Fig. 3.2), provided the mixture is nearly ideal. Least-squares analysis of the experimental vapor pressures for carbon tetrachloride + tetrachloroethylene mixtures gave Fried et al. (1969)

$$P(\text{torr}) = 95.1 + 348.3 X_{\text{CCl}_4}$$

D. Vapor–Liquid Equilibrium in an Ideal Solution

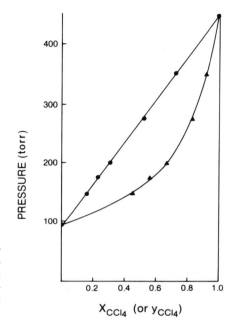

Fig. 3.2 Experimental liquid–vapor equilibrium data for carbon tetrachloride + tetrachloroethylene mixtures. The circles (●) and triangles (▲) refer to liquid and vapor compositions, respectively. Data were taken from the work of Fried et al. (1969).

which compares quite favorably to the Raoult's law expression

$$P(\text{torr}) = 94.9 + 349.3 X_{\text{CCl}_4}$$

based on the vapor pressures of the pure components $P^{\bullet}_{\text{CCl}_4} = 444.2$ torr, and $P^{\bullet}_{\text{TCE}} = 94.9$ torr, at 60°C, where TCE represents tetrachloroethylene.

To derive the relationship between the total pressure and vapor phase composition, a slightly different form of Dalton's law, $P_i = y_i P$, is more useful. For a binary mixture

$$P_1 = y_1 P \tag{3.66}$$

$$P_2 = y_2 P = (1 - y_1) P \tag{3.67}$$

and combination with Eq. (3.64) results in

$$P = P^{\bullet}_1 X_1 / y_1 \tag{3.68}$$

Through suitable manipulation of Eq. (3.65) and (3.68), it is found that the total pressure is

$$P = P^{\bullet}_1 P^{\bullet}_2 / [P^{\bullet}_1 - y_1(P^{\bullet}_2 - P^{\bullet}_1)] \tag{3.69}$$

not a linear function of vapor phase composition.

E. Behavior of Ideal Dilute Solutions

Now consider a special type of vapor–liquid equilibrium in which the temperature and pressure of the system lie above the critical state of one of the system's components. This component is a gas in its pure state and therefore, this type of equilibrium is that of gases dissolved in liquids.

At constant temperature the chemical potential of a real gas $\mu_2(g)$ is given by the equation

$$\mu_2(g) = \mu_2^{\bullet}(g, T, 1 \text{ atm}) + RT \ln f_2 \tag{3.70}$$

Similarly, the chemical potential of the gas dissolved in the ideal dilute solution $\mu_2(\ell)$ is expressed by

$$\mu_2(\ell) = (\mu_2^{\blacktriangle})_g(\ell, T, P) + RT \ln g_2 \tag{3.71}$$

where $(\mu_2^{\blacktriangle})_g(\ell, T, P)$ is the chemical potential referring to the infinitely dilute standard state, and $\mu_2^{\circ}(g, T, 1 \text{ atm})$ is the chemical potential referring to the pure gas at the temperature of the system and at a pressure of 1 atm. At equilibrium, $\mu_2(\ell) = \mu_2(g)$, and thus,

$$\mu_2^{\circ}(g, T, 1 \text{ atm}) + RT \ln f_2 = (\mu_2^{\blacktriangle})_g(\ell, T, P) + RT \ln g_2 \tag{3.72}$$

Consequently, the fugacity of component 2 can be written as

$$f_2 = g_2 \exp[(\mu_2^{\blacktriangle})_g(\ell, T, P) - \mu_2^{\circ}(g, T, 1 \text{ atm})]/RT \tag{3.73}$$

and

$$f_2 = g_2(k_H)_g \approx P_2 \tag{3.74}$$

because at low pressures the fugacity may be replaced, as a first approximation, by the partial pressure of the gas above the solution. The letter H here represents henry. As can be seen

$$(k_H)_g = \exp[(\mu_2^{\blacktriangle})_g(\ell, T, P) - \mu_2^{\circ}(g, T, 1 \text{ atm})]/RT \tag{3.75}$$

Equation (3.75) is a mathematical statement of Henry's law. According to this law, the solubility of a gas in a dilute solution is directly proportional to its partial pressure above the solution. The proportionality constant is mainly temperature dependent; it varies, however, with the solvent, the total pressure, and the chosen standard state. The standard states are generally chosen so the composition variable g_2 corresponds to either mole fraction, molality, or molarity.

Henry's law is not restricted to gas–liquid systems but is followed by a wide variety of dilute solutions and by all solutions in the limit of infinite

E. Behavior of Ideal Dilute Solutions

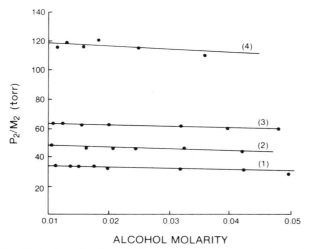

Fig. 3.3 Graphical determination of Henry's law constants for isomeric butanols in isooctane at 25°C. The various alcohols are: (1) 1-butanol, (2) isobutanol, (3) s-butanol and (4) t-butanol. Experimental data by Rytting et al. (1978).

dilution. Corrections, however, must be applied whenever the molecular structure of the solute undergoes a change in the solution (ionization, decomposition, or association).

The numerical value of the Henry's law constant $(k_H)_g$ may be obtained graphically by plotting the value of P_2/g_2 (or if one prefers, f_2/g_2) versus g_2, and extrapolating back to $g_2 = 0$. Such a plot is shown in Fig. 3.3 for various alcohols in isooctane. As is seen from the near horizontal slopes, the alcohols do obey Henry's law in this concentration region.

Examination of Eq. (3.74) reveals the constant $(k_H)_g$ cannot be identified with the vapor pressure of the pure solute $P_2^\bullet$ because the law applies only to dilute solutions of component 2 and not to the pure substance. In addition, if component 2 is a gas at the temperature and pressure of the system, it is unrealistic to assign a vapor pressure to it.

It may be recognized from the Gibbs–Duhem equation that if Raoult's law holds for one of the components of a binary solution (solvent), then Henry's law must hold in the same concentration region for the other component (solute). To prove this, suppose component 1 obeys Raoult's law over some range of concentration starting at $X_1 = 1$. In accordance with the Gibbs–Duhem equation, it can be asserted that

$$X_1(\partial \mu_1/\partial X_1) + X_2(\partial \mu_2/\partial X_1) = 0 \qquad (3.76)$$

Remembering that

$$\mu_1 = \mu_1^\circ(T, P) + RT \ln (f_1/f_1^\circ) \qquad (3.77)$$

$$\mu_2 = \mu_2^\circ(T, P) + RT \ln (f_2/f_2^\circ) \qquad (3.78)$$

permits Eq. (3.76) to be expressed in terms of fugacities

$$X_1(\partial \ln f_1/\partial X_1) + X_2(\partial \ln f_2/\partial X_1) = 0 \qquad (3.79)$$

It Raoult's law holds for component 1, then

$$(\partial \ln f_1/\partial X_1) = 1/X_1 \qquad (3.80)$$

and substitution of Eq. (3.80) into Eq. (3.79) gives, upon integration

$$\ln f_2 = \ln X_2 + W(T, P) \qquad (3.81)$$

where W is a function of temperature and pressure. Equation (3.81) expresses Henry's law in exponential form, provided $W(T, P) = \ln(k_H)_x$. Because the composition variables of a solute at infinite dilution are all co-linear, the mole fraction concentration can be readily converted to g_i with the proportionality constant incorporated into the Henry's law constant to give $(k_H)_g$.

The pressure dependence of $(k_H)_g$ is obtained by taking the partial derivative of Eq. (3.75) with respect to pressure at constant temperature:

$$(\partial \ln (k_H)_g/\partial P)_T = \frac{1}{RT}[(\partial(\mu_2^\blacktriangle)_g(\ell, T, P)/\partial P) - (\partial \mu_2^\circ(g, T, 1\,\text{atm})/\partial P)]$$

$$= \bar{V}_2^\blacktriangle(\ell, T, P)/RT \qquad (3.82)$$

The first term in the brackets is the partial molar volume of the dissolved gas in an infinitely dilute solution at the T and P of the system. The value of the second term in brackets is zero. Because of the small compressibility of liquids, the volume dependence on pressure is small, at least at low and moderate pressures. Thus, the variation of $(k_H)_g$ with pressure may be ignored.

The partial derivative of Eq. (3.75) with respect to T at constant pressure yields the temperature dependence of the Henry's law constant

$$(\partial \ln(k_H)_g/\partial T)_P = [\bar{H}_2^\circ(g, T, 1\,\text{atm}) - \bar{H}_2^\blacktriangle(\ell, T, P)]/RT^2$$

$$= -\bar{L}_2/RT^2 \qquad (3.83)$$

where $\bar{L}_2 = [\bar{H}_2^\blacktriangle(\ell, T, P) - \bar{H}_2^\circ(g, T, 1\,\text{atm})]$ is the latent differential heat of solution; that is, the change in enthalpy accompanying the transfer of 1 mole of gas into a nearly saturated solution. For exothermic reactions, the right-hand side of Eq. (3.83) is positive and thus, $(k_H)_g$ increases with increasing temperature; the solubility is reduced. In the case of an endothermic reaction, the solubility increases with increasing temperature. Although en-

F. Thermodynamic Excess Functions

dothermic solubility is not common for gases, several systems have been reported.

F. Thermodynamic Excess Functions

In a nonideal solution the chemical potential is related to the activity rather than to the mole fraction

$$\mu_i = \mu_i^{\bullet}(T, P) + RT \ln a_i$$

It can easily be proven from this equation that either ΔV^{mix} nor ΔH^{mix} is equal to zero in a nonideal solution. For the Gibbs free energy of mixing of a nonideal solution, it is written

$$\Delta G^{mix} = \sum_{i=1}^{N} n_i[\mu_i - \mu_i^{\bullet}(T, P)] = \Delta H^{mix} - T\Delta S^{mix}$$

$$= \sum_{i=1}^{N} n_i RT \ln a_i \qquad (3.84)$$

According to Eq. (3.24), $a_i = X_i \gamma_i$, and Eq. (3.84) assumes the form

$$\Delta G^{mix} = \sum_{i=1}^{N} n_i RT \ln X_i + \sum_{i=1}^{N} n_i RT \ln \gamma_i \qquad (3.85)$$

Comparison with Eq. (3.49) shows that the first term on the right-hand side of Eq. (3.85) is identical with the Gibbs free energy of mixing of an ideal solution (γ_i's = 1). The second term is responsible for the deviations from ideality and is called the excess Gibbs free energy, denoted by ΔG^{ex}. It is mathematically expressed as

$$\Delta G^{ex} = \Delta G^{mix}_{ac} - \Delta G^{mix}_{id} \qquad (3.86)$$

where ac and id are, respectively, the actual and ideal solution at T, P, and X_i. When $\Delta G^{ex} > 0$ or $\gamma_i > 1$, the system is said to exhibit positive deviations from ideal behavior. Conversely, when $\Delta G^{ex} < 0$ or $\gamma_i < 1$, the system exhibits negative deviations from ideal behavior. Finally, $\Delta G^{ex} = 0$ and $\gamma_i = 1$ defines an ideal solution.

Mathematical relationships between the various excess functions are identical to those between the total functions. That is,

$$\Delta H^{ex} = \Delta G^{ex} - T\Delta S^{ex} \qquad (3.87)$$

$$\Delta H^{ex} = \Delta U^{ex} + P\Delta V^{ex} \qquad (3.88)$$

$$\Delta A^{ex} = \Delta U^{ex} - T\Delta S^{ex} \qquad (3.89)$$

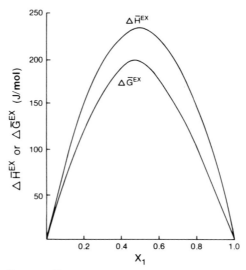

Fig. 3.4 Plot of $\Delta \bar{H}^{ex}$ and $\Delta \bar{G}^{ex}$ as a function of liquid phase composition for cyclopentane (1) + tetrachloroethylene (2) mixtures at 25°C. The experimental data were taken from a paper by Polak *et al.* (1970).

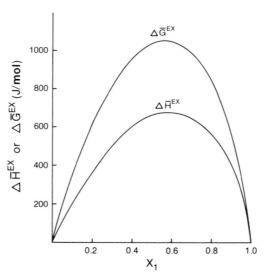

Fig. 3.5 Plot of $\Delta \bar{H}^{ex}$ and $\Delta \bar{G}^{ex}$ as a function of liquid phase composition for cyclohexane (1) + cyclopentanol (2) mixtures at 25°C. Experimental values were determined by Benson *et al.* (1974).

F. Thermodynamic Excess Functions

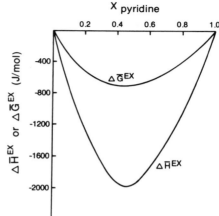

Fig. 3.6 Thermodynamic excess properties for binary liquid mixtures containing pyridine and chloroform. The $\Delta \bar{H}^{ex}$ data are at 25°C and were taken from a paper by Becker et al. (1970). The $\Delta \bar{G}^{ex}$ data are at 30°C and were taken from the work of Sharma and Singh (1975).

Partial molar excess functions are defined in a manner completely analogous to that used for partial molar properties

$$\Delta \bar{Z}_j^{ex} = (\partial \Delta Z^{ex}/\partial n_j)_{T,P,\,n_i \neq n_j} \tag{3.90}$$

Remember the subscript $n_i \neq n_j$ designates that the number of moles of all components other than j are held constant. From Euler's theorem, it follows

$$\Delta Z^{ex} = \sum_{i=1}^{N} n_i \Delta \bar{Z}_i^{ex} \tag{3.91}$$

$$\Delta \bar{Z}^{ex} = \sum_{i=1}^{N} X_i \Delta \bar{Z}_i^{ex} \tag{3.92}$$

Figures 3.4–3.7 show the results of some typical measurements of excess functions for mixtures of simple, complex, and polar molecules. Chapters 5–8 will take up the subject of the theoretical interpretation of such curves, in addition to the various predictive methods used for estimating the thermodynamic excess properties of mixtures from the properties of the pure components. For now, though, only the qualitative aspects will be examined. Deviations from ideality ($\Delta \bar{G}^{ex} \neq 0$, $\gamma_i \neq 1$) arise from a lack of balance in the intermolecular forces between the different species. If the molecules are similar in chemical nature (i.e., benzene + toluene), the mixture is nearly ideal. If they differ, the forces are usually greater between molecules of the same species. Thus, in cyclohexane (1) + cyclohexanol (2) mixtures, see Fig. 3.5, the mean 1–1 and 2–2 forces are, on the average, stronger than the 1–2 forces. This leads to positive values of the excess functions. Each molecule has a lower potential whenever it is preferentially surrounded by its own kind. If there is a specific complex between molecules of different species, as

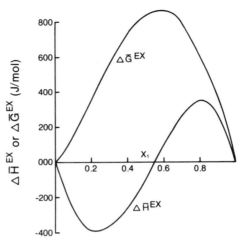

Fig. 3.7 Thermodynamic excess properties of chloroform (1) + methanol (2) mixtures at 303.15 K. Experimental values by Singh *et al.* (1979).

the hydrogen-bonded adduct of pyridine-chloroform, then $\Delta \bar{G}^{\text{ex}}$ and $\Delta \bar{H}^{\text{ex}}$ are negative. These crude arguments, as will be seen, cannot explain all of the details of the different curves.

G. Vapor–Liquid Equilibrium in Nonideal Solution

Consider a two component nonideal solution in equilibrium with an ideal vapor mixture of nonideal gases. If the same procedure that was employed for the ideal solution is followed, Eq. (3.93) may be derived to constant temperature[†]

$$P = X_1 \gamma_1 P_1^{\bullet} \exp[(\bar{V}_1^{\text{liq}} - B_{11})(P - P_1^{\bullet})/RT]$$
$$+ X_2 \gamma_2 P_2^{\bullet} \exp[(\bar{V}_2^{\text{liq}} - B_{22})(P - P_2^{\bullet})/RT] \qquad (3.93)$$

[†] Incorporation of vapor phase nonideality leads to a slightly more complex expression [see Barker (1953) and Scatchard and Raymond (1938)]:

$$P = X_1 \gamma_1 P_1^* + X_2 \gamma_2 P_2^*$$
$$P_1^* = P_1^{\bullet} \exp[\bar{V}_1^{\text{liq}} - B_{11})(P - P_1^{\bullet})/RT - (2B_{12} - B_{11} - B_{22})P y_2^2/RT]$$
$$P_2^* = P_2^{\bullet} \exp[\bar{V}_2^{\text{liq}} - B_{22})(P - P_2^{\bullet})/RT - (2B_{12} - B_{11} - B_{22})P y_1^2/RT]$$

requiring numerical values for three virial coefficients. Vapor phase nonideality will not be considered in the remaining chapters, which will focus instead on liquid phase nonidealities.

G. Vapor–Liquid Equilibrium in Nonideal Solution

On comparing this with Dalton's law, it is found that the partial pressure of component i can be written as

$$P_i = y_i P = X_i \gamma_i P_i^\bullet \exp[(\bar{V}_i^{\text{liq}} - B_{ii})(P - P_i^\bullet)/RT] \tag{3.94}$$

The value of the exponent is, in most cases, close to zero. Thus,

$$y_i = X_i P_i^\bullet \gamma_i / P \tag{3.95}$$

Equation (3.95) is the analog of Eq. (3.68), which describes the ideal solution. Formally, the two expressions differ only in that Eq. (3.95) contains a correction factor that is known as the activity coefficient γ_i. Knowledge of the total system pressure P, the component vapor pressure $P_i^\bullet$, and the appropriate activity coefficient γ_i permits the calculation of the vapor phase composition y_i, in equilibrium with the liquid solution of composition X_i.

The theoretical prediction of γ_i by means of classical and statistical thermodynamics is limited to simple systems. In the majority of systems the activity coefficients must be determined experimentally. For solutions of nonelectrolytes, vapor–liquid equilibrium data are often used for the evaluation of the activity coefficients, and the testing of empirical or semi-empirical predictive methods.

Quantitative methods to calculate phase equilibrium conditions can be formulated from the concepts just presented. The phase behavior of binary systems may also be represented by diagrams.

For a nonreacting PVT system containing two chemical species ($C = 2$), the phase rule ($P = C + 2 - \pi$) becomes $F = 4 - \pi$, and the maximum number of independent intensive variables required to specify the thermodynamic state of a stable system is *three*, corresponding to the case of a single equilibrium phase ($\pi = 1$). If P, T, and one of the mole fractions (or weight fraction) are chosen as the three intensive variables, then all equilibrium states of the system can be represented in three-dimensional P–T composition space. Within this space, the states of phase pairs, coexisting at equilibrium ($F = 4 - 2 = 2$), define surfaces; similarly, the states of three phases in equilibrium ($F = 4 - 3 = 1$) are represented as space curves.

Two-dimensional phase diagrams are obtained from intersections of the three-dimensional surfaces and curves with planes of constant pressure or constant temperature. By the former construction, one obtains a diagram having temperature and phase composition as the coordinates; this is called a T–X diagram. The second construction yields a diagram with the coordinates' pressure and composition; and is called a P–X diagram. The features of three diagrams are indicated by Figs. 3.8–3.13.

Figure 3.8 depicts a T–X diagram for the vapor–liquid equilibrium (VLE) of the carbon disulfide + benzene system at 760 torr total pressure, obtained from the intersection of the VLE surface with $P = 760$ torr plane. Figure 3.9

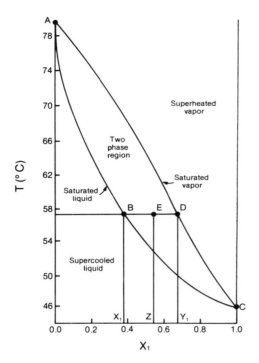

Fig. 3.8 Isobaric liquid–vapor equilibrium data at 760 torr for binary mixtures containing carbon disulfide (1) + benzene (2). Experimental data are from a paper by Golub et al. (1977).

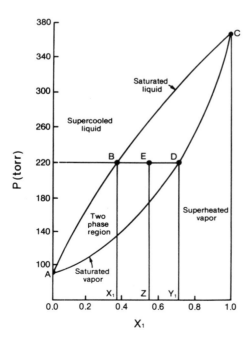

Fig. 3.9 Isothermal liquid–vapor equilibrium data for carbon disulfide (1) + benzene (2) mixtures at 25°C. Experimental results were taken from an article by Sameshima (1918).

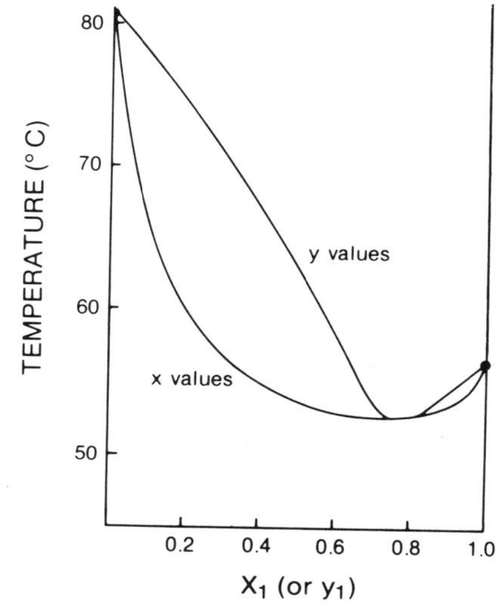

G. Vapor–Liquid Equilibrium in Nonideal Solution

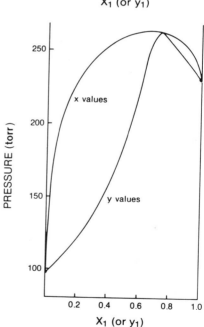

Fig. 3.10 Isobaric liquid–vapor equilibrium data for acetone (1) + cyclohexane (2) mixtures at 760 torr. Experimental data are taken from an article by Rao and Rao (1957).

Fig. 3.11 Isothermal liquid–vapor equilibrium data for acetone (1) + cyclohexane (2) mixtures at 25°C. Experimental values determined by Tasic, et al. (1978).

is a P–X diagram for the same system at 25°C, obtained from the intersection of the VLE surface with the $T = 25$°C plane. The "lens" shape of these figures is typical for many binary systems containing components of similar chemical nature, but having dissimilar vapor pressures.

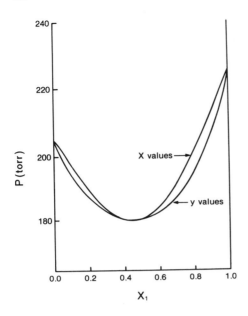

Fig. 3.12 Isothermal liquid–vapor equilibrium data for binary acetone (1) + chloroform (2) mixtures at 25.0°C. Experimental data are from Litvinov (1952).

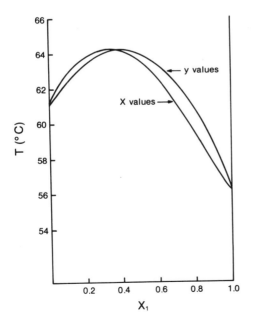

Fig. 3.13 Isobaric liquid–vapor equilibrium data for binary acetone (1) + chloroform (2) mixtures at 760 torr. Experimental values determined by Kudryavtseva and Susarev (1963).

G. Vapor–Liquid Equilibrium in Nonideal Solution

Curves ABC in Figs. 3.8 and 3.9 represent the states of saturated liquid mixtures, called the *bubble-point* curve. Curves ADC, the *dew-point* curve represent states of saturated vapor. The bubble- and dew-point curves converge to the pure component saturation values on the T- and P-axis at the composition extremes of $X_1 = 0$ and $X_1 = 1.0$. Thus, curves ABC and ADC in Fig. 3.8 intersect the T-axis at 46.2 and 80.1°C, the boiling points at 760 torr of carbon disulfide and benzene, respectively. Similarly, curves ABC and ADC in Fig. 3.9 intersect the P-axis at 361.1 and 93.9 torr, the vapor pressures of carbon disulfide and benzene at 25°C, respectively.

The region below ABC in Fig. 3.8 and above ABC in Fig. 3.9 corresponds to states of subcooled liquid; the region above ADC in Fig. 3.8 and below ADC in Fig. 3.9 is for superheated vapor. The area between ABC and ADC in both figures is the vapor–liquid two-phase region. Mixtures whose (T, X) or (P, X) coordinates fall within this area will spontaneously separate into a liquid and vapor phase. The equilibrium compositions of the phases formed in such a separation are determined from the phase diagrams by the intersections with the dew- and bubble-point curves of a horizontal straight line drawn through the point representing the overall state of the mixture. This construction derives from the requirement that T and P of coexisting phases must be the same. For example, a mixture of mole fraction z of carbon disulfide, when brought to a temperature of 57.8°C at 760 torr total pressure (point E in Figs. 3.8 and 3.9), forms a liquid containing mole fraction x_1 (point B) and a vapor of mole fraction y_1 (point D). Straight lines such as BD that connect states in phase equilibrium are called *tie* lines.

Tie lines have a useful stoichiometric property, derived as follows. Let n_t be the total number of moles of a mixture having a mole fraction z_1 of component 1, which separates into n^{liq} moles of liquid with a mole fraction x_1 and n^{vap} moles of vapor of mole fraction y_1. A mass balance on component 1 gives

$$x_1 n^{\text{liq}} + y_1 n^{\text{vap}} = z_1 n_t$$

and an overall mass balance (on both components) yields

$$n^{\text{liq}} + n^{\text{vap}} = n_t$$

Solving these equations for the mole number ratio $n^{\text{liq}}/n^{\text{vap}}$ gives

$$n^{\text{liq}}/n^{\text{vap}} = (y_1 - z_1)/(z_1 - x_1)$$

But from Figs. 3.8 and 3.9 (letting component 1 be carbon disulfide) it is seen that $y_1 - z_1$ and $z_1 - x_1$ are the lengths of segments ED and BE of the tie line BD. Thus, the last equation admits the geometrical interpretation

$$n^{\text{liq}}/n^{\text{vap}} = \text{ED}/\text{BE}$$

This is a statement of the *lever principle*, which asserts that the ratio of the number of moles of the equilibrium phases is inversely proportional to the ratio of the lengths of the corresponding segments of the tie line. The lever principle also holds for mass units if the analogous quantities are substituted for mole numbers and mole fractions in the previous equations.

A common variation on the type of VIE behavior shown in Figs. 3.8 and 3.9 is exemplified by the acetone + cyclohexane system. Figure 3.10 is a T–X diagram for this system at 760 torr, and Fig. 3.11 is a P–X diagram at 25°C. The significant feature of these figures is the occurrence of a state of intermediate composition for which the equilibrium liquid and vapor compositions are identical. Such a state is called an *azeotrope*.

The azeotropic state of a binary system is special in that it possesses only a single degree of freedom ($F = 1$), rather than the two degrees of freedom required for normal two-component, two-phase equilibrium. Thus, specification of any one of the coordinates T, P, or X_1 for a binary azeotrope fixes the other two provided the azeotrope actually exists. Binary azeotropes, therefore, are similar to the saturation states of the pure components.

An important characteristic of the azeotropic state, apparent in Figs. 3.10 and 3.11, is the occurrence of a minimum or maximum on the T–X and P–X diagrams at the azeotropic composition. The maximum or minimum occurs on both the bubble- and dew-point curves, and satisfies the appropriate pair of equations

$$(\partial T/\partial X_1)_{P,az} = (\partial T/\partial y_1)_{P,az} = 0$$

$$(\partial P/\partial X_1)_{T,az} = (\partial P/\partial y_1)_{T,az} = 0$$

Although the acetone + cyclohexane system has a *positive azeotrope* (characterized by a minimum boiling temperature under isobaric conditions and a maximum vapor pressure under isothermal conditions), *negative azeotropes* are also fairly common. Binary mixtures of acetone + chloroform (shown in Figs. 3.12 and 3.13) have a negative azeotrope that is characterized by a maximum boiling temperature under isobaric conditions and a minimum vapor pressure under isothermal conditions.

It can be proven mathematically that the composition of the two phases at the maximum or minimum point must be identical. For the sake of simplicity, assume the vapor may be treated as an ideal mixture. The total pressure is therefore the sum of the two partial pressures

$$P = P_1 + P_2$$

At the maximum or minimum

$$(\partial P/\partial X_1)_T = (\partial P_1/\partial X_1)_T + (\partial P_2/\partial X_1)_T = 0 \qquad (3.96)$$

G. Vapor–Liquid Equilibrium in Nonideal Solution

The Gibbs–Duhem equation for an ideal vapor may be written as

$$0 = X_1\left(\frac{\partial \ln P_1}{\partial X_1}\right)_T + X_2\left(\frac{\partial \ln P_2}{\partial X_1}\right)_T = \frac{X_1}{P_1}\left(\frac{\partial P_1}{\partial X_1}\right)_T + \frac{X_2}{P_2}\left(\frac{\partial P_2}{\partial X_1}\right)_T \quad (3.97)$$

Hence,

$$(\partial P_2/\partial X_1)_T = -(X_1 P_2/P_1 X_2)(\partial P_1/\partial X_1)_T \quad (3.98)$$

and Eq. (3.96) assumes the form

$$(\partial P_1/\partial X_1)_T [1 - X_1 P_2/X_2 P_1] = 0 \quad (3.99)$$

Because $(\partial P_1/\partial X_1)_T$ is never zero, the quantity contained within brackets must be equal to zero at the azeotropic composition. Because $P_1 = y_1 P$ and $P_2 = y_2 P$, we may also write

$$P_1/P_2 = y_1/y_2 \quad (3.100)$$

which, upon substitution into Eq. (3.99) yields

$$1 - (X_1 y_2)/(X_2 y_1) = 0 \quad (3.101)$$

One of the conditions set by this equation is $X_1 = y_1$ and $X_2 = y_2$.

Figures 3.8–3.13 illustrate the vapor–liquid equilibrium diagrams of two-component systems are, in general, uncomplicated. Nevertheless, their experimental determination may be tedious and time consuming. Thermodynamics provides several relationships one may use to reduce to a minimum the number of experimental observations.

The most important relationship in this connection is the Gibbs–Duhem equation

$$X_1(\partial \ln \gamma_1/\partial X_1)_{T,P} = X_2(\partial \ln \gamma_2/\partial X_2)_{T,P} \quad (3.102)$$

As is evident from this equation, if the dependence of the activity coefficient on composition is known for one of the components the dependence for the other component may be evaluated by integration.

The Gibbs–Duhem equation may serve also as a test for the thermodynamic consistency of the experimental vapor–liquid equilibrium data, provided the variation of both activity coefficients with liquid phase compositions are known over a wide range of concentration. Because the activity coefficient is related to the partial molar excess Gibbs free energy by the equation

$$\Delta \bar{G}_i^{ex} = \mu_i^{ex} = RT \ln \gamma_i \quad (3.103)$$

the Gibbs–Duhem equation may be expressed as

$$X_1(\partial \mu_i^{ex}/\partial X_1)_{T,P} = X_2(\partial \mu_2^{ex}/\partial X_2)_{T,P} \quad (3.104)$$

Integrating within the limits $X_1 = 0$, $X_1 = 1.0$; and $X_2 = 0$, $X_2 = 1.0$ gives

$$[X_1\mu_1^{ex}]_{X_1=0}^{X_1=1} - \int_{X_1=0}^{X_1=1} \mu_1^{ex}\,dX_1 = [X_2\mu_2^{ex}]_{X_2=0}^{X_2=1} - \int_{X_2=0}^{X_2=1} \mu_2^{ex}\,dX_2 \quad (3.105)$$

Because μ_1^{ex} is finite at $X_1 = 0$ and zero at $X_1 = 1.0$, and μ_2^{ex} is finite at $X_2 = 0$ and zero at $X_2 = 1.0$, Eq. (3.105) reduces to the form

$$\int_{X_1=0}^{X_1=1} (\mu_1^{ex} - \mu_2^{ex})\,dX_1 = 0 \quad (3.106)$$

Consequently,

$$\int_{X_1=0}^{X_1=1} \ln(\gamma_1/\gamma_2)\,dX_1 = 0 \quad (3.107)$$

which implies that, for thermodynamic consistency, the value of the integral in the given limits is equal to zero at constant T and P.[†]

Equation (3.107) provides an area test of phase-equilibrium data. A plot of $\ln(\gamma_1/\gamma_2)$ versus X_1 is prepared, an example is shown in Fig. 3.14 for the benzene + acetonitrile system. Because the integral on the left-hand side of Eq. (3.107) is given by the area under the curve shown in the figure, the requirement of thermodynamic consistency is met if that area is zero, i.e., the area above the x axis is equal to that below the x axis. These areas can be measured easily and accurately with a planimeter and thus, the area test is a particularly simple one to perform.

In performing the area test on real data, the experimentalist should realize the integral in Eq. (3.107) will rarely equal exactly zero. How, then, can one decide whether a particular set of data "passes" or "fails" the area test? There is no absolute answer to this question because all real data will have some error and thus, the answer depends on just how much error the experimentalist is willing to accept. As a practical guide, it is generally agreed that a

---

[†] Vapor–liquid equilibrium data for a binary system cannot be both isobaric and isothermal. It is a simple matter to write down rigorous expressions for thermodynamic consistency under nonisobaric and nonisothermal conditions. Isobaric but nonisothermal data:

$$\int_{X_1=0}^{X_1=1} \ln(\gamma_1/\gamma_2)\,dX_1 = \int_{X_1=0}^{X_1=1} \frac{\Delta \bar{H}_{12}^{ex}}{RT^2}\,dT$$

Isothermal but nonisobaric data:

$$\int_{X_1=0}^{X_1=1} \ln(\gamma_1/\gamma_2)\,dX_1 = -\int_{X_1=0}^{X_1=1} \frac{\Delta \bar{V}_{12}^{ex}}{RT}\,dP$$

In most cases, data required to evaluate the integrals on the right-hand sides are not available and the simplest approximation is to set the integrals equal to zero. In the latter case, this usually represents a good approximation. However, for isobaric nonisothermal data the right-hand integral often cannot be neglected.

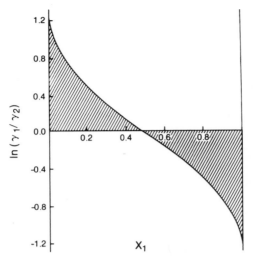

Fig. 3.14 Thermodynamic consistency test for the benzene (1) + acetonitrile (2) system. Experimental data from Monfort (1983).

set of data is thermodynamically consistent if

$$0.02 > \left|\frac{\text{area above } x \text{ axis} - \text{area below } x \text{ axis}}{\text{area above } x \text{ axis} + \text{area below } x \text{ axis}}\right|$$

The above criterion is necessarily arbitrary and the quantity on the left-hand side may be raised or lowered depending on how conservative or liberal the experimentalist wishes to be in interpreting the meaning of thermodynamic consistency.

Problems

3.1 Show that the adiabatic compressibility of an ideal solution is correctly given by the formula

$$(K_S)_{\text{id s}} = \sum_{i=1}^{N} \phi_i K_{T_i}^{\bullet} - \left[T\left(\sum_{i=1}^{N} X_i \bar{V}_i\right)\left(\sum_{i=1}^{N} \phi_i \alpha_i^{\bullet}\right)^2\right] \Big/ \sum_{i=1}^{N} X_i \bar{C}_{p_i}^{\bullet},$$

where the subscript id s signifies ideal solution and

$$K_S = -(1/V)(\partial V/\partial P)_S \quad \text{and} \quad K_T = -(1/V)(\partial V/\partial P)_T.$$

The ϕ is the volume fraction of constituent i calculated using the ideal molar volume approximation. The solid circles indicate that the properties refer to the pure component. Hint: The adiabatic compressibility K_S of any solution is related to the isothermal compressibility K_T by

$$K_S = K_T(C_v/C_p)$$

the ratio of heat capacities at constant volume and constant pressure, which are themselves related through

$$C_p - C_v = \alpha^2 VT/K_T$$

and the coefficient of thermal expansion $[\alpha = (\partial \ln V/\partial T)_P]$.

3.2 It is difficult to find experimental data reported in the literature according to the desired reference state. As a result, it becomes necessary mathematically to convert results from one reference state to another. For $(\mu_2^{\blacktriangle})_m$ and $(\mu_2^{\blacktriangle})_x$, show that this conversion takes the form

$$(\mu_2^{\blacktriangle})_m - (\mu_2^{\blacktriangle})_x = RT \ln(1000/\mathrm{MW}_{sol})$$

where MW_{sol} refers to the molecular weight of the solvent.

3.3 The fugacity of component A above mixtures of A and B obeys the equation for X_A between 0.0 and 0.3:

$$f_A = X_A \exp(5.70 - 3.50 X_A)$$

Calculate:
(a) The Henry's law constant for component A (mole fraction as composition variable).
(b) The activity coefficient of component A (relative to Henry's law) at $X_A = 0.10$.
(c) The activity coefficient of B at $X_A = 0.10$.

3.4 Show that the isochoric heat capacity of an ideal solution is given by

$$\bar{C}_v^{id\,s} = \sum_{i=1}^{N} X_i \bar{C}_{v_i}^{\bullet} + T \sum_{i=1}^{N} X_i \bar{V}_i (\alpha_i^{\bullet})^2/K_{t_i}^{\bullet}$$

$$- T \left[\sum_{i=1}^{N} X_i \bar{V}_i^{\bullet}\right] \left[\sum_{i=1}^{N} \phi_i \alpha_i^{\bullet}\right]^2 \bigg/ \left[\sum_{i=1}^{N} \phi_i K_{t_i}^{\bullet}\right]$$

For helpful hints see problem 3.1.

3.5 A solution containing four moles of A and eight moles of B is mixed with a solution containing eight moles of A and 14 moles of B in a reversible manner at 298.15 K. What is the change in the Gibbs free energy accompanying this mixing process? Assume all solutions to be ideal.

3.6 Morisue et al. (1972) proposed an expression for the mathematical representation of binary excess Gibbs free energy

$$\Delta \bar{G}_{12}^{ex} = RTX_1 X_2 [w_{12}/(X_2 + X_1 A_{12}) + w_{21}/(X_1 + X_2 A_{21})]$$

where

$$A_{12} = \exp(-W_{12}) \quad \text{and} \quad A_{21} = \exp(-W_{21})$$

and W_{ij} are binary parameters (i.e., they are numerical constants) independent of temperature. Using this form for $\Delta \bar{G}_{12}^{ex}$, derive expressions for the binary activity coefficients.

3.7 From the vapor–liquid equilibrium data in Table I, compute the Raoult's law activity coefficients of benzene and cyclohexane at the 11 binary compositions. Does the solution exhibit positive or negative deviations from ideality?

TABLE I
Vapor–Liquid Equilibrium Data for Benzene (1) + Cyclohexane Mixtures at 25°C.[a]

X_1	Y_1	P(mm Hg)
0.0000	0.0000	97.45
0.1035	0.1375	102.05
0.1750	0.2170	104.50
0.2760	0.3130	106.75
0.3770	0.4015	108.10
0.4330	0.4460	108.45
0.5090	0.5050	108.65
0.5830	0.5620	108.30
0.6940	0.6505	106.90
0.7945	0.7410	104.50
0.9005	0.8565	100.60
0.9500	0.9220	98.15
1.0000	1.0000	95.05

[a] Experimental values reproduced with permission from Tasic et al. (1978).

3.8 A naive approximation for the molar density of a mixture is provided by the expression

$$\rho^{mix} = \sum_{i=1}^{N} X_i \rho_i^{\bullet}$$

where $\rho_i^{\bullet}$ is the molar density of pure species i at the mixture temperature and pressure. Is this expression consistent with the ideal solution model as defined by Raoult's law? If not, derive an expression in density that is.

3.9 From the experimental data in Table II, compute the mol fraction compositions of acetone and acetonitrile in the vapor phase. Assume liquid phase ideality.

TABLE II
Isothermal Liquid–Vapor Equilibrium for Acetone (1) + Acetonitrile (2) Mixtures at 50°C.[a]

X_1	P(mm Hg)
0.0000	253.89
0.0824	283.94
0.1600	311.02
0.2531	346.15
0.3451	379.82
0.4314	411.59
0.4754	425.83
0.5077	438.83
0.5517	453.07
0.6350	482.21
0.7386	519.07
0.8138	546.56
0.8996	578.20
0.9581	599.34
1.0000	615.16

[a] Experimental values reproduced with permission from DiElsi et al. (1978). Copyright 1978 American Chemical Society.

3.10 Assume the following equation is valid:

$$\Delta \bar{G}_{12}^{ex} = -RT[X_1 \ln(X_1 + AX_2) + X_2 \ln(X_2 + BX_1)]$$

where A and B are numerical values independent of liquid phase composition. (As will later be seen, this is the general mathematical form of the Wilson equation.) Because this equation has the form of

$$\Delta \bar{Z}_{12}^{ex} = X_1 \Delta \bar{Z}_1^{ex} + X_2 \Delta \bar{Z}_2^{ex}$$

it is tempting to state

$$\Delta \bar{G}_1^{ex} = -RT \ln(X_1 + AX_2) \quad \text{and} \quad \Delta \bar{G}_2^{ex} = -RT \ln(X_2 + BX_1)$$

Prove whether or not this is true.

TABLE III
Experimental Data for the 2-Butanone (1) + n-Heptane (2) System at 45°C.[a]

X_1	Y_1	P(mm Hg)
0.0000	0.0000	115.09
0.0738	0.3130	157.40
0.1416	0.4348	182.05
0.2764	0.5542	210.26
0.3617	0.5971	220.23
0.4962	0.6513	230.95
0.5791	0.6807	234.96
0.7008	0.7289	238.93
0.8027	0.7819	238.98
0.9048	0.8608	233.71
0.9528	0.9185	227.86
1.0000	1.0000	218.26

[a] Experimental data is reproduced with permission from Takeo et al. (1979).

3.11 From the vapor–liquid equilibrium data in Table III, compute the Raoult's law activity coefficients of 2-butanone and n-heptane at the 10 binary compositions. Does the solution exhibit positive or negative deviations from ideality?

4

Empirical Expressions for Estimating Multicomponent Properties from Binary Data

For years the chemical industry has recognized the importance of the thermodynamic and physical properties of solution in design calculations involving chemical separations, fluid flow, and heat transfer. The development of flow calorimeters, continuous dilution dilatometers and vibrating-tube densimeters has enabled the experimental determination of excess enthalpies, heat capacities, and volumes of liquid mixtures with convenience and accuracy. Utilizing continuous dilution techniques has reduced the experimental time needed to determine Gibbs free energy through vapor pressure studies. But even with modern instrumentation, experimental measurements of thermodynamic properties has become progressively more expensive and time consuming with each additional component beyond binary mixtures. In the chemical literature, properties for binary systems are relatively abundant, properties for ternary systems are scarce and properties for higher order multicomponent systems are virtually nonexistent. Naturally, one of the primary goals of research in the area of solution thermodynamics has been the development of expressions for predicting the thermodynamic properties of multicomponent mixtures.

One of the earliest attempts to mathematically describe the changes in enthalpy of a mixture of two was that of van Laar (1906, 1910). Basing his development on the van der Waal's equation of state, van Laar expressed the enthalpy of a binary mixture as

$$\Delta \bar{U}^{ex}_{12} = \Delta \bar{H}^{ex}_{12} = [X_1 b_1 X_2 b_2/(X_1 b_1 + X_2 b_2)][(\sqrt{a_1}/b_1) - (\sqrt{a_2}/b_2)]^2 \quad (4.1)$$

The basic assumptions van Laar employed in his derivation were that:

(1) Both pure components and the binary mixture at all compositions obey the van der Waal's equation of state.

(2) The combinatorial relationships for the parameters a and b in fluid mixtures are those previously proposed by van der Waals (1890):

$$a_{mix} = X_1^2 a_1 + 2X_1 X_2 a_{12} + X_2^2 a_2$$

4 Empirical Expressions

and
$$b_{mix} = X_1 b_1 + X_2 b_2$$

(3) The excess volume of mixing is zero.
(4) The excess entropy of mixing is zero.
(5) The binary interaction parameter a_{12} is given by (Barthelot, 1898) as

$$a_{12} = (a_1 a_2)^{1/2}$$

The only parameters necessary to estimate mixing properties of a binary system are the van der Waal's parameters a and b, both of which are easily determined. Careful inspection of van Laar's approach reveals that only positive heats of mixing and only positive deviations from Raoult's law are permissible. In a later book, van Laar and Lorentz (1925) improved Eq. (4.1) by inserting the molar volumes of the pure components for the van der Waals bs.

Most equations of state have analogous "mixing rules" that enable estimation of multicomponent properties from pure component properties. Although these approaches provide reasonable estimates for several systems of practical importance, they do not address the problem of how to report experimental data. Vapor pressures, densities, and other solution properties are worthless unless they are transmitted from the experimentalist to the design engineer. Mathematical representation provides a convenient method to reduce extensive tables of experimental data into a few equations.

This chapter will be devoted to several of the empirical expressions that have been suggested for parametrizing and predicting mixture data. The more theoretical approaches, such as the Wilson equation, will be considered later.

Scatchard et al. (1925) expressed the excess enthalpy and free energy of binary systems containing benzene, cyclohexane, and carbon tetrachloride as

$$\Delta \bar{Z}^{ex}_{ij}/\bar{V} = \phi_i \phi_j (B_{ij} + C_{ij}\phi_j + D_{ij}\phi_j^2) \tag{4.2}$$

where $\bar{V}$ is the molar volume of the solution and Z the integral thermodynamic excess property referring to either enthalpy or Gibbs free energy. They proposed a predictive equation for the properties of a ternary mixture of nearly symmetrical binary solutions:

$$\Delta \bar{Z}^{ex}_{123}/\bar{V} = \sum_{i=1}^{2} \sum_{j>i}^{3} \phi_i \phi_j [B_{ij} + C_{ij}(1 - \phi_i - \phi_j)/2 \\ + D_{ij}(1 - \phi_i - \phi_j)^2/4] \tag{4.3}$$

using the coefficients determined from the binary systems.

Wohl (1946) published papers in which he considered the existing equations for representing the excess free energy of binary and ternary systems. Wohl developed expressions for the excess free energy for a multicomponent system involving weighted mole fractions (called q-fractions) and interaction terms up to a four-suffix (a_{hijk}) term. His development led directly to a combinatorial form for ternary systems that used ratios of weighting factors and interaction parameters determined from the contributing binary systems. Recognizing the advantages in requiring a normalized set of weighting factors for use in calculations, Wohl presented a rather complex definition of q-fractions.

Redlich and Kister (1948) proposed an expression for the excess free energy of a ternary mixture:

$$\Delta \bar{G}^{ex}_{123} = X_1 X_2 \sum_{v=0}^{r} (G_v)_{12}(X_1 - X_2)^v + X_1 X_3 \sum_{v=0}^{r} (G_v)_{13}(X_1 - X_3)^v$$
$$+ X_2 X_3 \sum_{v=0}^{r} (G_v)_{23}(X_2 - X_3)^v \quad (4.4)$$

with provisions for additional ternary parameters. The general form of Eq. (4.4) was also recommended for representing excess enthalpies and volume but with a different set of binary parameters, $(H_v)_{ij}$ and $(V_v)_{ij}$. The binary reduction of Eq. (4.4)

$$\Delta \bar{Z}^{ex}_{12} = X_1 X_2 \sum_{v=0}^{r} (Z_v)_{12}(X_1 - X_2)^v \quad (4.5)$$

is often used to mathematically represent thermodynamic excess properties and physical properties of binary mixtures. The initial popularity of the Redlich–Kister equation arose because the first parameter $(Z_0)_{12}$ could be determined conveniently from the experimental data at $X_1 = 0.50$, as 4 $\Delta \bar{Z}^{ex}_{12}$. If one looks at the binary form suggested by Scatchard, Wood, and Mochel in Eq. (4.3), it is seen the first parameter B_{12} is also easy to determine. Remember that computers were not available during the 1940s and the majority of experimental data were graphically presented in the literature. The Redlich–Kister, and Scatchard equations provided a means to transmit data from the experimentalist to the chemical engineer designing distillation columns.

For mixtures of one polar component (component 1) and two nonpolar components, Scatchard et al. (1932) suggested a modification of the Redlich–Kister equation

$$\Delta \bar{G}^{ex}_{123} = X_1 X_2 \sum_{v=0}^{r} (G_v)_{12}(2X_1 - 1)^v + X_1 X_3 \sum_{v=0}^{r} (G_v)_{13}(2X_1 - 1)^v$$
$$+ X_2 X_3 \sum_{v=0}^{r} (G_v)_{23}(X_2 - X_3)^v \quad (4.6)$$

Inspection of Eq. (4.6) reveals the 2–3 binary solution is treated differently than the two solutions containing the polar component.

Tsao and Smith (1953) proposed an equation for predicting the excess enthalpy of a ternary system

$$\Delta \bar{H}^{ex}_{123} = X_2(1 - X_1)^{-1} \Delta \bar{H}^{ex}_{12} + X_3(1 - X_1)^{-1} \Delta \bar{H}^{ex}_{13} + (1 - X_1)\Delta \bar{H}^{ex}_{23} \quad (4.7)$$

in which the $\Delta \bar{H}^{ex}_{ij}$ refer to the excess enthalpies for the binary mixtures at compositions (X_i°, X_j°), such that $X_i^\circ = X_1$ for the 1,2- and 1,3-binary systems, and $X_2^\circ = X_2/(X_2 + X_3)$ for the 2,3-binary system.

Equation (4.7) possesses a desirable mathematical form for the prediction of the ternary solutions' properties from those of the individual binary systems, in that the combinatorial equation is independent of binary parameterization. The three predictive expressions considered previously [Eqs. (4.3), (4.4), and (4.6)] require a specific form for data parameterization, as the binary coefficients appear in the predictive multicomponent (ternary) expression. The unsymmetrical nature of Eq. (4.7) is a highly undesirable feature and modifications suggested by Knobeloch and Schwartz (1962) are extremely complex, particularly for estimating partial molar quantities.

In studying the enthalpies of mixing for ternary hydrocarbon systems, Mathieson and Thynne (1956) employed an expression that takes the form

$$\Delta \bar{H}^{ex}_{123} = X_1 X_2 \sum_{v=0}^{r} (H_v)_{12}(X_1 - X_2 - X_3/2)^v$$

$$+ X_1 X_3 \sum_{v=0}^{r} (H_v)_{13}(X_1 - X_2 - X_3/2)^v$$

$$+ X_2 X_3 \sum_{v=0}^{r} (H_v)_{23}(X_2 - X_3)^v \quad (4.8)$$

Kohler (1960) proposed an equation for the excess Gibbs free energy of mixing of a ternary solution

$$\Delta \bar{G}^{ex}_{123} = (X_1 + X_2)\Delta \bar{G}^{ex}_{12} + (X_1 + X_3)^2 \Delta \bar{G}^{ex}_{13} + (X_2 + X_3)^2 \Delta \bar{G}^{ex}_{23} \quad (4.9)$$

in which $\Delta \bar{G}^{ex}_{ij}$ refers to the excess free energies of the binary mixtures at a composition (X_i°, X_j°), such that

$$X_i^\circ = 1 - X_j^\circ = X_i/(X_i + X_j)$$

Unlike Eq. (4.7), Kohler's equation is symmetrical in that all three binary systems are treated identically. Its numerical predictions do not depend on the arbitrary designation of component numbering.

Toop (1965) developed an equation for the excess Gibbs free energy of mixing similar to that suggested previously by Tsao and Smith for excess enthalpies

$$\Delta \bar{G}^{ex}_{123} = X_2(1 - X_1)^{-1}\Delta \bar{G}^{ex}_{12} + X_3(1 - X_1)^{-1}\Delta \bar{G}^{ex}_{13} + (1 - X_1)^2 \Delta \bar{G}^{ex}_{23} \quad (4.10)$$

The $\Delta \bar{G}^{ex}_{ij}$ refers to the excess free energy of the binary mixture calculated in a method analogous to the $\Delta \bar{H}^{ex}_{ij}$ of Tsao and Smith.

Colinet (1967) established a slightly more complex relationship for expressing the thermodynamic excess properties of multicomponent systems,

$$\Delta \bar{G}^{ex}_{12\cdots n} = 0.5 \sum_{i=1}^{n-1} \sum_{j>i}^{n} X_i(1 - X_j)^{-1}[\Delta \bar{G}^{ex}_{ij}]_{X_j} \quad (4.11)$$

in which $[\Delta \bar{G}^{ex}_{ij}]_{X_j}$'s are calculated from the binary data at constant mole fraction X_j. This equation, although perfectly symmetrical, requires the addition of the thermodynamic properties at six different binary compositions for a ternary solution.

In their studies of the enthalpies of mixing of polyoxyethylene glycol in dioxane–water and ethanol–water mixtures, Lakhanpal et al., (1975, 1976) expressed the enthalpy of a substance (component 3) with a binary solvent (mixture of components 1 and 2) in terms of the enthalpies of mixing of the contributing binary mixtures,

$$\Delta \bar{H}^{ex}_{3-12} = \phi_3 \left\{ \frac{(X_1 + X_3)}{\phi_{31}} \Delta \bar{H}^{ex}_{13} + \frac{(X_2 + X_3)}{\phi_{32}} \Delta \bar{H}^{ex}_{23} \right.$$
$$\left. - (X_1 + X_2)\Delta \bar{H}^{ex}_{23} \right\} \quad (4.12)$$

in which the $\Delta \bar{H}^{ex}_{ij}$ are enthalpies of the binary mixtures at mole fraction compositions $X_i/(X_i + X_j)$ and $X_j/(X_i + X_j)$; and ϕ_{31} and ϕ_{32} are the volume fractions of component 3 in the 3–1 and 3–2 binary mixtures respectively. Mathematical manipulations of Eq. (4.12) yields an expression for the total enthalpy of mixing of three *pure* components

$$\Delta \bar{H}^{ex}_{123} = (X_1 + X_2)(\phi_1 + \phi_2)\Delta \bar{H}^{ex}_{12} + (X_1 + X_3)(\phi_1 + \phi_3)\Delta \bar{H}^{ex}_{13}$$
$$+ (X_2 + X_3)(\phi_2 + \phi_3)\Delta \bar{H}^{ex}_{23} \quad (4.13)$$

which is applicable for the entire composition range.

Jacob and Fitzner (1977) suggested an equation for estimating the properties of a ternary solution based on the binary data at compositions nearest the ternary composition, taking the form

$$\Delta \bar{G}^{ex}_{123} = \frac{X_1 X_2 \Delta \bar{G}^{ex}_{12}}{(X_1 + X_3/2)(X_2 + X_3/2)} + \frac{X_1 X_3 \Delta \bar{G}^{ex}_{13}}{(X_1 + X_2/2)(X_3 + X_2/2)}$$
$$+ \frac{X_2 X_3 \Delta \bar{G}^{ex}_{23}}{(X_2 + X_1/2)(X_3 + X_1/2)} \quad (4.14)$$

4 Empirical Expressions

The $\Delta \bar{G}_{ij}^{ex}$ is the excess Gibbs free energy of the binary mixture at compositions (X_i°, X_j°), such that $X_i - X_j = X_i^\circ - X_j^\circ$. Essentially, Meschel and Kleppa (1968) employed this equation in predicting the enthalpies of mixing in ternary fused-nitrate salts in their consideration of the compositional dependence upon the binary interaction parameters λ_{ij}.

Rastogi et al. (1977a,b) suggested a form for predicting the excess volumes of a ternary solution:

$$\Delta \bar{V}_{123}^{ex} = \tfrac{1}{2}[(X_1 + X_2)\Delta \bar{V}_{12}^{ex} + (X_1 + X_3)\Delta \bar{V}_{13}^{ex} + (X_2 + X_3)\Delta \bar{V}_{23}^{ex}] \quad (4.15)$$

in which $\Delta \bar{V}_{ij}^{ex}$ represents the excess molar volume of the binary mixture at compositions (X_i°, X_j°), such that

$$X_i^\circ = 1 - X_j^\circ = X_i/(X_i + X_j)$$

Failure to reduce to a correct description of the binary systems as the composition of the third component approaches zero makes this expression inappropriate for describing the system near infinite dilution. For example, when $X_1 = 0.0$, Eq. (4.15) gives only one-half of the 2–3 binary value.

This proliferation of similar expressions becomes confusing, especially when the equations are encountered for the first time. Figure 4.1 illustrates the characteristics of the five predictive expressions that employ numerical

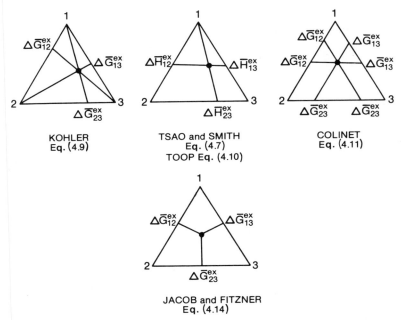

Fig. 4.1 Graphical representation of $\Delta \bar{G}_{ij}^{ex}$ (or $\Delta \bar{H}_{ij}^{ex}$) for the five predictive methods.

data ($\Delta \bar{Z}_{ij}^{ex}$) calculated at specific binary compositions. There is no preferred way of knowing which method will provide the best predictions for a given system. The fact that so many empirical equations have been developed suggests that no single equation can describe all types of systems encountered.

The 14 expressions mentioned in this chapter can be used to predict any intensive thermodynamic excess property, such as excess free energies, excess volumes, and excess enthalpies. From a historical perspective, it is perferable to present each approach in terms of its original applications. This does not imply that the expression can only be used for this single thermodynamic property.

To illustrate several of the predictive methods let us assume one wishes to use the Redlich–Kister, Kohler, Tsao–Smith, and Jacob–Fitzner equations to predict the excess volumes of methyl ethyl ketone (1) + 1-butanol (2) + n-heptane (3) mixtures at $X_1 = 0.50$, $X_2 = 0.25$ and 25°C.

The binary data Naidu and Naidu (1981) are parametrized as

$\Delta \bar{V}_{13}^{ex}(\text{cm}^3/\text{mol}) = X_1 X_3 [2.1829 - 0.22109(X_1 - X_3) + 0.20465(X_1 - X_3)^2]$

$\Delta \bar{V}_{12}^{ex}(\text{cm}^3/\text{mol}) = X_1 X_2 [0.05098 + 0.0283(X_1 - X_2) + 0.0156(X_1 - X_2)^2]$

$\Delta \bar{V}_{23}^{ex}(\text{cm}^3/\text{mol}) = X_2 X_3 [0.87729 - 0.43338(X_2 - X_3) + 0.14511(X_2 - X_3)^2]$

REDLICH–KISTER EQUATION

$\Delta \bar{V}_{123}^{ex} = (0.50)(0.25)[2.1829 - 0.22109(0.50 - 0.25) + 0.20465(0.50 - 0.25)^2]$
$\qquad + (0.50)(0.25)[0.0598 + 0.0283(0.50 - 0.25) + 0.0156(0.50 - 0.25)^2]$
$\qquad + (0.25)(0.25)[0.87729]$
$\qquad = 0.331 \text{ cm}^3/\text{mol}$

KOHLER EQUATION

The excess volumes of the binary systems at X_i°, X_j° are first calculated:

$\Delta \bar{V}_{13}^{ex} = (0.67)(0.33)[2.1829 - 0.22109(0.67 - 0.33) + 0.20465(0.67 - 0.33)^2]$
$\qquad = 0.471 \text{ cm}^3/\text{mol}$

$\Delta \bar{V}_{12}^{ex} = (0.67)(0.33)[0.0598 + 0.0283(0.67 - 0.33) + 0.0156(0.67 - 0.33)^2]$
$\qquad = 0.016 \text{ cm}^3/\text{mol}$

$\Delta \bar{V}_{23}^{ex} = (0.50)(0.50)[0.87729]$
$\qquad = 0.219 \text{ cm}^3/\text{mol}$

and these quantities are then added in accordance to Eq. (4.9)

$\Delta \bar{V}_{123}^{ex} = (0.50 + 0.25)^2(0.016) + (0.50 + 0.25)^2(0.471) + (0.25 + 0.25)^2(0.219)$
$\qquad = 0.329 \text{ cm}^3/\text{mol}$

4 Empirical Expressions

TSAO–SMITH EQUATION

Because the Tsao–Smith equation is unsymmetrical, there are three different predicted values. For the purposes here, methyl ethyl ketone is component 1:

$$\Delta \bar{V}^{ex}_{13} = (0.50)(0.50)[2.1829] = 0.546 \text{ cm}^3/\text{mol}$$

$$\Delta \bar{V}^{ex}_{12} = (0.50)(0.50)[0.0598] = 0.015 \text{ cm}^3/\text{mol}$$

$$\Delta \bar{V}^{ex}_{23} = (0.50)(0.50)[0.87729] = 0.219 \text{ cm}^3/\text{mol}$$

The binary $\Delta \bar{V}^{ex}_{ij}$ is then combined in accordance with Eq. (4.7)

$$\bar{V}^{ex}_{123} = (0.25)(0.015)/(0.50) + (0.25)(0.546)/(0.50) + (0.50)(0.219)$$
$$= 0.390 \text{ cm}^3/\text{m}$$

JACOB–FITZNER EQUATION

$$\Delta \bar{V}^{ex}_{13} = (0.625)(0.375)[2.1829 - 0.22109(0.625 - 0.375) + 0.20465(0.625 - 0.375)^2]$$
$$= 0.502 \text{ cm}^3/\text{mol}$$

$$\Delta \bar{V}^{ex}_{12} = (0.625)(0.375)[0.0598 + 0.0283(0.625 - 0.375) + 0.0156(0.625 - 0.375)^2]$$
$$= 0.016 \text{ cm}^3/\text{mol}$$

$$\Delta \bar{V}^{ex}_{23} = (0.50)(0.50)[0.87729]$$
$$= 0.219 \text{ cm}^3/\text{mol}$$

The binary $\Delta \bar{V}^{ex}_{ij}$ is then combined in accordance with Eq. (4.14):

$$\Delta \bar{V}^{ex}_{123} = \frac{(0.50)(0.25)(0.502)}{(0.625)(0.375)} + \frac{(0.50)(0.25)(0.016)}{(0.625)(0.375)}$$
$$+ \frac{(0.25)(0.25)(0.219)}{(0.50)(0.50)} = 0.331 \text{ cm}^3/\text{mol}$$

Mathematically, it can be shown that the predictions of Eq. (4.4) must be equal to the predictions of Eq. (4.14) whenever the three binary systems are parametrized with the Redlich–Kister equation.

Up to this point the primary emphasis has been on predicting multicomponent properties from binary data; several of the empirical expressions proposed during the past 40 years were summarized. These expressions also served as the point-of-departure for the mathematical representation of multicomponent excess properties. Differences between the predicted values and experimentally determined values are expressed as

$$(\Delta \bar{Z}_{123}^{ex})^{obs} - (\Delta \bar{Z}_{123}^{ex})^{pre} = X_1 X_2 X_3 Q_{123} \tag{4.16}$$

with Q-functions of varying complexity. The abbreviations obs and pre indicate observed and predicted, respectively. For most systems commonly encountered, the experimental data can be adequately represented by the power series expansion

$$Q_{123} = A_{123} + \sum_{i=1}^{r} B_{12}^i (X_1 - X_2)^i + \sum_{j=1}^{r} B_{13}^j (X_1 - X_3)^j$$

$$+ \sum_{k=1}^{r} B_{23}^k (X_2 - X_3)^k \tag{4.17}$$

though it is unlikely data for multicomponent systems will be obtained with sufficient precision to warrant more than a few parameters.

Problems[†]

4.1 Naidu and Naidu reported the excess volumes of methyl ethyl ketone + n-heptane, methyl ethyl ketone + 1-propanol, and 1-propanol + n-heptane mixtures in terms of Redlich–Kister coefficients.

Methyl ethyl ketone (1) + n-heptane (3)
$\Delta \bar{V}_{13}^{ex}(cm^3/mol)$
$= X_1 X_3 [2.1829 - 0.22109(X_1 - X_3) + 0.20465(X_1 - X_3)^2]$

Methyl ethyl ketone (1) + 1-propanol (2)
$\Delta \bar{V}_{12}^{ex}(cm^3/mol)$
$= X_1 X_2 [-0.1709 + 0.0122(X_1 - X_2) + 0.0118(X_1 - X_2)^2]$

1-Propanol (2) + n-heptane (3)
$\Delta \bar{V}_{23}^{ex}(cm^3/mol)$
$= X_2 X_3 [1.3472 - 0.43715(X_2 - X_3) - 0.15407(X_2 - X_3)^2]$

Estimate the ternary $\Delta \bar{V}_{123}^{ex}$ values at the mole fraction compositions listed in Table I. The actual experimental values are listed for comparison.

4.2 Shatas et al. (1975) parametrized the excess enthalpies of chloroform + ethanol, chloroform + n-heptane, and ethanol + n-heptane mixtures at

---
[†] The numerical values to the problems will depend to a large extent upon which predictive expression is used. Unless otherwise instructed, estimate the ternary properties with the most convenient method.

TABLE I
Experimental Excess Molar Volumes of Methyl Ethyl
Ketone (1) + 1-Propanol (2) + n-Heptane (3)
Mixtures at 303.15 K.[a]

X_1	X_2	$\Delta \bar{V}^{ex}_{123}$ (cm^3/mol)
0.7538	0.1278	0.294
0.7211	0.1556	0.328
0.6078	0.2912	0.238
0.4794	0.4105	0.228
0.4579	0.4126	0.282
0.3235	0.5486	0.246
0.2183	0.6865	0.158
0.1108	0.7873	0.151

[a] Reproduced with permission from Naidu and Naidu (1981). Copyright 1981 American Chemical Society.

30°C according to the formula

Chloroform (1) + Ethanol (2)

$$\frac{\Delta \bar{H}^{ex}_{12}}{X_1 X_2 RT} = 3.7860 X_1 - 2.4322 X_2 - (15.8145 X_1 - 0.4649 X_2) X_1 X_2$$
$$+ (33.2820 X_1 + 1.2813 X_2) X_1^2 X_2^2$$

Chloroform (1) + n-Heptane (3)

$$\frac{\Delta \bar{H}^{ex}_{13}}{X_1 X_3 RT} = 1.5413 X_1 + 1.1350 X_3 - (0.4862 X_1 + 0.1661 X_3) X_1 X_3$$

Ethanol (2) + n-Heptane (3)

$$\frac{\Delta \bar{H}^{ex}_{23}}{X_2 X_3 RT} = 1.2478 X_2 + 10.6488 X_3$$
$$- \frac{(591.2876)(11.0298) X_2 X_3}{591.2876 X_2 + 11.0298 X_3 + (151.9309 X_2 + 90.0913 X_3) X_2 X_3}$$

Estimate the excess enthalpies for the ternary chloroform + ethanol + n-heptane mixtures at the five mole fraction compositions $X_1 = 0.25$, $X_2 = 0.25$; $X_1 = 0.33$, $X_2 = 0.33$; $X_1 = 0.25$, $X_2 = 0.50$; $X_1 = 0.50$, $X_2 = 0.25$; and $X_1 = 0.33$, $X_2 = 0.50$.

4.3 Mathieson and Thynne reported the excess enthalpies for benzene + cyclohexane, benzene + n-hexane, and cyclohexane + n-hexane mixtures at 20°C.

TABLE II
Experimental Values for the Excess Enthalpies of
Benzene (1) + Cyclohexane (2) + n-Hexane (3)
Mixtures at 20°C.[a]

X_1	X_2	$\Delta \bar{H}^{ex}_{123}$ (J/mol)
0.266	0.398	745
0.383	0.335	874
0.455	0.230	870
0.300	0.578	749
0.578	0.250	883
0.424	0.282	866

[a] Reproduced with permission from Mathieson and Thynne (1956).

Benzene (1) + Cyclohexane (2)
$$\Delta H^{ex}_{12}(\text{J/ml}) = \phi_1\phi_2[33.53 + 3.81(\phi_1 - \phi_2) - 2.0(\phi_1 - \phi_2)^2]$$
Benzene (1) + n-Hexane (3)
$$\Delta H^{ex}_{13}(\text{J/ml}) = \phi_1\phi_3[32.60 + 1.12(\phi_1 - \phi_3) + 1.12(\phi_1 - \phi_3)^2]$$
Cyclohexane (2) + n-Hexane (3)
$$\Delta H^{ex}_{23}(\text{J/ml}) = \phi_2\phi_3[7.36 + 1.34(\phi_2 - \phi_3) - 1.76(\phi_2 - \phi_3)^2]$$

Calculate the $\Delta \bar{H}^{ex}_{123}$ values at the mole fraction compositions listed in Table II. The experimentally determined values are given comparisons.

4.4 Li (1974) mathematically represented the excess free energies of n-hexane (1) + cyclohexane (2) + benzene (3) mixtures at 343.15 K by a Redlich–Kister expression

$$\frac{\Delta \bar{G}^{ex}_{123}}{RT} = X_1X_2[0.02859 - 0.02700(X_1 - X_2) - 0.08193(X_1 - X_2)^2]$$
$$+ X_2X_3[0.32472 - 0.05915(X_2 - X_3) - 0.09896(X_2 - X_3)^2]$$
$$+ X_1X_3[0.38993 + 0.03172(X_1 - X_2) - 0.02215(X_1 - X_3)^2]$$
$$+ X_1X_2X_3[0.50877 - 1.45605(X_2 - X_3) + 0.99648(X_1 - X_3)]$$

Using this equation, calculate $\Delta \bar{G}^{ex}_{123}$ at the mole fraction compositions $X_1 = 0.33$, $X_2 = 0.33$; $X_1 = 0.50$, $X_2 = 0.50$; $X_1 = 0.25$, $X_2 = 0.25$; $X_1 = 0.25$, $X_2 = 0.50$; $X_1 = 0.50$, $X_2 = 0.25$; and $X_1 = 0.33$, $X_2 = 0.50$.

4.5 Brynestad (1981) demonstrated that the Kohler method is incompatible with the subregular model

$$\Delta \bar{G}^{ex}_{ij} = X_iX_j(\psi_{ij}X_i + \psi_{ji}X_j)$$

where ψ_{ij} and ψ_{ji} are binary parameters because it gives rise to a dichotomy in the predicted behavior of the partial molar excess properties of the solute in the terminal regions of the ternary system. By differentiating

$$\Delta \bar{G}_{123}^{ex} = \frac{n_1 n_2}{n_1 + n_2}(\psi_{12}X_1 + \psi_{21}X_2) + \frac{n_1 n_3}{n_1 + n_3}(\psi_{13}X_1 + \psi_{31}X_3)$$
$$+ \frac{n_2 n_3}{n_2 + n_3}(\psi_{23}X_2 + \psi_{32}X_3)$$

show that the partial molar excess free energy $\Delta \bar{G}_3^{ex}$ is given by

$$\Delta \bar{G}_3^{ex} = \frac{X_1}{1 - X_2}\left[\left(\frac{X_1}{1 - X_2} - X_3\right)(X_1\psi_{13} + X_3\psi_{31}) + X_3\psi_{31}\right]$$
$$+ \frac{X_2}{1 - X_1}\left[\left(\frac{X_2}{1 - X_1} - X_3\right)(X_2\psi_{23} + X_3\psi_{32}) + X_3\psi_{32}\right]$$
$$- \frac{X_1 X_2}{X_1 + X_2}[X_1\psi_{12} + X_2\psi_{21}]$$

Furthermore, show that the slope of $\Delta \bar{G}_3^{ex}$ (with respect to X_3) changes discontinuously when the solute is added to a pure solvent. (Hint: After using the chain rule to differentiate $\Delta \bar{G}_3^{ex}$ with respect to X_3, allow $X_3 \to 0$ and then $X_2 \to 0$. Next, allow $X_2 \to 0$ and then $X_3 \to 0$. Are the two results identical?)

4.6 Assume the heat of mixing of a ternary mixture can be estimated via

$$\Delta \bar{H}_{123}^{ex} = X_1 X_2(\psi_{12}X_1 + \psi_{21}X_2) + X_1 X_3(\psi_{13}X_1 + \psi_{31}X_3)$$
$$+ X_2 X_3(\psi_{23}X_2 + \psi_{32}X_3)$$

Using this expression, develop a predictive method independent of binary parametrization. How does this newly derived method differ from the Kohler equation?

5
Binary and Ternary Mixtures Containing Only Nonspecific Interactions

Solutions of nonelectrolytes are classified according to their thermodynamic or molecular properties. The thermodynamic classification is based on a consideration of the excess functions of mixing. The simplest classification can be made by considering the signs of $\Delta \bar{G}^{ex}$, $\Delta \bar{H}^{ex}$, $\Delta \bar{S}^{ex}$, $\Delta \bar{C}_p^{ex}$, $\Delta \bar{V}^{ex}$, etc. As was shown in a previous chapter, knowledge of $\Delta \bar{G}^{ex}(T, P, X)$ enables the remaining thermodynamic functions to be determined via the appropriate differentiation.

Alternatively, the solutions may be classified according to their molecular properties by considering the kind of internal forces acting between like and unlike molecules. Such a division of intermolecular forces leads to classification of the pure liquids into *simple* and *complex* liquids. The interactions in simple liquids result exclusively from the dispersion forces, or nonspecific interactions, as they are often called. On the other hand, in complex liquids the molecules have a permanent nonuniform distribution of charge (polar liquids) so they interact through electrostatic forces in addition to the dispersion forces. The electrostatic interactions (primarily dipole–dipole interactions) lead to some degree of specific interactions resulting in a specific orientation of one molecule with respect to an adjacent molecule. For this reason it is sometimes called the *orientation effect*. If these interactions are strong enough, stable dimers or larger complexes may be formed and the liquid is then said to be associated. The boundary between the weakly polar liquids and the associated liquids is nebulous, and there have been numerous disagreements in the literature involving the separation of specific and nonspecific interactions.

A mixture composed of simple liquids is called a simple mixture or a simple solution. A mixture containing at least one complex liquid is called a complex mixture. The molecular interactions in mixtures containing at least two complex liquids may lead to a homogeneous A–A type association, a heterogeneous A–B type association, or both simultaneously. Interactions of the type A–B of increasing interaction strength may finally lead to the formation of stable intermolecular complexes.

A second factor affecting the properties of a solution is the shape and relative size of the molecules. Whereas the influence of shape has not been well established, the influence of size is better known. The limiting case is that of polymer solutions, amongst which the simplest groups are those forming athermal solutions. The Flory–Huggins model of solution ideality (to be discussed in Section C) provides a reasonable description for thermodynamic properties of polymer solutions.

An extensive survey and discussions of the functions $\Delta \bar{G}^{ex}$, $\Delta \bar{H}^{ex}$ and $\Delta \bar{S}^{ex}$ for binary mixtures is given in two books by Rowlinson (1959, 1969). Numerous examples cited by Rowlinson show that mixtures of simple liquids, in which the deviations from ideality depend primarily on the differences in nonspecific interactions, have positive (endothermic) enthalpies of mixing. Conversely, it is typical for mixtures of molecules differing greatly in size but only slightly in dispersion forces to show negative deviations from Raoult's law. These mixtures have positive entropies of mixing, i.e., $\Delta \bar{S}^{ex} > 0$. Thus, it is possible to distinguish between two types of simple mixtures with the transition between the two types being continuous. For mixtures of similar size molecules the dominant factor is the enthalpy of mixing; with the (positive) excess entropy of mixing increasing as the relative size of the molecules increases. That is, the value of $\Delta \bar{S}^{ex}$ increases gradually, compensates for the enthalpic effect; becomes equal to it for a pseudo-ideal mixture, and finally surpasses it, producing negative deviations from Raoult's law. Deviations from the Flory–Huggins model, though, are still positive.

As examples of simple mixtures in which positive $\Delta \bar{H}^{ex}$ leads to positive deviations from ideality, the following systems can be cited: n-hexane + cyclohexane, benzene + n-hexane, benzene + cyclohexane, toluene + methylcyclohexane, toluene + cyclohexane, benzene + carbon tetrachloride, and perfluoromethylcyclohexane + methylcyclohexane; all at room temperature. As examples of simple mixtures showing negative deviations because of a positive value of $\Delta \bar{S}^{ex}$, these systems can be listed: n-hexane + n-hexadecane, n-hexadecane + n-heptane, n-dodecane + n-hexane, and benzene + diphenyl.

There are some systems for the signs of $\Delta \bar{S}^{ex}$ and $\Delta \bar{H}^{ex}$ can be accounted for in terms of the differences in molecular shape Rowlinson (1959). For instance, $\Delta \bar{H}^{ex}$ and $\Delta \bar{S}^{ex}$ are both negative in the n-octane + tetraethylmethane system.

The question of the enthalpy of mixing sign is relatively clear in the case of complex mixtures. Upon dilution of a polar liquid with a nonpolar (inert) component, a part of the electrostatic interactions is lost, so this process is endothermic in nature, i.e., $\Delta \bar{H}^{ex} < 0$. Mixing of two polar liquids may produce various effects. If the interactions between unlike molecules are stronger than those acting between like molecules, the enthalpy of mixing is negative. If, however, the interaction of the type A–B are weaker, the enthalpy of mixing is positive. Of course, the nonspecific interactions are acting and their contribution in systems of weakly polar molecules may be a dominating one. It may also happen that the A–B interactions are comparable to the A–A and B–B interactions, this occurs in mixtures of homologs such as 1-butanol and 1-pentanol mixtures. When the components do not belong to one homologous series the sign of $\Delta \bar{H}^{ex}$ cannot be predicted, although partial compensation of the dissociation of the A–A and B–B complexes by the A–B complexes must occur.

It is much more difficult to predict the sign of the entropy of mixing for complex mixtures. When the components are isomers or adjacent homologs, as in systems formed by 1-propanol + 2-propanol, phenol + cresol, or propionic acid + butanoic acid, the entropy of mixing is almost identical to that for an ideal solution. This is probably because of the compensation of two effects, dissociation and association. The first increases the number of configurations, whereas the second decreases the number of possible configurations.

This compensation of the gain and loss of orientational degrees of freedom is, in general, only partial in mixtures of complex liquids other than homologs or isomers. The excess entropy of mixing is positive when the prevailing effects lead to more orientational freedom of the molecules. Conversely, if the mutual association of the components predominates, the entropic contribution will have a negative value.

Examples of systems of two polar liquids with positive values of $\Delta \bar{S}^{ex}$ are mixtures of acetone with either ethanol or 2-propanol. The enthalpy of mixing is positive for both systems, confirming that the prevailing effect is that of dissociation of complexes of like molecules, in all likelihood alcohol complexes. For these two systems $\Delta \bar{H}^{ex} > T \Delta \bar{S}^{ex}$, and consequently $\Delta \bar{G}^{ex} > 0$. Examples of systems with negative $\Delta \bar{S}^{ex}$ values are water + ethanol, and water + 1-propanol. The sign of the enthalpy of mixing varies depending on the composition and the temperature. The excess Gibbs free energy of mixing is positive at all mixture compositions because of the large negative value of $\Delta \bar{S}^{ex}$. The enthalpy of mixing and excess entropy of mixing are negative for systems such as nitromethane + p-dioxane, and water + pyridine, where the A–B type interactions become more apparent. The excess Gibbs free energy for these systems is still positive because $|T\Delta \bar{S}^{ex}| > |\Delta \bar{H}^{ex}|$. Finally, it is possible to list systems such as acetone + chloroform, diethyl ether + chloroform, and benzene + chloroform, for which $\Delta \bar{S}^{ex}$ and $\Delta \bar{H}^{ex}$ similarly take on a negative sign, but because $|T\Delta \bar{S}^{ex}| < |\Delta \bar{H}^{ex}|$, the excess Gibbs free energy is negative.

Consider now the sign of $\Delta \bar{S}^{ex}$ for a binary mixture containing of polar and nonpolar molecule such as n-alkane + alcohol, acetonitrile + carbon tetrachloride, and methanol + carbon tetrachloride system. Here, $\Delta \bar{S}^{ex}$

assumes negative values, whereas $\Delta \bar{H}^{ex}$ is positive. As a result, the excess free energies are particularly large and positive. The entropy of mixing of these systems is negative because the strong orientational effects persist in the solution. This is to be expected because the presence of the *inert* cosolvent gives rise to a slight dissociation of A–A complexes, resulting in an endothermic enthalpy of mixing, the nonpolar nature of the *inert* cosolvent does not permit the formation of heterogeneous A–B-type complexes.

Solutions of polar liquids in nonpolar solvents can also exhibit positive excess entropies. Examples reported in literature include acetone + carbon disulfide, nitromethane + benzene, and carbon disulfide + nitromethane. The positive values of $\Delta \bar{S}^{ex}$ are because of the pronounced destruction of orientational effects by the nonpolar solvent.

The interpretation of solution nonideality in terms of molecular interactions is a fascinating area and many valuable contributions have been made in the past 50 years. A large number of semi-theoretical models have been developed to explain the thermodynamic properties of binary mixtures on a quantitative basis. This chapter will discuss several of the more successful approaches proposed for describing the properties of noncomplexing mixtures. Liquid mixtures having a heterogeneous A–B-type association, or having one self-associating component (A–A-type complexes) will be discussed at length in Chapters 7 and 8, respectively.

A. The Theory of Van Laar

One of the essential requirements of any successful solution model is judicious simplification. If one tried to describe all the properties of a liquid mixture it would become a hopelessly complicated situation. To make progress it is often necessary to neglect certain aspects of the physical situation. This can make the difference between a solution model that endures the "test of time" and one that merely becomes an academic exercise. Many of van Laar's contributions to science are still recognized today, primarily because he chose good simplifying assumptions that made the problem solvable, and yet did not greatly violate physical reality.

Van Laar (1906) assumed the formation of the binary mixture from the pure components

X_1 Component 1(ℓ, T, P) + X_2 Component 2(ℓ, T, P) $\longrightarrow$

$\qquad\qquad\qquad\qquad$ [X_1 Component 1 + X_2 Component 2](ℓ, T, P)

occurred in such a manner that

(a) the volume change upon mixing is zero, i.e., $\Delta V^{mix} = 0$

(b) the entropy of mixing is given by that corresponding to an ideal solution, i.e.,

$$\Delta S^{mix} = -n_1 \ln X_1 - n_2 \ln X_2$$

Because

$$\Delta G^{ex} = \Delta U^{ex} + P\Delta V^{ex} - T\Delta S^{ex} \tag{5.1}$$

it follows that van Laar's simplifying assumption is

$$\Delta G^{ex} \approx \Delta H^{ex} \approx \Delta U^{ex} \tag{5.2}$$

To calculate the energy change, van Laar constructed a three step thermodynamic cycle

X_1 Component 1(ℓ, T, P) + X_2 Component 2(ℓ, T, P) $\xrightarrow{\Delta U_I}$
$\qquad X_1$ Component 1(g, T, $P^* \to 0$) + X_2 Component 2(g, T, $P^* \to 0$)

X_1 Component 1(g, T, $P^* \to 0$) + X_2 Component 2(g, T, $P^* \to 0$) $\xrightarrow{\Delta U_{II}}$
$\qquad (X_1$ Component 1 + X_2 Component 2)(g, T, X_1, $P^* \to 0$)

$(X_1$ Component 1 + X_2 Component 2)(g, T, X_1, $P^* \to 0$) $\xrightarrow{\Delta U_{III}}$
$\qquad (X_1$ Component 1 + X_2 Component 2)(g or ℓ, T, X_1, P)

the overall process given by

X_1 Component 1(ℓ, T, P) + X_2 Component 2(ℓ, T, P) $\xrightarrow{\Delta U_{IV}}$
$\qquad (X_1$ Component 1 + X_2 Component 2)(g or ℓ, T, X_1, P)

with

$$\Delta U_{IV} = \Delta U_I + \Delta U_{II} + \Delta U_{III} \tag{5.3}$$

In the first step the liquids are vaporized to pure vapors at some very low pressure, such that they behave as ideal gases. The energy accompanying this process is calculated from the thermodynamic relationship

$$(\partial U/\partial V)_T = T(\partial P/\partial T)_V - P \tag{5.4}$$

At this point van Laar assumed the volumetric properties of the pure fluids could be described by the van der Waals equation of state. In which case

$$(\partial U/\partial V)_T = a/V \tag{5.5}$$

where a is the constant appearing in the van der Waals equation. By mixing X_1 moles of liquid 1 with X_2 moles of liquid 2, one obtains 1 mole of the mixture. Then

$$X_1(U_1^{id} - U_1) = \int_{\bar{V}_1^\ell}^{\infty} (a_1 X_1/V^2)\, dV = a_1 X_1/\bar{V}_1^\ell \tag{5.6}$$

A. The Theory of Van Laar

$$X_2(U_2^{id} - U_2) = \int_{\bar{V}_2^\ell}^{\infty} (a_2 X_2/V^2) dV = a_2 X_2/\bar{V}_2^\ell \tag{5.7}$$

where U_i^{id} is the energy of the ideal gas and $\bar{V}_i^\ell$ the molar volume of liquid. In accordance with the van der Waals theory approximate the molar volume of the liquid (well below its critical temperature) by constant b. Thus,

$$\Delta U_I = X_1(a_1/b_1) + X_2(a_2/b_2) \tag{5.8}$$

Step II is simply the mixing of two ideal gases at constant temperature and pressure

$$\Delta U_{II} = 0 \tag{5.9}$$

Step III is the isothermal compression of the mixture from a very low pressure to the original pressure P. The thermodynamic Eq. (5.4) also applies for the binary mixture. Because van Laar assumed the volumetric properties could also be described by the van der Waals equation, the energy change is

$$\Delta U_{III} = -a_{mix}/b_{mix} \tag{5.10}$$

With the additional assumptions

$$\sqrt{a_{mix}} = X_1\sqrt{a_1} + X_2\sqrt{a_2} \tag{5.11}$$

$$b_{mix} = X_1 b_1 + X_2 b_2 \tag{5.12}$$

the combination of Eqs. (5.3) and (5.6–5.10) yields

$$\Delta \bar{G}_{12}^{ex} \approx \Delta \bar{H}_{12}^{ex} \approx \Delta \bar{U}_{12}^{ex} = \frac{X_1 b_1 X_2 b_2}{X_1 b_1 + X_2 b_2}\left(\frac{\sqrt{a_1}}{b_1} - \frac{\sqrt{a_2}}{b_2}\right)^2 \tag{5.13}$$

The activity coefficient of component 1 is obtained via

$$RT \ln \gamma_1 = \Delta \bar{G}_1^{ex} = \Delta \bar{G}_{12}^{ex} + X_2 \left(\frac{\partial \Delta \bar{G}_{12}^{ex}}{\partial X_1}\right)_{T,P}$$

$$= \frac{X_1 b_1 X_2 b_2}{X_1 b_1 + X_2 b_2}\left(\frac{\sqrt{a_1}}{b_1} - \frac{\sqrt{a_2}}{b_2}\right)^2$$

$$+ X_2 \left[\frac{X_2 b_1 b_2^2}{(X_1 b_1 + X_2 b_2)^2}\left(\frac{\sqrt{a_1}}{b_1} - \frac{\sqrt{a_2}}{b_2}\right)^2 \right.$$

$$\left. - \frac{X_1 b_1 b_2}{X_1 b_1 + X_2 b_2}\left(\frac{\sqrt{a_1}}{b_1} - \frac{\sqrt{a_2}}{b_2}\right)^2 \right]$$

$$= \frac{X_2^2 b_2^2 b_1}{(X_1 b_1 + X_2 b_2)^2}\left(\frac{\sqrt{a_1}}{b_1} - \frac{\sqrt{a_2}}{b_2}\right)^2 \tag{5.14}$$

Similarly

$$RT \ln \gamma_2 = \frac{X_1^2 b_1^2 b_2}{(X_1 b_1 + X_2 b_2)^2} \left(\frac{\sqrt{a_1}}{b_1} - \frac{\sqrt{a_2}}{b_2}\right)^2 \qquad (5.15)$$

Equations (5.14) and (5.15) relate the activity coefficients to the temperature, composition, and properties of the pure components, i.e., (a_1, b_1), and (a_2, b_2).[†] Careful inspection of the van Laar equations reveals the activity coefficients of both components are never less than unity; hence this theory always predicts positive deviations from Raoult's law. This result follows naturally from Eq. (5.11), which states

$$a_{\text{mix}} < X_1 a_1 + X_2 a_2 \qquad (5.16)$$

for $a_1 \neq a_2$. Because the van der Waals constant a is directly proportional to the attractive forces between the molecules, Eq. (5.11) (or Eq. (5.16)) implies that the attractive forces between the molecules in the binary mixture are less than they would be if they were additive on a molar basis. If van Laar had assumed

$$a_{\text{mix}} > X_1 a_1 + X_2 a_2$$

he would have obtained activity coefficients less than one, i.e., $\gamma_i < 1$, for all mole fraction compositions and hence, negative deviations from ideality. The particular combining rules used to describe the mixture constants in terms of the pure component constants have a large influence on the predicted results.

As might be expected from such a simple solution model, quantitative agreement between van Laar's equations and the experimental data is not good. This poor agreement is not as much a result of van Laar's simplifica-

---

[†] For the mathematical representation of experimental data, a slightly different form is preferred

$$\ln \gamma_1 = A'/[1 + (A'X_1/B'X_2)]^2 \quad \text{and} \quad \ln \gamma_2 = B'/[1 + (B'X_2/A'X_1)]^2$$

where

$$A' = (b_1/RT)[(\sqrt{a_1}/b_1) - (\sqrt{a_2}/b_2)]^2$$

and

$$B' = (b_2/RT)[(\sqrt{a_1}/b_1) - (\sqrt{a_2}/b_2)]^2$$

A' and B' are treated as adjustable parameters with their numerical values determined through at least squares analysis of the binary data. These two-parameter expressions can correlate experimental activity coefficients for many binary systems, including some that show large deviations from ideal behavior.

B. The Scatchard–Hildebrand Model

Hildebrand and Scott (1930b) and Scatchard (1931a, 1932, 1934, 1937) realized van Laar's theory could be greatly improved if it could be freed from the limitations of the van der Waal's equation of state. This can be accomplished by defining a new parameter c according to

$$c_i = \Delta \bar{U}_i^{\text{vap}}/\bar{V}_i \tag{5.17}$$

where $\Delta \bar{U}_i^{\text{vap}}$ is the energy of complete vaporization, that is, the energy change accompanying the isothermal vaporization of the saturated liquid to the ideal gas state. The newly defined parameter is referred to as the cohesive energy density.

Generalizing Eq. (5.17) to one mole of a binary liquid mixture, Hildebrand and Scatchard wrote

$$-(\bar{U}_{\text{liq}} - \bar{U}_{\text{id g}})_{\text{bm}} = \frac{c_1 \bar{V}_1^{\bullet 2} X_1^2 + 2c_{12} X_1 \bar{V}_1^\bullet X_2 \bar{V}_2^\bullet + c_2 \bar{V}_2^{\bullet 2} X_2^2}{X_1^\bullet \bar{V}_1^\bullet + X_2 \bar{V}_2^\bullet} \tag{5.18}$$

where bm denotes the binary mixture and $\bar{V}_i^\bullet$ is the molar volume of a pure liquid.

Equation (5.18) assumes that the energy of the binary liquid mixture (relative to the ideal gas at the same temperature and composition) can be expressed as a quadratic function of the volume fractions and also implies that the volume of the binary liquid mixture is given by the ideal molar volume approximation (i.e., $X_1 \bar{V}_1^\bullet + X_2 \bar{V}_2^\bullet$). The constants c_1 and c_2 refer to binary interactions between like molecules, whereas the constant c_{12} refers to interactions between unlike molecules.

To simplify the notation the symbols ϕ_1 and ϕ_2, which designate the ideal volume fractions of components 1 and 2, are introduced

$$\phi_1 = X_1 \bar{V}_1^\bullet /(X_1 \bar{V}_1^\bullet + X_2 \bar{V}_2^\bullet) \tag{5.19}$$

$$\phi_2 = X_2 \bar{V}_2^\bullet /(X_1 \bar{V}_1^\bullet + X_2 \bar{V}_2^\bullet) \tag{5.20}$$

Making these substitutions, Eq. (5.18) becomes

$$-(\bar{U}_{\text{liq}} - \bar{U}_{\text{id g}})_{\text{bm}} = (X_1 \bar{V}_1^\bullet + X_2 \bar{V}_2^\bullet)[\phi_1^2 c_1 + 2c_{12} \phi_1 \phi_2 + \phi_2^2 c_2] \tag{5.21}$$

The molar energy change of mixing, which is also the excess energy of mixing, is defined by

$$\Delta \bar{U}_{12}^{\text{ex}} = \bar{U}_{\text{bm}} - X_1 \bar{U}_1^\bullet - X_2 \bar{U}_2^\bullet \tag{5.22}$$

Utilizing the relation for ideal gases

$$\Delta \bar{U}_{id}^{ex} = \bar{U}_{id\,g} - X_1 \bar{U}_1^\bullet - X_2 \bar{U}_2^\bullet = 0 \tag{5.23}$$

combine Eq. (5.17) for each component and Eq. (5.21) to give

$$\Delta \bar{U}_{12}^{ex} = (\bar{U}_{bm} - X_1 \bar{U}_1^\bullet - X_2 \bar{U}_2^\bullet) - (\bar{U}_{id\,g} - X_1 \bar{U}_1^\bullet - X_2 \bar{U}_2^\bullet)$$
$$= (c_1 + c_2 - 2c_{12})\phi_1 \phi_2 (X_1 \bar{V}_1^\bullet + X_2 \bar{V}_2^\bullet) \tag{5.24}$$

Scatchard and Hildebrand make what is probably the most important assumption in their treatment. They assume that for molecules whose forces of attraction result primarily from dispersion forces there is a simple relationship between c_1, c_2, and c_{12}:

$$c_{12} = (c_1 c_2)^{1/2} \tag{5.25}$$

Substituting Eq. (5.25) into Eq. (5.24) gives

$$\Delta \bar{U}_{12}^{ex} = \phi_1 \phi_2 (X_1 \bar{V}_1^\bullet + X_2 \bar{V}_2^\bullet)(\delta_1 - \delta_2)^2 \tag{5.26}$$

where

$$\delta_1 = c_1^{1/2} = (\Delta \bar{U}_1^{vap}/\bar{V}_1^\bullet)^{1/2} \quad \text{and} \quad \delta_2 = c_2^{1/2} = (\Delta \bar{U}_2^{vap}/\bar{V}_2^\bullet)^{1/2}$$

The positive square root of c is given the special symbol δ, which is called the solubility parameter.

To complete their theory of solutions, Scatchard and Hildebrand made one additional assumption: At constant temperature and pressure the excess entropy of mixing vanishes. This assumption is consistent with Hildebrand's definition of regular solutions because in the treatment outlined previously it was already assumed there was no volume change upon mixing. Eliminating the excess entropy and volume gives

$$\Delta \bar{G}_{12}^{ex} = \Delta \bar{H}_{12}^{ex} = \Delta \bar{U}_{12}^{ex} = \phi_1 \phi_2 (X_1 \bar{V}_1^\bullet + X_2 \bar{V}_2^\bullet)(\delta_1 - \delta_2)^2 \tag{5.27}$$

The corresponding activity coefficients are obtained through the appropriate differentiation of Eq (5.27)

$$RT \ln \gamma_1 = \bar{V}_1^\bullet \phi_2^2 (\delta_1 - \delta_2)^2 \tag{5.28}$$

and

$$RT \ln \gamma_2 = \bar{V}_2^\bullet \phi_1^2 (\delta_1 - \delta_2)^2 \tag{5.29}$$

Equations (5.28) and (5.29) are known as the *Regular Solution* equations and they have much in common with the van Laar relationships developed in the previous section of this chapter. In fact, the Regular Solution equa-

B. The Scatchard–Hildebrand Model

tions can be rearranged into the van Laar form merely by writing for the parameters

$$A' = (\bar{V}_1^{\bullet}/RT)(\delta_1 - \delta_2)^2 \tag{5.30}$$

$$B' = (\bar{V}_2^{\bullet}/RT)(\delta_1 - \delta_2)^2 \tag{5.31}$$

Just as the van Laar equations do, the Regular Solution equations always predict positive deviations from Raoult's law. This result is again a direct consequence of the geometric mean assumption, and follows from Eq. (5.25), wherein the cohesive energy density corresponding to the interactions between dissimilar molecules is given by the geometric average of the cohesive energy densities corresponding to interactions between similar molecules. Had Scatchard and Hildebrand assumed an arithmetic average for the cohesive energy density for the interactions between unlike molecules

$$c_{12} = \tfrac{1}{2}(c_1 + c_2)$$

then the energy change upon mixing would have corresponded to that of an ideal solution, namely $\Delta \bar{U}^{\text{ex}} = 0$.

The solubility parameters δ_1 and δ_2 are functions of temperature but the difference between these parameters, $\delta_1 - \delta_2$, is often nearly independent of temperature. Because the Regular Solution model assumes the excess entropy is zero, it follows that at constant composition the logarithm of each activity coefficient must be inversely proportional to the absolute temperature. As a result, the model assumes the enthalpy of mixing remains constant as the temperature is varied.

For many solutions of nonpolar liquids this is a reasonable approximation, provided the temperature range is not too wide and the solution is remote from critical conditions.

Table I gives liquid molar volumes and solubility parameters for some typical nonpolar liquids at 25°C. A more complete list can be found in Appendix A. Inspection of the solubility parameters of different liquids can be useful in making some qualitative statements regarding the nonideality of certain mixtures. Remembering that the logarithm of the activity coefficient varies directly as the square of the difference in solubility parameters, one can see, for instance, a mixture of carbon disulfide with isooctane exhibits large deviations from ideality, whereas a mixture of benzene with carbon tetrachloride is very nearly ideal.

Expressions derived from the Regular Solution model give a semiquantitative description of activity coefficients, excess free energies, and enthalpies of mixing for many solutions containing only nonpolar components. Because of the simplifying assumptions that have been made in their derivation,

TABLE I
Selected Values of Solubility Parameters and Molar Volumes at 25°C.[a]

Substance	δ_i (cal/cm³)$^{1/2}$	$\bar{V}_i^\bullet$ (ml/mol)
Acetone	9.62	74.00
Acetonitrile	12.11	52.88
Benzene	9.16	89.41
n-Butylcyclohexane	7.53	176.32
Carbon Disulfide	9.92	60.62
Carbon Tetrachloride	8.55	97.08
Chloroform	9.16	80.64
Cyclohexane	8.19	108.76
Cyclopentane	8.10	94.71
Decane	7.74	195.91
2,2-Dimethylbutane	6.71	133.70
2,3-Dimethylbutane	6.97	131.14
Dodecane	7.92	228.54
Ethylbenzene	8.84	123.10
Heptane	7.50	147.48
Hexane	7.27	131.51
Octane	7.54	163.48
Toluene	8.92	106.84
1,2,3-Trimethylbenzene	9.18	134.97
1,2,4-Trimethylbenzene	8.99	137.86
1,3,5-Trimethylbenzene	8.88	139.57
2,2,4-Trimethylpentane	6.86	166.09
m-xylene	8.88	123.48
o-xylene	9.06	121.21
p-xylene	8.82	124.00

[a] Numerical values of the solubility parameters were taken from a tabulation by K. L. Hoy (1970).

the expressions cannot be expected to provide complete quantitative agreement between calculated and experimental results. For approximate work, the solubility parameter expressions provide reasonable predictions, and in many design calculations a reasonable estimate will more than suffice.

Figures 5.1–5.4 compare the Regular Solution predictions for $\Delta \bar{G}^{ex}$ and $\Delta \bar{H}^{ex}$ to the experimentally observed values for benzene + n-heptane, benzene + carbon tetrachloride, cyclohexane n-heptane, and cyclohexane + carbon disulfide mixtures. Numerical values of δ_i and $\bar{V}_i^\bullet$ were taken from Table I. It can be seen from the four figures that Eq. (5.27) generally gives a better description for the excess free energy than for the excess enthalpy. It is now known that the assumptions of regularity ($\Delta \bar{S}^{ex} = 0$) and isometric

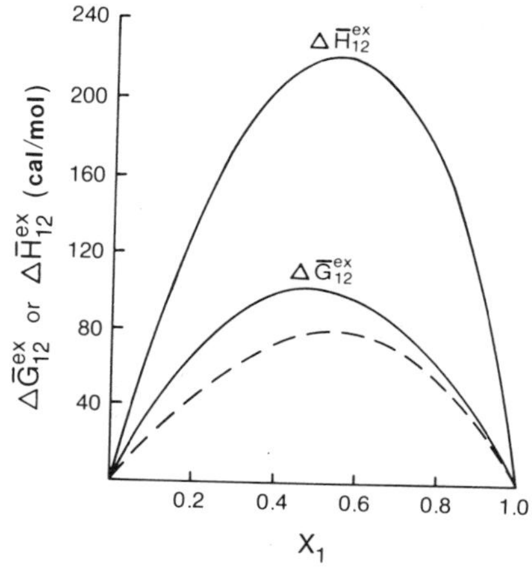

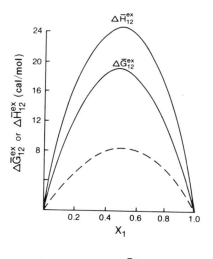

Fig. 5.1 Comparison between the experimental thermodynamic excess properties and the values calculated with Eq. (5.27) for benzene (1) + n-heptane (2) mixtures. The calculated values are depicted with a dashed line whereas experimental values are shown using solid lines. Experimental free energies are taken from Jain et al. (1973) and enthalpies from Lunderg (1964).

Fig. 5.2 Comparison between the experimental thermodynamic excess properties and the values calculated with Eq. (5.27) for benzene (1) + carbon tetrachloride (2) mixtures. The calculated values are depicted with a dashed line whereas experimental values are shown using solid lines. Excess free energies and enthalpies of mixing for the binary system were determined by Goates et al. (1959).

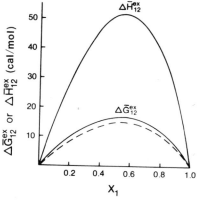

Fig. 5.3 Comparison between the experimental thermodynamic excess properties and the values calculated with Eq. (5.27) for cyclohexane (1) + n-heptane (2) mixtures. The calculated values are depicted with a dashed line whereas experimental values are shown using solid lines. Experimental values of $\Delta \bar{G}_{12}^{ex}$ were determined by Young et al. (1977) and values of $\Delta \bar{H}_{12}^{ex}$ by Lunderg (1964).

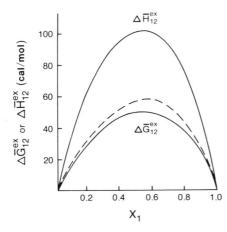

Fig. 5.4 Comparison between the experimental properties and the values calculated with Eq. (5.27) for carbon disulfide (1) + cyclohexane (2) mixtures. The calculated values are depicted with a dashed line whereas the experimental values are shown using solid lines. Experimental values of $\Delta \bar{G}_{12}^{ex}$ from Bernatova and Boublik (1977) and $\Delta \bar{H}_{12}^{ex}$ values from Swinton (1980).

mixing ($\Delta V^{ex} = 0$) at constant temperature and pressure are not correct even for simple mixtures, but because of the cancellation of errors, these assumptions frequently do not seriously affect calculations of the excess free energy. When Regular Solution theory is used to calculate excess enthalpies, however, the results are usually much worse, as seen in Figs. 5.1–5.4.

The most serious defect of the Regular Solution model involves the geometric mean approximation and the evaluation of solubility parameters from energies of vaporization. If $(\delta_1 - \delta_2)^2$ is treated as an empirically determined interaction parameter A_{12}

$$A_{12} = (\delta_1 - \delta_2)^2$$

then Eq. (5.27) is remarkably successful for excess enthalpies and free energies. Few one-parameter equations do as well. This liberal application permits positive and negative deviations from Raoult's law, as there are no restrictions placed on the numerical value of A_{12}. Looking at Eq. (5.24), it is found the general mathematical form of the Regular Solution model

$$\Delta \bar{U}_{12}^{ex} = \Delta \bar{H}_{12}^{ex} = \Delta \bar{G}_{12}^{ex} = \phi_1 \phi_2 (X_1 \bar{V}_1^{\bullet} + X_2 \bar{V}_2^{\bullet}) A_{12} \quad (5.32)$$

is obtainable by setting $A_{12} = (c_1 + c_2 - 2c_{12})$ and does not require the geometric mean approximation.

One of the main advantages of the Regular Solution equations is their simplicity, and this simplicity is retained even when the model is extended to liquid solutions having more than two components. The derivation for the multicomponent case is completely analogous to that given previously for a binary system. To illustrate how this is done, the molar energy of

B. The Scatchard–Hildebrand Model

a ternary mixture is written as

$$-(\bar{U}_{\text{liq}} - \bar{U}_{i\,g})_{\text{tm}}$$

$$= \frac{X_1^2 \bar{V}_1^{\bullet 2} c_1 + X_2^2 \bar{V}_2^{\bullet 2} c_2 + X_3^2 \bar{V}_3^{\bullet 2} c_3 + 2X_1 X_2 \bar{V}_1^{\bullet} \bar{V}_2^{\bullet} c_{12}}{(X_1 \bar{V}_1^{\bullet} + X_2 \bar{V}_2^{\bullet} + X_3 \bar{V}_3^{\bullet})}$$

$$+ \frac{2X_1 X_3 \bar{V}_1^{\bullet} \bar{V}_3^{\bullet} c_{13} + 2X_2 X_3 \bar{V}_2^{\bullet} \bar{V}_3^{\bullet} c_{23}}{(X_1 \bar{V}_1^{\bullet} + X_2 \bar{V}_2^{\bullet} + X_3 \bar{V}_3^{\bullet})}$$

$$= (X_1 \bar{V}_1^{\bullet} + X_2 \bar{V}_2^{\bullet} + X_3 \bar{V}_3^{\bullet})(\phi_1^2 c_1 + 2\phi_1 \phi_2 c_{12} + \phi_2^2 c_2$$
$$+ \phi_3^2 c_3 + 2\phi_1 \phi_3 c_{13} + 2\phi_2 \phi_3 c_{23}) \tag{5.33}$$

where tm is the ternary mixture. The volume fraction of component i is now defined by

$$\phi_i = X_i \bar{V}_i^{\bullet} / (X_1 \bar{V}_1^{\bullet} + X_2 \bar{V}_2^{\bullet} + X_3 \bar{V}_3^{\bullet}) \tag{5.34}$$

and the excess energy of mixing is defined by

$$\Delta \bar{U}_{123}^{\text{ex}} = \bar{U}_{\text{tm}} - X_1 \bar{U}_1^{\bullet} - X_2 \bar{U}_2^{\bullet} - X_3 \bar{U}_3^{\bullet} \tag{5.35}$$

Combining Eqs. (5.33–5.35), the excess energy of mixing in terms of the cohesive energy densities is written as

$$\Delta \bar{U}_{123}^{\text{ex}} = (X_1 \bar{V}_1^{\bullet} + X_2 \bar{V}_2^{\bullet} + X_3 \bar{V}_3^{\bullet})$$
$$\times [\phi_1 \phi_2 (c_1 + c_2 - 2c_{12}) + \phi_1 \phi_3 (c_1 + c_3 - 2c_{13})$$
$$+ \phi_2 \phi_3 (c_2 + c_3 - 2c_{23})] \tag{5.36}$$

Assuming the cohesive energy density c_{ij} is given by the geometric mean

$$c_{ij} = (c_i c_j)^{1/2}$$

and that

$$\Delta \bar{S}_{123}^{\text{ex}} = \Delta \bar{V}_{123}^{\text{ex}} = 0$$

the excess Gibbs free energy (or the excess enthalpy) of a ternary liquid mixture can be expressed in terms of solubility parameters

$$\Delta \bar{U}_{123}^{\text{ex}} = \Delta \bar{G}_{123}^{\text{ex}} = \Delta \bar{H}_{123}^{\text{ex}}$$
$$= (X_1 \bar{V}_1^{\bullet} + X_2 \bar{V}_2^{\bullet} + X_3 \bar{V}_3^{\bullet})[\phi_1 \phi_2 (\delta_1 - \delta_2)^2 + \phi_1 \phi_3 (\delta_1 - \delta_3)^2$$
$$+ \phi_2 \phi_3 (\delta_2 - \delta_3)^2] \tag{5.37}$$

or in terms of the more general binary interaction parameters A_{ij}

$$\Delta \bar{U}_{123}^{\text{ex}} = \Delta \bar{G}_{123}^{\text{ex}} = \Delta \bar{H}_{123}^{\text{ex}} = (X_1 \bar{V}_1^{\bullet} + X_2 \bar{V}_1^{\bullet} + X_3 \bar{V}_3^{\bullet})$$
$$\times [\phi_1 \phi_2 A_{12} + \phi_1 \phi_3 A_{13} + \phi_2 \phi_3 A_{23}] \tag{5.38}$$

Inspection of Eq. (5.38) reveals that for regular ternary solutions obeying this model equation, the properties of the contributive binary systems would obey (per mole of solution)

$$\Delta \bar{G}_{ij}^{ex} = \Delta \bar{H}_{ij}^{ex} = \Delta \bar{U}_{ij}^{ex} = (X_i^{\circ} \bar{V}_i^{\bullet} + X_j^{\circ} \bar{V}_j^{\bullet})\phi_i \phi_j A_{ij} \qquad (5.39)$$

where the mole fraction compositions $(X_i^{\circ}, X_j^{\circ})$ are now calculated as if the third component were not present

$$X_i^{\circ} = 1 - X_j^{\circ} = X_i/(X_i + X_j)$$

Suitable mathematical manipulation of Eqs. (5.38) and (5.39) yields the following expression for the excess Gibbs free energy (or excess enthalpy) of a ternary mixture

$$\Delta \bar{G}_{123}^{ex} = (X_1 + X_2)(\phi_1 + \phi_2)\Delta \bar{G}_{12}^{ex} + (X_1 + X_3)(\phi_1 + \phi_3)\Delta \bar{G}_{13}^{ex}$$
$$+ (X_2 + X_3)(\phi_2 + \phi_3)\Delta \bar{G}_{23}^{ex} \qquad (5.40)$$

which is identical to the predictive expression of Lakhanpal *et al.* (1975, 1976), discussed in Chapter 4.

C. The Flory–Huggins Model

In the theory of Regular Solutions for molecules of similar size it was assumed the entropy terms corresponds to that of an ideal solution and attention was focused on the enthalpy of mixing. When considering solutions containing molecules of differing sizes, however, it has been found advantageous to assume, at least as a first approximation, that the enthalpy of mixing is zero and concentrate on the entropic contributions. Solutions having zero enthalpy of mixing are referred to as *athermal solutions* because their components mix with no liberation or absorption of heat. True athermal behavior is rarely observed, but is approximated by mixtures having components similar in their chemical characteristics even is their sizes differ.

Using the concept of a crystalline lattice as a model for the liquid state, Flory (1942) and Huggins (1941, 1942) independently derived an expression for the entropy of mixing in an athermal solution. Because the derivation is clearly presented in several references (for example, see Flory 1953) and requires some familiarity with statistical mechanics, a brief discussion of the essential ideas will be presented.

Flory and Huggins assumed a polymer molecule in solution behaves like a chain with a large number of equal sized segments identical in size to the solvent molecule. It was also assumed that each segment occupies a single site in the crystalline lattice and adjacent segments must occupy adjacent

C. The Flory–Huggins Model

sites. Assume there are n_1 moles of solvent, n_2 moles of polymer, and m segments in a polymer molecule. The total number of sites is then $(n_1 + mn_2)N_A$ is Avogadro's number. The *volume* fraction of the solvent and polymer are given by

$$\phi_1 = n_1/(n_1 + n_2 m) \quad \text{and} \quad \phi_2 = n_2 m/(n_1 + n_2 m) \quad (5.41)$$

Flory and Huggins went on to show that if the amorphous polymer and the solvent mix without any energetic effects, the change in Gibbs free energy and entropy upon mixing are given by the remarkably simple expression

$$\Delta G_{12}^{\text{mix}} = -T \Delta S_{12}^{\text{mix}} = RT(n_1 \ln \phi_1 + n_2 \ln \phi_2) \quad (5.42)$$

The entropy change described by Eq. (5.42) is similar in form to that of Eq. (3.49) for an ideal solution, except volume fractions are used rather than mole fractions. For the special case when the solute and solvent molecules are identical in size ($m = 1$), the change in entropy given by Eq. (5.42) reduces to that of Eq. (3.49) as expected.

The expression of Flory and Huggins immediately leads to an equation for the excess entropy, which is, per mole of mixture,

$$\Delta \bar{S}_{12}^{\text{ex}} = -RX_1 \ln\left[1 - \phi_2(m-1)/m\right] - RX_2 \ln\left[m - \phi_2(m-1)\right] \quad (5.43)$$

Mathematically, it can be shown that Eq. (5.43) is positive for all values of $m > 1$. Therefore, for an athermal solution of components whose molecules differ in size, the Flory–Huggins solution model predicts negative deviations from Raoult's law

$$\Delta \bar{G}^{\text{ex}}/RT = (\Delta \bar{H}^{\text{ex}}/RT) - (\Delta \bar{S}^{\text{ex}}/R) = 0 - (\Delta \bar{S}^{\text{ex}}/R) < 0 \quad (5.44)$$

The activity coefficient of the solvent follows from differentiation of Eq. (5.44)[†]

$$\ln \gamma_1 = \ln[1 - \phi_2(m-1)/m] + \phi_2(m-1)/m \quad (5.45)$$

[†] In many theoretical interpretations of solution nonideality, it is useful to calculate the excess Gibbs free energies relative to the Flory–Huggins model, rather than Raoult's law. The Flory–Huggins based excess free energy is defined by

$$\Delta \bar{G}^{\text{fh}} = \Delta \bar{G}^{\text{mix}} - RT \sum_{i=1}^{N} X_i \ln \phi_i$$

For a binary liquid mixture, the excess molar Gibbs free energy over the predictions of the Flory–Huggins equation is related to the normal excess free energy by

$$\Delta \bar{G}_{12}^{\text{fh}} = \Delta \bar{G}_{12}^{\text{ex}} + RT[\ln(X_1 \bar{V}_1^\bullet + X_2 \bar{V}_2^\bullet) - X_1 \ln \bar{V}_1^\bullet - X_2 \ln \bar{V}_2^\bullet]$$

The quantity m has been set equal to the ratio of molar volumes of polymer and solvent so the volume fractions ϕ_1 and ϕ_2 are given by Eqs. (5.19) and (5.20).

To apply the theoretical result of Flory and Huggins to real polymer solutions it becomes necessary to add to Eq. (5.42) a semi-empirical term for the enthalpy of mixing. The form of this term is the same as that used in the development of the Scatchard–Hildebrand theory of solutions; the excess enthalpy is set proportional to the volume of the solution and to the product of the volume fractions. The Flory–Huggins equation for real polymer solutions then becomes

$$\Delta \bar{G}^{mix}_{12}/RT = X_1 \ln \phi_1 + X_2 \ln \phi_2 + \phi_1 \phi_2 (X_1 + X_2 m)\chi \qquad (5.46)$$

and the activity coefficient of component 1 is

$$\ln \gamma_1 = \ln\{1 - [(m-1)/m]\phi_2\} + [(m-1)/m]\phi_2 + \phi_2^2 \chi \qquad (5.47)$$

where χ, called Flory interaction parameter, is determined by intermolecular forces between the molecules in solution. If m is set equal to the ratio of molar volumes, as is usually done, the definition of volume fraction ϕ is identical to those used in the Scatchard–Hildebrand theory.

The dimensionless parameter χ is determined by the energies that characterize the interactions between the pairs of polymer segments, pairs of solvent molecules, and between one polymer segment and one solvent molecule. For athermal solutions, χ is zero; for mixtures of components that are chemically similar, χ is small.

The addition of an enthalpic term to the theoretical result for athermal mixtures is essentially an empirical modification required to obtain a reasonable expression for the Gibbs free energy of mixing. According to the theory, χ should be independent of polymer concentration and of polymer molecular weight, but in many systems χ is found to vary with both variables. Furthermore, the theory erroneously assumes the enthalpy of mixing should be described by the last term in Eq. (5.46). Calorimetric studies indicate that the numerical value of χ needed to describe experimental enthalpies of mixing is different from the value obtained from experimental Gibbs free energies reduced through Eq. (5.46) or (5.47). The Flory–Huggins equation for real polymer solutions is not a perfect description of the thermodynamic properties of such systems, but there is little doubt this relatively simple theory contains most of the essential features that distinguish solutions of large molecules from those containing molecules of approximately equal sizes.

D. The Wilson Model

For mixtures of molecules that are chemically similar (athermal solutions) and differ only in size, Flory and Huggins derived a simple mathematical expression for the excess Gibbs free energy. According to Eq. (5.42), ideal

D. The Wilson Model

behavior is expected whenever the mixture components have identical molar volumes that are taken as measures of the molecule's size.

About 20 years later, Wilson (1964) considered the case where the components in a mixture differ not only in molecular size but also in their intermolecular forces. Wilson's point of departure was Eq. (5.42), which he used as the basis for his semitheoretical description of liquid mixtures.

To derive the Wilson equation for a binary mixture, start with the following two assumptions:

(a) The free energy of mixing is given by a relation similar to the athermal Flory–Huggins equation

$$\Delta \bar{G}_{12}^{\text{mix}} = RT[X_1 \ln \xi_1 + X_2 \ln \xi_2] \tag{5.48}$$

where ξ_i is the "local" volume fraction of component i about a central molecule of the same type.

(b) The probability of finding a molecule of type 2 relative to finding a molecule of type 1 around a central type 1 molecule is assumed to be expressible in terms of Boltzmann factors

$$\frac{X_{21}}{X_{11}} = \frac{X_2 \exp(-g_{12}/RT)}{X_1 \exp(-g_{11}/RT)} \tag{5.49}$$

Similarly, the probability of finding a molecule of type 1 relative to finding a molecule of type 2 about a central molecule 2 is

$$\frac{X_{12}}{X_{22}} = \frac{X_1 \exp(-g_{12}/RT)}{X_2 \exp(-g_{22}/RT)} \tag{5.50}$$

where g_{ij} is proportional to the interaction energy between molecules i and j.

According to the second assumption the local volume fraction of components 1 and 2 are given by

$$\begin{aligned}\xi_1 &= \frac{X_{11}\bar{V}_1^\bullet}{X_{11}\bar{V}_1^\bullet + X_{21}\bar{V}_2^\bullet} \\ &= \frac{X_1\bar{V}_1^\bullet \exp(-g_{11}/RT)}{X_1\bar{V}_1^\bullet \exp(-g_{11}/RT) + X_2\bar{V}_2^\bullet \exp(-g_{12}/RT)}\end{aligned} \tag{5.51}$$

and

$$\begin{aligned}\xi_2 &= \frac{X_{22}\bar{V}_2^\bullet}{X_{12}\bar{V}_1^\bullet + X_{22}\bar{V}_2^\bullet} \\ &= \frac{X_2\bar{V}_2^\bullet \exp(-g_{22}/RT)}{X_1\bar{V}_1^\bullet \exp(-g_{12}/RT) + X_2\bar{V}_2^\bullet \exp(-g_{22}/RT)}\end{aligned} \tag{5.52}$$

respectively. It is seen from the definitions that if $g_{11} = g_{12} = g_{22}$, the *local* volume fraction is equivalent to the ideal volume fraction ϕ_i and Eq. (5.48) reduces to the Flory–Huggins equation.

Substituting Eqs. (5.51) and (5.52) into Eq. (5.48) gives the relation

$$\Delta \bar{G}_{12}^{mix} = RT \left\{ X_1 \ln \frac{X_1 \bar{V}_1^\bullet}{X_1 \bar{V}_1^\bullet + X_2 \bar{V}_2^\bullet \exp[-(g_{12} - g_{11})/RT]} \right.$$
$$\left. + X_2 \ln \frac{X_2 \bar{V}_2^\bullet}{X_2 \bar{V}_2^\bullet + X_1 \bar{V}_1^\bullet \exp[-(g_{12} - g_{22})/RT]} \right\} \quad (5.53)$$

from which the excess free energy of mixing is calculated to be

$$\Delta \bar{G}_{12}^{ex} = RT[X_1 \ln(\xi_1/X_1) + X_2 \ln(\xi_2/X_2)]$$
$$= -RT(X_1 \ln\{X_1 + X_2(\bar{V}_2^\bullet/\bar{V}_1^\bullet) \exp[-(g_{12} - g_{11})/RT]\}$$
$$+ X_2 \ln\{X_2 + X_1(\bar{V}_1^\bullet/\bar{V}_2^\bullet) \exp[-(g_{12} - g_{22})/RT]\}) \quad (5.54)$$

To simplify notation it is convenient for us to define to new parameters Λ_{12} and Λ_{12} in terms of the molar volumes $\bar{V}_1^\bullet$ and $\bar{V}_2^\bullet$, and the energies g_{11}, g_{12}, and g_{22}. Using these definitions

$$\Lambda_{12} = (\bar{V}_2^\bullet/\bar{V}_1^\bullet) \exp[-(g_{12} - g_{11})] \quad (5.55)$$
$$\Lambda_{21} = (\bar{V}_1^\bullet/\bar{V}_2^\bullet) \exp[-(g_{12} - g_{22})] \quad (5.56)$$

the Wilson equation becomes

$$\Delta \bar{G}_{12}^{ex} = -RT[X_1 \ln(X_1 + X_2 \Lambda_{12}) + X_2 \ln(X_2 + X_1 \Lambda_{21})] \quad (5.57)$$

The activity coefficients of components 1 and 2 are obtained through the appropriate differentiation of Eq. (5.57)

$$\ln \gamma_1 = -\ln(X_1 + X_2 \Lambda_{12}) + X_2 \left[\frac{\Lambda_{12}}{X_1 + \Lambda_{12} X_2} - \frac{\Lambda_{21}}{\Lambda_{21} X_1 + X_2} \right] \quad (5.58)$$

$$\ln \gamma_2 = -\ln(X_2 + X_1 \Lambda_{21}) - X_1 \left[\frac{\Lambda_{12}}{X_1 + \Lambda_{12} X_2} - \frac{\Lambda_{21}}{\Lambda_{21} X_1 + X_2} \right] \quad (5.59)$$

Wilson's mathematical description of binary liquid mixtures contains two adjustable parameters, Λ_{12} and Λ_{21}. In Wilson's derivation, these are related to the pure component molar volumes and characteristic energy differences via Eq. (5.55) and (5.56). To a fair approximation, the differences in the characteristic energies are temperature independent, at least over a modest interval. As a result, Wilson's equation gives not only an expression for the

D. The Wilson Model

activity coefficients as a function of composition, but also an expression for the variation of the activity coefficients with temperature. This is an important practical advantage in isobaric conditions where the temperature varies as the composition changes. For accurate work, $(g_{12} - g_{11})$ and $(g_{12} - g_{22})$ should be considered temperature dependent. Ratkovics and Rehim (1970), and Nagata and Yamada (1972, 1973b) assumed the energy parameter differences varied linearly with temperature. Nagata and Yamada (1973a) and Nagata et al. (1973) further demonstrated the assumption of a quadratic function of temperature for the energy parameter differences is suitable for the simultaneous correlation of excess Gibbs free energy and enthalpy of mixing data over a moderate temperature range, it is also successful for the prediction of excess heat capacities for binary alcohol–hydrocarbon and hydrocarbon–hydrocarbon mixtures. For the purposes of this book, however, temperature independent energy differences will be assumed. Interested readers are encouraged to review the articles listed previously.

A study of the Wilson equation by Orye and Prausnitz (1965) showed that for approximately 100 miscible binary mixtures of various chemical types, activity coefficients could be adequately described by the Wilson equation. To demonstrate the applicability of Eq. (5.57), Tables II and III, compare calculated and experimental excess free energies for methanol + benzene,

TABLE II
Comparison of the Descriptive Abilities of the Wilson Equation and the Redlich–Kister Equation for the Methanol (1) + Benzene (2) System.[a]

X_1	$(\Delta \bar{G}_{12}^{ex})_{35°C}^{meas}$	$(\Delta \bar{G}_{12}^{ex})$ See Eq. (5.57)	$(\Delta \bar{G}_{12}^{ex})$ See Eq. (5.60) (5 Parameters)
0.0242	47.15	38.30	40.15
0.0254	40.67	44.21	41.96
0.1302	173.40	173.34	170.97
0.3107	281.08	284.87	282.23
0.4989	306.06	308.49	306.01
0.5191	304.24	306.39	304.16
0.6305	278.46	279.76	278.54
0.7965	192.65	192.53	192.55
0.9197	89.15	88.46	88.87

$\Lambda_{12} = 0.8774 \qquad \Lambda_{21} = 0.3421$

[a] Calculations in calories per mole were taken from Wilson (1964). Reprinted with permission, copyright American Chemical Society.

TABLE III

Comparison of the Descriptive Abilities of the
Wilson Equation and the Redlich–Kister Equation
for the Carbon Tetrachloride (1) + Acetonitrile (2) System[a]

X_1	$(\Delta \bar{G}_{12}^{ex})_{45°C}^{meas}$	$(\Delta \bar{G}_{12}^{ex})$ See Eq. (5.57)	$(\Delta \bar{G}_{12}^{ex})$ See Eq. (5.60) (4 parameters)
0.1914	168.0	169.9	165.9
0.2887	226.9	227.2	227.0
0.3752	261.0	261.3	261.0
0.4567	279.5	280.0	279.2
0.5060	284.3	284.8	283.9
0.6049	279.2	279.6	279.0
0.7164	248.0	247.7	248.7
0.8069	200.1	199.0	201.2
0.8959	129.2	126.5	128.7
0.9609	56.7	54.0	54.8
	$\Lambda_{21} = 0.1713$	$\Lambda_{12} = 0.3882$	

[a] Calculations in calories per mole were taken from Wilson (1964). Reprinted with permission, copyright American Chemical Society.

and carbon tetrachloride + acetonitrile mixtures. Included in these comparisons is the Redlich–Kister equation

$$\Delta \bar{G}_{12}^{ex} = X_1 X_2 \sum_{v=0}^{r} G_v (X_1 - X_2)^v \tag{5.60}$$

which is one of the most popular expressions used in the mathematical representation of thermodynamic excess properties. As can be seen, the two-parameter Wilson equation describes the experimental data with approximately the same degree of precision as the four or five Redlich–Kister equations.

Despite the success of Eq. (5.57), it can not be used to predict phase separation. According to the Wilson model, a system will be close to separation into two phases when the parameters Λ_{ij} are close to zero. But in this limit the excess free energy becomes equal and opposite in sign to the ideal free energy of mixing, thus requiring the free energy of mixing to equal zero at all binary compositions.[†]

[†] For partially miscible systems, Wilson suggested the right-hand side of Eq. (5.57) be multiplied by a constant.

D. The Wilson Model

TABLE IV

Experimental and Predicted Enthalpies of Mixing for Carbon Tetrachloride (1) + Acetonitrile (2) Mixtures at 45°C

X_2	$(\Delta \bar{H}^{ex}_{12})$ exp	$(\Delta \bar{H}^{ex}_{12})$ See Eq. (5.61)
	(cal/mol)	
0.218	183	140
0.252	193	146
0.353	213	151
0.399	215	149
0.450	220	145
0.697	176	102
0.835	126	64

$\Lambda_{12} = 0.3882$, $\quad g_{12} - g_{11} = 214.1$ cal/mol

$\Lambda_{21} = 0.1713$, $\quad g_{21} - g_{22} = 1499.3$ cal/mol

The Wilson expression for the excess enthalpy of mixing is easily obtained using the Gibbs–Helmholtz relationship

$$\Delta \bar{H}^{ex}_{12} = \left(\frac{\partial \Delta \bar{G}^{ex}_{12}/T}{\partial 1/T} \right)_{P, X_1, X_2}$$

$$\Delta \bar{H}^{ex}_{12} = X_1 X_2 \left[\frac{\Lambda_{12}(g_{12} - g_{11})}{X_1 + X_2 \Lambda_{12}} + \frac{\Lambda_{21}(g_{12} - g_{22})}{X_2 + X_1 \Lambda_{21}} \right] \quad (5.61)$$

Calorimetrically determined enthalpies of mixing for the carbon tetrachloride + acetonitrile system Lien and Missen (1974) provide us with the opportunity to test Eq. (5.61). Table IV compares the experimental data to values calculated via Eq. (5.61). Essentially this represents a predictive application of the Wilson equation rather than a curve-fitting application, as the binary parameters used in the calculations were evaluated from free energy data listed in Table III. Deviations between the calculated and observed values diminish whenever $g_{12} - g_{11}$ and $g_{12} - g_{22}$ are determined from the excess enthalpies.[†] These calculations further support earlier contentions that the energy parameter differences may vary with temperature. Personal experience using the Wilson equation indicate it gives a good representation of $\Delta \bar{G}^{ex}$ and $\Delta \bar{H}^{ex}$ for nonideal systems whose enthalpy of mixing values are

---

[†] Better mathematical representation of $\Delta \bar{G}^{ex}_{12}$ and $\Delta \bar{H}^{ex}_{12}$ using a single set of binary parameters can often be obtained by evaluating $g_{12} - g_{11}$ and $g_{12} - g_{22}$ from $\Delta \bar{H}^{ex}_{12}$, as Eq. (5.61) is more sensitive to the numerical values of the Wilson parameters.

less than 500 J/mol. Simultaneous representation of both sets of excess properties for systems having enthalpies of mixing greater than 500 J/mol generally requires temperature dependent parameters.

One of the advantages of the Wilson equation is that it may be extended to as many components as desired without any additional assumptions and without introducing any constants other than those obtained from the binary data. For a solution of n components Wilson's equation for $\Delta \bar{G}_{12\cdots n}^{ex}$ is written as

$$\Delta \bar{G}_{12\cdots n}^{ex} = -RT \sum_{i=1}^{N} X_i \ln \left[\sum_{j=1}^{N} X_j \Lambda_{ij} \right] \tag{5.62}$$

and the activity coefficient of any component k as

$$\ln \gamma_k = -\ln \left[\sum_{j=1}^{N} X_j \Lambda_{kj} \right] + 1 - \sum_{i=1}^{N} \left(X_i \Lambda_{ik} \Big/ \sum_{j=1}^{N} X_j \Lambda_{ij} \right) \tag{5.63}$$

As mentioned earlier, Eq. (5.62) requires only parameters that can be obtained from binary data; two parameters being determined from each possible binary pair in the multicomponent solution. Certain restraining conditions between binary constants, however, must be fulfilled if the expressions for the multicomponent system are to be consistent with the theoretical model. According to Eqs. (5.55) and (5.56) a ternary system can be written as

$$\ln(\Lambda_{12}/\Lambda_{21}) - 2 \ln(\bar{V}_2^{\bullet}/\bar{V}_1^{\bullet}) = g_{11} - g_{22} \tag{5.64}$$

and similarly

$$\ln(\Lambda_{13}/\Lambda_{31}) - 2 \ln(\bar{V}_3^{\bullet}/\bar{V}_1^{\bullet}) = g_{11} - g_{33} \tag{5.65}$$

$$\ln(\Lambda_{23}/\Lambda_{32}) - 2 \ln(\bar{V}_3^{\bullet}/\bar{V}_2^{\bullet}) = g_{22} - g_{33} \tag{5.66}$$

It is evident from Eqs. (5.64–5.66) the Wilson constants in a ternary system are related by the equation

$$(\Lambda_{23}/\Lambda_{32}) = (\Lambda_{13}/\Lambda_{31})(\Lambda_{21}/\Lambda_{12}) \tag{5.67}$$

As one might expect from the mathematical form of the Wilson expressions, parametrization of binary in accordance to Eqs. (5.57) and (5.61) generally involves reiterative calculations. Rarely do these calculations give Λ_{ij} parameters that satisfy the normalization conditions described by Eq. (5.67). Although the difficulty in obtaining a normalized set of binary parameters distorts to some extent the physical significance of the energy difference terms, the Wilson equation still provides fairly good predictions for the properties of multicomponent systems based on binary data. For example, Govindaswamy et al. (1976) compared the predictions of the Wilson equation

E. The Nonrandom Two-Liquids (NRTL) Model

TABLE V

Isobaric Vapor-Liquid Equilibrium Data for n-Hexane (1) + Benzene (2) + 1-Butanol at 760 Torr[a]

Temperature (°C)	Liquid composition		Experimental vapor composition		Calculated temperature (°C)	Calculated values vapor composition	
	X_1	X_2	Y_1	Y_2		Y_1	Y_2
87.60	0.0375	0.3050	0.1432	0.5870	89.31	0.1466	0.6189
80.20	0.0833	0.3768	0.2850	0.5620	83.62	0.2484	0.5874
77.70	0.1433	0.3967	0.3655	0.5120	79.89	0.3500	0.5215
77.90	0.1934	0.1934	0.5590	0.2934	80.54	0.5364	0.3079
73.70	0.2834	0.2750	0.5734	0.3300	75.75	0.5675	0.3266
72.80	0.3867	0.1532	0.7200	0.1968	73.85	0.7181	0.1813
71.10	0.4667	0.2000	0.7165	0.1940	72.07	0.7155	0.2028
70.20	0.5690	0.1500	0.7733	0.1468	70.80	0.7812	0.1445
68.30	0.7167	0.1450	0.8030	0.1432	69.53	0.8148	0.1268
68.70	0.8100	0.0715	0.8567	0.0700	68.96	0.8800	0.0630

Wilson binary energy parameters (cal/mol):

$g_{12} - g_{11} = 30.567$, $\quad g_{21} - g_{22} = 256.45$, $\quad g_{23} - g_{22} = 138.61$,

$g_{32} - g_{33} = 902.61$, $\quad g_{31} - g_{33} = 1968.86$ and $g_{13} - g_{11} = -60.30$

[a] Reprinted in part with permission from Govindaswamy et al. (1976), copyright 1976 American Chemical Society.

to the experimental vapor–liquid equilibrium data for the ternary mixture n-hexane, benzene, and 1-butanol. The results of this comparison are summarized in Table V for 10 mole-fraction compositions.

E. The Nonrandom Two-Liquids (NRTL) Model

Renon and Prausnitz (1968) further developed Wilson's concept of local concentrations with the view of obtaining a general equation applicable to miscible and partially miscible liquids. Consider a binary mixture as illustrated in Fig. 5.5. Each molecule is closely surrounded by other molecules; the immediate region around any central molecule is referred to as that molecule's cell. In a binary mixture of components 1 and 2 there are two types of cells, one type contains molecule 1 at its center, the other contains molecule 2 at its center. The chemical nature (1 or 2) of the molecules surrounding a central molecule depends on the mole fractions X_1 and X_2.

Let $G^{(1)}$ be the residual Gibbs free energy of a hypothetical fluid containing only type 1 cells; similarly, let $G^{(2)}$ be the same residual property of a hypothetical fluid containing only type 2 cells. The two-liquid theory states that

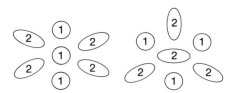

MOLECULE 1 AT CENTER MOLECULE 2 AT CENTER
Fig. 5.5 Two types of cells according to the two-liquid cell model.

the extensive Gibbs free energy of the mixture is given by

$$G^{\text{mix}} = X_1 G^{(1)} + X_2 G^{(2)} \tag{5.68}$$

The quantities $G^{(1)}$ and $G^{(2)}$ are assumed to be related to the local mole fraction compositions through

$$G^{(1)} = X_{11} g_{11} + X_{21} g_{21} \tag{5.69}$$

$$G^{(2)} = X_{12} g_{12} + X_{22} g_{22} \tag{5.70}$$

where $g_{ij}\,(=g_{ji})$ represents a parameter characteristic of binary i–j interactions. The local mole fraction X_{ij} refers to the mole fraction of component i in the immediate vicinity of a molecule j. The local mole fractions must obey the conservation equations

$$X_{21} + X_{11} = 1 \tag{5.71}$$

$$X_{12} + X_{22} = 1 \tag{5.72}$$

For pure component 1, $X_{21} = 0$ and therefore

$$G^{(1)}_{\text{pure}} = g_{11} \tag{5.73}$$

Similarly, for pure component 2

$$G^{(2)}_{\text{pure}} = g_{22} \tag{5.74}$$

The excess molar Gibbs free energy can now be found by considering first the change in residual Gibbs free energy that results when X_1 moles of component 1 are transferred from cells containing only component 1 to their cells in the binary mixture, and then the similar change for X_2 moles of component 2

$$\Delta \bar{G}^{\text{ex}}_{12} = X_1 [G^{(1)} - G^{(1)}_{\text{pure}}] + X_2 [G^{(2)} - G^{(2)}_{\text{pure}}] \tag{5.75}$$

Equation (5.75) gives the excess Gibbs free energy rather than the Gibbs free energy of mixing because a residual property is defined as that property relative to the ideal gas at the same temperature, pressure, and composition.

E. The Nonrandom Two-Liquids (NRTL) Model

Hence, the ideal solution contribution to the Gibbs free energy of mixing cancels.

Combining Eqs. (5.69–5.75) results in

$$\Delta \bar{G}_{12}^{ex} = X_1 X_{21}(g_{21} - g_{11}) + X_2 X_{12}(g_{12} - g_{22}) \quad (5.76)$$

In analogy to Eqs. (5.49) and (5.50), Renon and Prausnitz relate the local mole fractions to the overall mole fractions through Boltzmann factors

$$X_{21}/X_{11} = [X_2 \exp(-\alpha_{12} g_{21}/RT)/X_1 \exp(-\alpha_{12} g_{11}/RT)] \quad (5.77)$$

$$X_{12}/X_{22} = [X_1 \exp(-\alpha_{21} g_{12}/RT)/X_2 \exp(-\alpha_{21} g_{22}/RT)] \quad (5.78)$$

The parameter α_{12} ($=\alpha_{21}$) is a constant, characterizing the tendency of the components to mix in a nonrandom manner. When α_{12} equals zero, the local mole fractions are equal to the overall mole fractions and mixing is completely random.

Substituting Eqs. (5.77) and (5.78) into Eq. (5.76) enables the excess Gibbs free energy to be expressed in terms of the overall mole fractions X_1 and X_2 and the three binary parameters

$$\Delta \bar{G}_{12}^{ex} = RT X_1 X_2 \{[\tau_{21} G_{21}/(X_1 + X_2 G_{21})] + [\tau_{12} G_{12}/(X_2 + X_1 G_{12})]\} \quad (5.79)$$

where

$$\tau_{12} = (g_{12} - g_{22})/RT \quad \text{and} \quad \tau_{21} = (g_{21} - g_{11})/RT$$

$$G_{12} = \exp(-\alpha_{12} \tau_{12}) \quad \text{and} \quad G_{21} = \exp(-\alpha_{21} \tau_{21})$$

The working equations for the activity coefficients are obtained through the appropriate differentiation of Eq. (5.79)

$$\ln \gamma_1 = X_2^2 \{\tau_{21} [G_{21}/(X_1 + X_2 G_{21})]^2 + [\tau_{12} G_{12}/(X_2 + X_1 G_{12})^2]\} \quad (5.80)$$

$$\ln \gamma_2 = X_1^2 \{\tau_{12} [G_{12}/(X_2 + X_1 G_{12})]^2 + [\tau_{21} G_{21}/(X_1 + X_2 G_{21})^2]\} \quad (5.81)$$

For a solution containing n components, the NRTL expression for $\Delta \bar{G}_{12\cdots n}^{ex}$ is

$$\Delta \bar{G}_{12\cdots n}^{ex} = RT \sum_{i=1}^{N} X_i \frac{\sum_{j=1}^{N} X_j \tau_{ji} G_{ji}}{\sum_{m=1}^{N} G_{mi} X_m} \quad (5.82)$$

and the activity coefficient for a component i is

$$\ln \gamma_i = \frac{\sum_{j=1}^{N} X_j \tau_{ji} G_{ji}}{\sum_{m=1}^{N} X_m G_{mi}} + \sum_{j=1}^{N} \frac{X_j G_{ij}}{\sum_{m=1}^{N} G_{mj} X_m} \left(\tau_{ij} - \frac{\sum_{r=1}^{N} X_r \tau_{rj} G_{rj}}{\sum_{m=1}^{N} G_{mj} X_m} \right) \quad (5.83)$$

TABLE VI
Isobaric Vapor–Liquid Equilibrium Data for n-Hexane (1) + Benzene (2) + tert-Butyl Alcohol (3)[a]

Temperature (°C)	Liquid composition		Vapor composition		Calculated values Wilson			Calculated values NRTL		
	X_1	X_2	Y_1	Y_2	Temperature (°C)	Y_1	Y_2	Temperature (°C)	Y_1	Y_2
74.80	0.0234	0.2200	0.0634	0.3217	75.46	0.0656	0.3384	74.91	0.0665	0.3413
73.60	0.0350	0.2750	0.0934	0.3717	74.24	0.0885	0.3781	73.79	0.0897	0.3806
72.15	0.0500	0.4236	0.1036	0.4700	72.70	0.1045	0.4668	72.41	0.1055	0.4675
72.10	0.0584	0.3305	0.1400	0.4000	72.86	0.1306	0.4011	72.55	0.1323	0.4024
72.50	0.0734	0.2334	0.1720	0.3095	73.04	0.1759	0.3142	72.73	0.1768	0.3153
71.00	0.0800	0.4320	0.1834	0.4450	71.79	0.1549	0.4483	71.60	0.1566	0.4477
71.70	0.0968	0.1985	0.2300	0.2568	72.47	0.2277	0.2683	72.21	0.2319	0.2687
70.20	0.1200	0.4067	0.2080	0.4115	70.80	0.2175	0.4073	70.71	0.2202	0.4056
69.90	0.1367	0.2500	0.2817	0.2867	70.80	0.2757	0.2921	70.69	0.2807	0.2912
69.10	0.1585	0.3334	0.2880	0.3417	70.00	0.2820	0.3409	69.98	0.2863	0.3388
68.75	0.1767	0.3415	0.3020	0.3317	69.61	0.3011	0.3375	69.63	0.3056	0.3348
68.15	0.2210	0.3667	0.3350	0.3380	68.85	0.3403	0.3345	68.95	0.3448	0.3306
65.10	0.5734	0.2140	0.5967	0.1750	65.71	0.6057	0.1683	66.09	0.6086	0.1636
66.85	0.9640	0.0167	0.9150	0.0117	67.34	0.9253	0.0154	67.65	0.9231	0.0151
74.30	0.0717	0.8516	0.1385	0.7250	74.77	0.1203	0.7246	74.35	0.1239	0.7253
72.80	0.1400	0.7933	0.2166	0.6500	73.46	0.2139	0.6522	73.17	0.2198	0.6515

Wilson parameters (cal/mol):

$g_{12} - g_{11} = 30.567$, $g_{21} - g_{22} = 256.45$, $g_{23} - g_{22} = 210.556$,

$g_{32} - g_{33} = 706.139$, $g_{31} - g_{33} = 1313.188$,   and $g_{13} - g_{11} = -7.3445$

NRTL parameters (cal/mol):

$g_{12} - g_{22} = -115.16$, $g_{21} - g_{11} = 410.00$, $g_{23} - g_{33} = 804.25$,

$g_{32} - g_{22} = 50.91$, $g_{31} - g_{11} = 53.31$ and $g_{13} - g_{33} = 1130.68$

[a] Reprinted in part with permission from Govindaswamy et al. (1977). Copyright 1977 American Chemical Society.

F. The UNIQUAC and Effective UNIQUAC Models   101

As mentioned earlier, the NRTL model contains three constants for each binary system that are adjustable in fitting data. To simplify calculations, Renon and Prausnitz recommended values of α for broad classes of mixture systems. The values of α range from 0.20 to 0.50, and generally increase with the complexity of the i–j molecular interaction. When a value of α is selected in this manner, only two parameters τ_{12} and τ_{21} need to be determined from binary data.

An example of the applicability of the NRTL model is provided by the calculations of Govindaswamy *et al* (1977) for the system n-hexane + benzene + tert-butyl alcohol at one atmosphere. The NRTL parameters were calculated from the binary data; vapor compositions of the ternary system were then calculated and compared to experimental results, as shown in Table VI. The authors also calculated ternary equilibrium with the Wilson equation. Both the NRTL and Wilson model provide a good description for this ternary system.

F. The UNIQUAC and Effective UNIQUAC Models

The Universal Quasi Chemical (UNIQUAC) model, proposed by Prausnitz and co-workers (Abrams and Prausnitz, 1975; Maurer and Prausnitz, 1978), is based on a relatively simple two-fluid model, shown in Fig. 5.6. The binary mixture is considered to consist of X_1 moles of hypothetical fluid (1) and X_2 moles of hypothetical fluid (2). Each molecule in the mixture is divided into attached segments such that the molecules of type i have r_i segments. Although all segments have identical sizes, they differ in their external contact areas. For a molecule of type i, the number of nearest neighbors is given by zq_i, where z is the coordination number of the molecule surrounded by $z^{(1)}\theta_{11}q_1$ segments belonging to other molecules of type 1 and $z^{(1)}\theta_{21}q_1$ segments belonging to molecules of type 2.

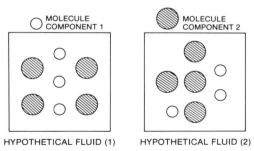

Fig. 5.6 Essential idea of the two-fluid model.

Therefore, $\theta_{11} + \theta_{21} = 1$, where θ_{11} is the local composition of component 1 (about a central molecule 1) and θ_{21} the local composition of component 2 (about a central molecule 1) in hypothetical fluid (1). Assuming $z^{(1)}$ in fluid (1) is identical to $z^{(\bullet)}$ in the pure fluid, it is found that the energy released by this condensation process is given by

$$U^{\text{cond}} = \tfrac{1}{2}z[\theta_{11}U_{11}^{(1)} + \theta_{21}U_{21}^{(1)}] \tag{5.84}$$

The superscript on z has been dropped for notational simplicity, and cond refers to condensation.

As also indicated in Fig. 5.6, a similar transfer can be made for a molecule 2 from the pure liquid denoted by the superscript ($\bullet$), to a hypothetical fluid denoted by the superscript (2).

The total change in energy accompanying the transfer of X_1 moles of species 1 from the pure liquid 1, and X_2 moles of species 2 from pure liquid 2 to the two-liquid mixture can be written as

$$\Delta U_{12}^{\text{mix}} = \tfrac{1}{2}zX_1 N_A[\theta_{11}q_1 U_{11}^{(1)} + \theta_{21}q_1 U_{11}^{(1)} - q_1 U_{11}^{(\bullet)}]$$
$$+ \tfrac{1}{2}zX_2 N_A[\theta_{22}q_2 U_{22}^{(2)} + \theta_{12}q_2 U_{12}^{(2)} - q_2 U_{22}^{(\bullet)}] \tag{5.85}$$

where N_A is Avogadro's number.

Remembering that the various U_{ij} terms represent interaction parameters characterizing the potential energies between molecules i and j, it is reasonable to expect as a first approximation that the interactions will be independent of other adjacent molecules, i.e., $U_{11}^{(\bullet)} = U_{11}^{(1)}$. These assumptions enable Eq. (5.85) to be rewritten as

$$\Delta \bar{U}_{12}^{\text{mix}} = \tfrac{1}{2}zN_A[X_1 q_1 \theta_{21}(U_{21} - U_{11}) + X_2 q_2 \theta_{12}(U_{12} - U_{22})] \tag{5.86}$$

in terms of potential energy differences $U_{21} - U_{11}$ and $U_{12} - U_{22}$. As will be later discovered, these differences in potential energies become the UNIQUAC parameters.

In analogy to Eqs. (5.49) and (5.50), Prausnitz et al. now relate the local area fractions to the surface compositions through Boltzmann factors

$$\theta_{21}/\theta_{11} = (\theta_2/\theta_1)\exp[-\tfrac{1}{2}z(U_{21} - U_{11})/kT] \tag{5.87}$$

$$\theta_{12}/\theta_{22} = (\theta_1/\theta_2)\exp[-\tfrac{1}{2}z(U_{12} - U_{22})/kT] \tag{5.88}$$

with the surface fraction θ_i calculated via

$$\theta_i = X_i q_i/(X_1 q_1 + X_2 q_2) \tag{5.89}$$

Combination of Eqs. (5.86–5.89) yields the following expression for the excess energy of mixing per mole of solution:

$$\Delta \bar{U}_{12}^{\text{mix}} = \Delta \bar{U}_{12}^{\text{ex}} = X_1 q_1 \theta_{21} \Delta u_{21} + X_2 q_2 \theta_{12} \Delta u_{12} \tag{5.90}$$

F. The UNIQUAC and Effective UNIQUAC Models

where

$$\Delta u_{21} = \tfrac{1}{2} z N_A (U_{21} - U_{11}) \quad \text{and} \quad \Delta u_{12} = \tfrac{1}{2} z N_A (U_{12} - U_{22})$$

$$\theta_{21} = \theta_2 \exp(-\Delta u_{21}/RT)/[\theta_1 + \theta_2 \exp(-\Delta u_{21}/RT)] \quad (5.91)$$

$$\theta_{12} = \theta_1 \exp(-\Delta u_{12}/RT)/[\theta_2 + \theta_1 \exp(-\Delta u_{12}/RT)] \quad (5.92)$$

Equation (5.90) is the basic relation based on the two-fluid theory, utilizing the notation of local composition.

To obtain an expression for the excess molar Gibbs free energy, one integrates $\Delta \bar{U}_{12}^{ex}$ from $1/T_o$ to $1/T$

$$\frac{\Delta \bar{G}_{12}^{ex}}{T} \approx \frac{\Delta \bar{A}_{12}^{ex}}{T} = \int_{1/T_o}^{1/T} \Delta \bar{U}_{12}^{ex} \, d(1/T) + \text{constant of integration} \quad (5.93)$$

The constant of integration can be evaluated by letting $1/T_o \to 0$. At very high temperatures, Prausnitz et al. assumed the binary mixture obeyed the equation of Guggenheim (1952) and Staverman (1950) for athermal mixtures of molecules of arbitrary shape and size

$$(\Delta \bar{G}_{12}^{ex}/RT)_{ath} = -(\Delta \bar{S}_{12}^{ex}/R)_{ath} = X_1 \ln(\phi_1/X_1) + X_2 \ln(\phi_2/X_2)$$
$$+ \tfrac{1}{2} z [q_1 X_1 \ln(\theta_1/\phi_1) + q_2 X_2 \ln(\theta_2/\phi_2)] \quad (5.94)$$

where ath denotes athermal. The UNIQUAC equation is derived by integrating Eq. (5.93)

$$\Delta \bar{G}_{12}^{ex}/RT = X_1 \ln(\phi_1/X_1) + X_2 \ln(\phi_2/X_2)$$
$$+ \tfrac{1}{2} z [q_1 X_1 \ln(\theta_1/\phi_1) + q_2 X_2 \ln(\theta_2/\phi_2)]$$
$$- X_1 q_1 \ln[\theta_1 + \theta_2 \exp(-\Delta u_{21}/RT)]$$
$$- X_2 q_2 \ln[\theta_2 + \theta_1 \exp(-\Delta u_{12}/RT)] \quad (5.95)$$

The activity coefficients of the individual components are

$$\ln \gamma_1 = \ln(\phi_1/X_1) + \tfrac{1}{2} z q_1 \ln(\theta_1/\phi_1) + \phi_2(\ell_1 - (r_1/r_2)\ell_2)$$
$$- q_1 \ln(\theta_1 + \theta_2 \tau_{21}) + \theta_2 q_1 \{[\tau_{21}/(\theta_1 + \theta_2 \tau_{21})] - [\tau_{12}/(\theta_2 + \theta_1 \tau_{12})]\} \quad (5.96)$$

and

$$\ln \gamma_2 = \ln(\phi_2/X_2) + \tfrac{1}{2} z q_2 \ln(\theta_2/\phi_2) + \phi_1[\ell_2 - (r_2\ell_1/r_1)]$$
$$- q_2 \ln(\theta_2 + \theta_1 \tau_{12}) + \theta_1 q_2 \{[\tau_{12}/(\theta_2 + \theta_1 \tau_{12})] - [\tau_{21}/(\theta_1 + \theta_2 \tau_{21})]\} \quad (5.97)$$

where

$$\ell_i = \tfrac{1}{2} z (r_i - q_i) - (r_i - 1)$$
$$\tau_{ij} = \exp(-\Delta u_{ij}/RT)$$

TABLE VII
Comparison between the UNIQUAC Representation and Experimental Vapor–Liquid Equilibria Data for the Ethanol (1) + 2-Butanone (2) System at 298.15 K

X_1	Experimental values			UNIQUAC model		
	γ_1	γ_2	P (kPa)	γ_1	γ_2	P (kPa)
0.073	2.103	1.013	12.63	2.19	1.00	12.51
0.138	1.917	1.021	12.80	1.95	1.01	12.71
0.273	1.604	1.083	13.05	1.58	1.08	12.96
0.328	1.481	1.100	12.85	1.46	1.11	12.86
0.482	1.251	1.251	12.68	1.26	1.24	12.63
0.505	1.240	1.261	12.57	1.25	1.24	12.48
0.570	1.176	1.351	12.40	1.18	1.33	12.30
0.659	1.096	1.478	11.88	1.11	1.46	11.87
0.770	1.042	1.648	11.00	1.05	1.68	11.13
0.839	1.028	1.839	10.47	1.02	1.85	10.43
0.918	1.019	2.152	9.60	1.01	2.18	9.56
0.939	1.004	2.334	9.24	1.00	2.27	9.16

UNIQUAC parameters: $\Delta u_{12}/R = -44.05$ and $\Delta u_{21}/R = 195.45$

TABLE VIII
Comparison between the UNIQUAC Representation and Experimental Vapor–Liquid Equilibria Data for the Ethanol (1) + Benzene (3) System at 298.15 K

X_1	Experimental values			UNIQUAC model		
	γ_1	γ_3	P (kPa)	γ_1	γ_3	P (kPa)
0.065	7.209	1.039	16.16	7.00	1.02	15.80
0.187	3.119	1.162	16.73	3.40	1.12	16.70
0.257	2.396	1.254	16.83	2.53	1.22	16.75
0.326	2.017	1.355	16.93	2.04	1.34	16.83
0.483	1.440	1.668	16.57	1.42	1.70	16.69
0.545	1.318	1.848	16.48	1.33	1.86	16.58
0.613	1.200	2.067	16.09	1.18	2.12	16.24
0.734	1.084	2.617	15.24	1.08	2.54	14.95
0.770	1.066	2.787	14.73	1.05	2.78	14.61
0.795	1.055	2.961	14.44	1.02	2.92	14.10
0.943	0.996	4.232	10.56	1.00	3.94	10.38

UNIQUAC parameters: $\Delta u_{13}/R = -78.66$ and $\Delta u_{31}/R = 454.39$

F. The UNIQUAC and Effective UNIQUAC Models

TABLE IX
Comparison between the UNIQUAC Representation and Experimental Vapor–Liquid Equilibria Data for the 2-Butanone (2) + Benzene (3) System at 298.15 K

X_2	Experimental values			UNIQUAC model		
	γ_2	γ_3	P (kPa)	γ_2	γ_3	P (kPa)
0.083	1.235	1.022	13.21	1.26	1.01	13.02
0.110	1.216	1.025	13.28	1.20	1.01	13.00
0.227	1.143	1.037	13.39	1.16	1.02	13.18
0.303	1.086	1.055	13.39	1.08	1.03	13.06
0.400	1.064	1.067	13.33	1.05	1.05	13.06
0.518	1.042	1.072	13.15	1.02	1.09	13.04
0.679	1.020	1.095	12.89	1.02	1.13	12.97
0.762	1.010	1.104	12.69	1.01	1.14	12.74
0.854	1.003	1.104	12.45	1.01	1.14	12.53
0.964	0.998	1.116	12.19	1.00	1.15	12.17

UNIQUAC parameters: $\Delta u_{23}/R = -175.82$ and $\Delta u_{32}/R = 258.87$.

is readily obtained through the appropriate differentiation of Eq. (5.95). The quantities Δu_{12} and Δu_{21} represent the two adjustable UNIQUAC parameters whose values are determined from the parametrization of binary data in accordance with Eqs. (5.95–5.96).

Like most of the two-parameter models presented previously, the UNIQUAC model can mathematically represent thermodynamic excess properties with a reasonable degree of accuracy. Tables VII–IX compare the descriptive ability of the UNIQUAC model to the liquid–vapor pressure measurements of Ohta et al. (1981) for ethanol + benzene, ethanol + 2-butanone, and benzene + 2-butanone mixtures. The calculated values are primarily within 6% of the experimental values, with the exceptions occurring in the ethanol + benzene system. While these deviations may seem large at first, it must be remembered that this system is nonideal, as is indicated by the large activity coefficients for ethanol.

The UNIQUAC model fails to give a satisfactory representation of vapor–liquid equilibrium for some highly nonideal systems. Highly nonideal refers to mixtures with limiting activity coefficients of 10 or more. The nonidealities in these systems often arise from strong, specific interactions, such as association and solvation. Thermodynamic properties of systems having strong specific interactions are more appropriately described by expressions containing equilibrium constants.

The derivation of Eq. (5.95) is readily extended to mixtures containing three or more components without additional assumptions. For a multicomponent mixture the UNIQUAC equations becomes

$$\Delta \bar{G}_{12\cdots n}^{ex}/RT = \sum_{i=1}^{N} X_i \ln(\phi_i/X_i) + \tfrac{1}{2} z \sum_{i=1}^{N} q_i X_i \ln(\theta_i/\phi_i)$$

$$- \sum_{i=1}^{N} q_i X_i \ln\left(\sum_j \theta_j \tau_{ji}\right) \qquad (5.98)$$

where the average area fraction θ_i and the average segment fraction ϕ_i are defined by

$$\theta_i = q_i X_i \bigg/ \sum_{j=1}^{N} X_j q_j$$

$$\phi_i = r_i X_i \bigg/ \sum_{j=1}^{N} X_j r_j$$

For any component i the activity coefficient is given by

$$\ln \gamma_i = \ln(\phi_i/X_i) + \tfrac{1}{2} z q_i \ln(\theta_i/\phi_i) + \ell_i - (\phi_i/X_i) \sum_j X_j \ell_j$$

$$+ q_i - q_i \ln\left(\sum_j^N \theta_j \tau_{ji}\right) - q_i \sum_j \frac{\theta_j \tau_{ij}}{\sum_k \theta_k \tau_{kj}} \qquad (5.99)$$

Inspection of Eqs. (5.98) and (5.99) reveals they contain only pure component and binary parameters. A prior knowledge of these parameters enable the estimation of liquid–vapor equilibrium in multicomponent mixtures. Table X compares the predictions of the UNIQUAC model to the experimental data of Ohta *et al.* (1981) for the ethanol + benzene + 2-butanone system. The interaction parameters used in the calculations were determined from contributing binary systems and are given in Tables VII–IX. From the comparison given in Table X it can be concluded the UNIQUAC model can provide reasonable estimates of multicomponent properties based on parameters determined from binary data.

In attempting to improve the predictive ability of the UNIQUAC model, Nagata and Katoh (1980) assumed the energy of mixing $\Delta \bar{U}_{12}^{mix}$ for a binary mixture was given by

$$\Delta \bar{U}_{12}^{mix} = X_1 \theta_{21} \Delta u_{21} + X_2 \theta_{12} \Delta u_{12} \qquad (5.100)$$

where θ_{ij} is the local area fraction of molecule i about a central molecule j

$$\theta_{21} = \theta_2 \exp(-\Delta u_{21}/CRT)/[\theta_1 + \theta_2 \exp(-\Delta u_{21}/CRT)] \qquad (5.101)$$

$$\theta_{12} = \theta_1 \exp(-\Delta u_{12}/CRT)/[\theta_2 + \theta_1 \exp(-\Delta u_{12}/CRT)] \qquad (5.102)$$

As before, Δu_{21} and Δu_{12} represent the binary interaction parameters, and C is a proportionality constant with a value near unity.[†]

[†] For the 24 different binary systems parametrized by Nagata and Katoh (1980), the numerical value of C ranged between 1.0 and 1.2.

F. The UNIQUAC and Effective UNIQUAC Models

TABLE X

Comparison between the UNIQUAC Representation and Experimental Activity Coefficients for Ethanol (1) + 2-Butanone (2) + Benzene (3) System at 298.15 K

			Experimental values			UNIQUAC model		
X_1	X_2	X_3	γ_1	γ_2	γ_3	γ_1	γ_2	γ_3
0.397	0.092	0.511	1.635	0.995	1.508	1.61	1.06	1.51
0.212	0.201	0.587	2.418	0.979	1.249	2.69	0.95	1.23
0.703	0.194	0.103	1.104	1.494	2.593	1.09	1.51	2.40
0.310	0.302	0.388	1.768	1.051	1.417	1.72	1.00	1.39
0.178	0.474	0.348	2.234	1.006	1.259	2.16	1.00	1.60
0.095	0.807	0.098	2.209	1.010	1.205	2.14	1.00	1.24
0.150	0.101	0.749	3.332	1.002	1.147	3.55	0.98	1.13
0.223	0.493	0.284	1.941	1.022	1.297	1.95	1.01	1.31
0.109	0.695	0.196	2.263.	1.008	1.203	2.26	1.01	1.19
0.121	0.415	0.464	2.696	0.997	1.165	2.81	0.99	1.16
0.816	0.108	0.076	1.046	1.747	3.159	1.02	1.69	2.93
0.502	0.195	0.303	1.360	1.153	1.834	1.26	1.20	1.81

The molar excess Gibbs free energy is

$$\Delta \bar{G}_{12}^{ex}/RT = X_1 \ln(\phi_1/X_1) + X_2 \ln(\phi_2/X_2)$$
$$+ \tfrac{1}{2}z[X_1 q_1 \ln(\theta_1/\phi_1) + X_2 q_2 \ln(\theta_2/\phi_2)]$$
$$- CX_1 \ln(\theta_1 + \theta_2 \tau'_{21}) - CX_2 \ln(\theta_2 + \theta_1 \tau'_{12}) \quad (5.103)$$

and

$$\tau'_{12} = \exp(-\Delta u_{12}/CRT) = \exp(-\alpha_{12}/CT)$$

obtained by integrating Eq. (5.93), the Guggenheim–Staverman equation again serving as the boundary condition. This newly derived excess Gibbs function, Eq. (5.103), is called the *Effective UNIQUAC Equation*.

The three-parameter effective UNIQUAC equation can be extended to multicomponent mixtures under the additional assumptions that the constant C is identical for all contributing binary systems. For a multicomponent system, Eq. (5.103) becomes

$$\Delta \bar{G}_{12\cdots n}^{ex}/RT = \sum_{i=1}^{N} X_i \ln(\phi_i/X_i) + \tfrac{1}{2}z \sum_{i=1}^{N} q_i X_i \ln(\theta_i/\phi_i)$$
$$- C \sum_{i=1}^{N} X_i \ln\left(\sum_j \theta_j \tau'_{ji}\right) \quad (5.104)$$

with the activity coefficients given by

$$\ln \gamma_i = \ln(\phi_i/X_i) + \tfrac{1}{2}zq_i \ln(\theta_i/\phi_i) + \ell_i - (\phi_i/X_i)\sum_j X_j \ell_j$$

$$+ C\left[-\ln\left(\sum_j X_j G_{ji}\right) + 1 - \sum_k \frac{X_k G_{ik}}{\sum_j X_j G_{jk}} - \ln\left(\frac{\theta_i}{X_i}\right) - 1 + \left(\frac{\theta_i}{X_i}\right)\right]$$

(5.105)

where

$$G_{ji} = (q_j/q_i)\tau'_{ji}$$

Whereas there have been few comparisons between the predictive abilities of the UNIQUAC and Effective UNIQUAC models, Nagata and Katoh showed the Effective UNIQUAC model gave a closed solubility curve for the acetonitrile + ethanol + cyclohexane system, whereas the original UNIQUAC equation erroneously predicted phase separation for the ethanol + cyclohexane system.

G. Summary

In experimental thermodynamics, it is frequently found that the excess free energy calculated from a simple theoretical model agrees well with experimental results, whereas the agreement for excess enthalpy and excess entropy, taken separately, is generally poor. The prediction of excess volume, if possible, is the worst. Similarly, if a solution theory is improved by choosing a better model or by introducing new parameters, the modifications for predicting excess entropy, enthalpy, and volume are much greater in magnitude than those in the free energy. Hildebrand and Scott (1950a) explained this observation based on the following mathematical argument:

> In order to calculate the thermodynamic properties of a particular system, we assume, as a first approximation, an oversimplified model. Thermodynamic functions calculated from such a model we denote by G_o, H_o, S_o, etc.
>
> We know, however, than in a particular respect, our system does not correspond to the assumed model. With respect to a particular property (for example, volume or distribution of like and unlike pairs), designated by a parameter α, the system has the value of α' rather than α_o. We may now now express the free energy of the actual α' in the form of a Taylor series expansion around α_o.
>
> $$G' = G_o + (\partial G/\partial \alpha)_{\alpha=\alpha_o}(\alpha' - \alpha_o) + \tfrac{1}{2}(\partial^2 G/\partial \alpha^2)_{\alpha=\alpha_o}(\alpha - \alpha_o)^2 + \cdots \quad (5.106)$$

G. Summary

We may equally well, however, expand around α'.

$$G_o = G' + (\partial G/\partial \alpha)_{\alpha=\alpha'}(\alpha_o - \alpha') + \tfrac{1}{2}(\partial^2 G/\partial \alpha^2)_{\alpha=\alpha'}(\alpha_o - \alpha')^2 + \cdots \quad (5.107)$$

The system, at equilibrium, must have chosen the value of α which minimizes G. Hence we may set $(\partial G/\partial \alpha)_{\alpha=\alpha'}$ equal to zero. Rearranging, we obtain:

$$G' - G_o = \delta G = -\tfrac{1}{2}(\partial^2 G/\partial \alpha^2)_{\alpha=\alpha'}(\alpha' - \alpha_o)^2 + \cdots \quad (5.108)$$

We see that we have lost the term in $\Delta\alpha$ and that the correction to be applied to the free energy depends only on the square of the quantity representing the deviation of the actual system from the model.

On the contrary, the entropy and enthalpy are not minimized, and applying the same reasoning to them, we obtain:

$$S' - S_o = \delta S = (\partial S/\partial \alpha)_{\alpha=\alpha'}(\alpha' - \alpha_o) - \tfrac{1}{2}(\partial^2 S/\partial \alpha^2)_{\alpha=\alpha'}(\alpha' - \alpha_o)^2 + \cdots \quad (5.109)$$

$$H' - H_o = \delta H = (\partial H/\partial \alpha)_{\alpha=\alpha'}(\alpha' - \alpha_o) - \tfrac{1}{2}(\partial^2 H/\partial \alpha^2)_{\alpha=\alpha'}(\alpha' - \alpha_o)^2 + \cdots \quad (5.110)$$

We may calculate $\Delta\alpha$ from the relation

$$(\partial G/\partial \alpha)_{\alpha=\alpha'} = 0 \quad (5.111)$$

or equivalently:

$$T(\partial s/\partial \alpha)_{\alpha=\alpha'} = (\partial H/\partial \alpha)_{\alpha=\alpha'} \quad (5.112)$$

Provided that our initial choice of model was not completely unreasonable, $\Delta\alpha$ will be small, and the series will rapidly converge. Since δG depends only upon $(\Delta\alpha)^2$ and higher powers, while the δS and δH series begin with $\Delta\alpha$, we conclude that an oversimplified model introduced far more serious errors in the calculation of the enthalpy and the entropy separately than in in that of the free energy (pp. 135–136)

Because Gibbs free energy, enthalpy, and entropy can all be expressed as a power series of deviation, $\Delta\alpha = \alpha' - \alpha_o$, only one of these thermodynamic properties can be minimized to determine the parameter α. Usually, excess Gibbs free energy data are applied, and the calculated parameter value generally gives large errors for both excess enthalpy and excess entropy. If a better model is chosen, the improvements in predicting excess enthalpy and entropy are proportional to $\Delta\alpha$, and therefore larger than the improvements in predicting the excess free energy that is proportional to $(\Delta\alpha)^2$. As a result, a comparison of excess enthalpy and entropy in place of excess free energy provides a more sensitive test of a particular model.

Alternatively, if α is selected such that $(\partial H/\partial \alpha)$ equals zero, the model will give better results for enthalpy than for entropy or free energy. Therefore, in order to provide a good description of the solution's thermodynamic properties, the model must be able to predict two of these three quantities: G, H and S, with the third quantity calculated from the definition of the Gibbs free energy.

A consequence of the mathematics presented here is the familiar linear free energy relationships in kinetic studies [see Eyring et al. (1961), Frost and Pearson (1961), and Moore and Pearson (1981)], the Hammett (1935) equation, the Taft (1952) equation, and the Bronsted equation (Bronstead and Penderson, 1924). The observed phenomena indicate that a slight change in the enthalpy term is usually compensated for by a corresponding change in the entropy term, and the combined effect is a small change in the free energy. The Hammett equation (Eyring et al. 1961; Hammett, 1935) represents the change in reaction rates of different benzene derivatives because of inductive effects. The Taft equation (Eyring et al. 1961; Taft, 1952) correlates the reaction rates of aliphatic acid and their steric effects. The Bronsted equation (Eyring et al. 1961; Bronstead and Penderson, 1924) relates the reaction rate of an acid to its acid dissociation constant.

In terms of transition state theory, the reaction rate depends on the free energy of activation. The correlation of reaction rates can be considered as a correlation of activational free energies. These correlations arise as a natural consequence of the linear relationship between an entropic and enthalpic term. For example, when the molecular environment of a molecule changes because of an inductive or substituent effect, both entropy and enthalpy of activation change proportionally but not necessarily equally. The entropy and enthalpy terms tend to cancel each other, resulting in a much smaller variation for the free energy term.

Linear free energy relationships are also observed in solution processes (see Bronstead and Penderson, 1924; Butler, 1937; and Langer and Purnell, 1963, 1966). A solution process can be envisioned as a change in the environment of a solute. The α in Eq. (5.106) is the parameter that characterizes the degree of change during a mixing process. The enthalpy of mixing and the associated orientational entropy for solutes are proportional to $\Delta\alpha$, and thus a linear relationship can be expected:

$$\Delta \bar{S}^{ex} = \Delta \bar{H}^{ex} (\partial \bar{S}^{ex}/\partial \bar{H}^{ex})_{\alpha=\alpha_o} \tag{5.113}$$

Although Eq. (5.113) has a zero intercept when $\Delta \bar{H}^{ex} = 0$, experimental observations (Langer and Purnell, 1963, 1966) for excess thermodynamic properties of solutions do not indicate this. The apparent discrepancy arises because a combinatorial entropy term exists even when a solution process occurs athermally.

It is sufficient to state that excess free energy, enthalpy, entropy are not equally weighted in the mathematical structure of solution theory. Excess entropy and enthalpy, separately, will give more insight into solution properties. Excess free energies, however, are more important from an engineering point of view and it is for this reason the "success" of a solution model depends on its ability to predict activity coefficients and phase equilibrium.

Problems

Problems

5.1 Treating the energy differences as temperature independent quantities, show that the excess isobaric heat capacity ($\Delta \bar{C}^{ex}_{p_{12}}$) of the Wilson model is given by:

$$\Delta \bar{C}^{ex}_{p_{12}} = \frac{X_1 X_2}{RT^2} \left[\frac{X_1 \Lambda_{12}(g_{12} - g_{11})^2}{(X_1 + X_2 \Lambda_{12})^2} + \frac{X_2 \Lambda_{21}(g_{12} - g_{22})^2}{(X_2 + X_1 \Lambda_{21})^2} \right]$$

5.2 Listed in Table XI are the excess Gibbs free energies (base on Raoult's law) of cyclohexane (1) + octamethylcyclotetrasiloxane (2) mixtures. Convert these excess free energies to values based on the Flory–Huggins model of solution ideality. The molar volumes of cyclohexane and OMCTS are 108.76 and 314.00 cm³/mol, respectively.

5.3 Derive expressions for the activity coefficients of a binary mixture based on the nonathermal Flory–Huggins solution model. What would the activity coefficients be for the athermal model?

5.4 The NRTL equation for the excess free energy of mixing is:

$$\Delta \bar{G}^{ex}_{12} = X_1 X_2 RT \left[\frac{\tau_{21} \exp(-\alpha \tau_{21})}{X_1 + X_2 \exp(-\alpha \tau_{21})} + \frac{\tau_{12} \exp(-\alpha \tau_{12})}{X_2 + X_1 \exp(-\alpha \tau_{12})} \right]$$

$$\tau_{21} = (g_{21} - g_{11})/RT \quad \text{and} \quad \tau_{12} = (g_{12} - g_{22})/RT$$

where α and the various g_{ij} terms are assumed to be independent of temperature. For the nitromethane (1) + isooctane (2) system calculate

TABLE XI
Experimental Excess Gibbs Free Energies for Cyclohexane (1) + OMCTS (2) at 308.15 K[a]

X_2	$\Delta \bar{G}^{ex}_{12}$ (J/mol)
0.10103	−20.8
0.20720	−40.7
0.28156	−52.1
0.36325	−61.0
0.44613	−65.6
0.53953	−64.9
0.62624	−58.6
0.72897	−45.1
0.78479	−42.9
0.90591	−19.8

[a] Table reproduced with permission from Tomlins and Marsh (1976). Copyright Academic Press, New York.

TABLE XII
Vapor-Liquid Equilibria of Mixtures of 1-Hexanol (1) + n-Hexane (2) Mixtures at 293.15 K[a]

X_1	γ_1	γ_2	P (kPa)
0.0000	23.76	1.000	16.17
0.1074	4.537	1.075	15.77
0.1990	2.696	1.178	15.33
0.2967	1.937	1.311	14.93
0.4010	1.537	1.483	14.29
0.4942	1.329	1.667	13.49
0.5976	1.184	1.915	12.39
0.7080	1.088	2.242	10.77
0.8242	1.030	2.683	7.64
0.9176	1.006	3.132	4.24
1.0000	1.000	3.622	0.06

[a] Reproduced with permission from Wolff and Shadiakly (1981).

TABLE XIII
Vapor Pressures Above Binary Mixtures of n-Heptane (1) + 1-Propanol (2) at 298.15 K[a]

X_1	P (kPa)
0.0000	2.849
0.0699	5.169
0.1523	6.490
0.2570	7.261
0.3602	7.590
0.4789	7.783
0.6489	7.893
0.7164	7.914
0.7725	7.915
0.8374	7.915
0.8746	7.881
0.9493	7.686
1.0000	6.101

[a] Reproduced with permission from Sayegh et al. (1979).

$\Delta \bar{G}_{12}^{ex}$ and $\Delta \bar{H}_{12}^{ex}$ at 25°C for $X_1 = 0.25$, 0.50, and 0.75. Numerical values of the NRTL parameters are $\alpha = 0.20$, $g_{12} - g_{11} = 4{,}180$ J/mol, and $g_{21} - g_{22} = 2{,}090$ J/mol.

5.5 In reporting the vapor-pressure behavior of the 1-hexanol (1) + n-hexane (2) system at 293.15 K, Wolff and Shakiakly elected to parametrize the experimental data according to the Wilson equation. Using their values of $g_{12} - g_{11} = 7{,}127$ J/mol and $g_{12} - g_{22} = 723$ J/mol, calculate the activity coefficients of both components and the pressure above the binary mixture at the nine compositions listed in Table XII. Compare your calculated values to the experimental values given in Table XII. Does the Wilson equation provide a satisfactory mathematical representation of the binary data?

5.6 Sayegh et al. (1979) parameterized the vapor-liquid equilibrium data of n-heptane (1) + 1-propanol (2) mixture with the Wilson equation. Using their binary parameters, $g_{12} - g_{11} = 8883$ J/mol and $g_{12} - g_{22} = 1147$ J/mol, calculate the vapor pressure at the 11 binary compositions listed in Table XIII. Does the Wilson equation adequately describe the experimental vapor pressures?

5.7 Using the Wilson equation and binary parameters from problem 5.5, predict the excess enthalpy of n-heptane (1) + 1-propanol (2) mixtures. Compare your predictions to the experimental values of Savini et al.

TABLE XIV
Calorimetric Data for Binary Mixtures of n-Heptane (1) + 1-Propanol (2) at 30°C[a]

X_2	$\Delta \bar{H}_{12}^{ex}$ (J/mol)
0.100	529
0.200	654
0.300	708
0.400	708
0.500	668
0.600	590
0.700	485
0.800	354
0.900	194

[a] Reprinted in part with permission from Savini, Winterhalter and Van Ness (1965). Copyright 1965 American Chemical Society.

TABLE XV
Experimental Vapor–Liquid Equilibria Data for the 1-Butanol (1) + n-Hexane (2) System at 332.53 K[a]

X_1	P (kPa)
0.0000	76.284
0.1291	76.323
0.2152	75.054
0.3160	73.251
0.3905	71.857
0.5114	68.501
0.5928	65.350
0.6911	60.018
0.7810	52.066
0.9125	31.623
1.0000	8.099

[a] Reprinted with permission from Berro et al. (1982). Copyright 1982 American Chemical Society.

listed in Table XIV. Briefly discuss reasons for any discrepancies between the predicted and experimental values.

5.8 Although the Wilson parameters (Λ_{12} and Λ_{21}) are generally determined from binary data via reiterative calculations, it is possible to estimate these parameters from infinite dilution activity coefficients. Develop a numerical procedure for calculating the parameters Λ_{12} and Λ_{21} from known values of γ_1^∞ and γ_2^∞. (Hint: Let $X_i \to 0$ in the Wilson equation for the acitivity coefficients.)

5.9 In reporting the vapor–liquid equilibrium of the binary 1-butanol (1) + n-hexane (2) system, Berro et al. (1982) parametrized the experimental data to the Wilson equation. Using their values of $\Lambda_{12} = 0.117451$ and $\Lambda_{21} = 0.516085$, calculate the vapor pressure at the nine binary compositions listed in Table XV. Does the Wilson equation adequately describe the experimental vapor pressures?

5.10 From a theoretical standpoint, the Wilson parameters in a ternary system are related by

$$(\Lambda_{23}/\Lambda_{32}) = (\Lambda_{13}/\Lambda_{31})(\Lambda_{21}/\Lambda_{12}).$$

Does the NRTL equation have a similar restraining condition? If a restraint exists, what is it?

TABLE XVI
Experimental Data for the Methyl Acetate (1) + 1-Hexene (2) System at 50°C[a]

X_1	P (mm Hg)
0.0000	485.18
0.1087	568.95
0.2285	630.00
0.2948	654.56
0.4147	683.74
0.5417	696.75
0.6159	698.63
0.7611	690.45
0.8163	680.18
0.8904	660.86
1.0000	595.05

[a] Reprinted in part with permission from Gmehling (1983). Copyright 1983 American Chemical Society.

TABLE XVII
Experimental Liquid-Vapor Equilibrium Data for n-Heptane (1) + 1-Propanol (2) + Chlorobutane (3) Mixtures at 318.15 K[a]

X_1	X_2	P (kPa)
0.0000	0.0000	31.53
1.0000	0.0000	15.30
0.0000	1.0000	9.34
0.1095	0.3895	31.25
0.1650	0.5884	27.43
0.2041	0.7296	22.10
0.2535	0.2511	30.51
0.3981	0.3946	26.25
0.4764	0.4724	22.42
0.6902	0.2594	22.34
0.8163	0.0604	22.57
0.8547	0.0929	21.69

[a] Reproduced with permission from Ashraf and Vera (1981).

5.11 Gmehling parametrized the liquid–vapor equilibrium data of methyl acetate (1) + 1-hexene (2) mixtures with the NRTL equation. Using his binary parameters $g_{21} - g_{11} = 80.0781$ J/mol, $g_{12} - g_{22} = 619.9473$ J/mol, and $\alpha_{12} = 0.298$, calculate the vapor pressure at the nine binary compositions listed in Table XVI.

5.12 Predict the vapor pressure above n-heptane (1) + 1-propanol (2) + chlorobutane (3) mixtures at 318.15K using the Wilson equation. Compare your predictions to the experimental values of Ashraf and Vera listed in Table XVII. Values needed in the calculations include $g_{12} - g_{11} = 1122$ J/mol, $g_{13} - g_{33} = 1286$ J/mol, $g_{12} - g_{22} = 7985$ J/mol, $g_{23} - g_{22} = 615.2$ J/mol, $g_{13} - g_{11} = -229.5$ J/mol, $g_{23} - g_{33} = 5420$ J/mol, and $\bar{V}_3^{\bullet} = 105.8$ cm^3/mol.

6

Prediction of Thermodynamic Excess Properties of Liquid Mixtures Based on Group Contribution Methods

In designing chemical processes for the separation of fluid mixtures, chemical engineers must frequently estimate the liquid-phase activity coefficients of multicomponent systems. As seen in earlier chapters, there are many methods for estimating multicomponent properties, provided the binary data is available either in the form of actual numerical values or in the form of binary coefficients determined from a least-squares analysis of binary data. These predictive methods, however, are virtually useless in many instances as experimental measurements were made on only a small fraction of all possible binary systems. To address this need, various individuals[†] developed predictive expressions based on a group contribution approach.

The rationale behind this idea is that whereas there are thousands of chemical compounds of interest in chemical technology, the number of functional groups that constitute these compounds is much smaller. Therefore, if one assumes that a physical property of liquid mixtures is the sum of contributions made by the molecules' functional groups, one obtains a possible technique for correlating and predicting the properties of a large number of compounds in terms of a much smaller number of parameters that characterize the contributions of individual groups.

The most important discussion in early studies of group contribution methods for excess properties was made by Pierotti, Deal et al. (1962). These authors discussed the extensive measurement and phenomenological correlations for infinite dilution activity coefficients determined by gas–liquid chromatography. They noted that the activity coefficients for infinitely dilute solutions are summed functions of the various groups. Six different kinds of

---

[†] See Derr and Deal (1969), Fredenslund et al. (1975, 1977a,b), Gmehling et al. (1978, 1982), Lai et al. (1978), Maripuri and Ratcliff (1971a,b), Nguyen and Ratcliff (1971a,b, 1974), (1975), Ratcliff and Chao (1969), Ronc and Ratcliff (1975), Sayegh and Ratcliff (1976), Siman and Vera (1979), Skjøld–Jorgensen et al. (1979), and Wilson and Deal (1962).

interactions were needed to describe the functional group and skeletal interactions in homologous monofunctional compounds.

Deal, *et al.* (1962) considered the activity coefficients of hydrocarbons in infinitely dilute solutions of some polar solvents and found it was possible to correlate data by assuming olefins, naphthenes and aromatics to be pseudo-saturated hydrocarbons with few parameters.

It was not until the work of Wilson and Deal (1962) that the group method was used to describe thermodynamic excess properties at finite concentrations. Wilson and Deal's description of liquid mixtures is based on the following four assumptions:

(1) The partial molar excess free energy, or simply, the logarithm of the activity coefficient is separated into two contributions, one associated with differences in molecular size and the second associated with interactions between structural "groups." For molecule i in any solution

$$\log \gamma_i = \log \gamma_i^s + \log \gamma_i^g. \tag{6.1}$$

(2) The contribution associated with molecular size differences is given by a Flory–Huggins relation expressed in terms of the numbers of constituent atoms other than hydrogen. For molecule i this becomes

$$\log \gamma_i^s = \log\left(v_i \bigg/ \sum_{j=1}^{N} X_j v_j\right) + 0.4343\left[1 - v_i \bigg/ \left(\sum_{j=1}^{N} X_j v_j\right)\right] \tag{6.2}$$

where v_i is the number of atoms (other than hydrogen) in molecular component i, and X_i the mole fraction of component i in the liquid mixture and the summations extend over all the components present in the solution.

(3) The contribution from interactions between molecular groups can be expressed as the sum of the individual contributions of each solute group in the solution minus the sum of the individual contributions in the conventional standard state environment. For molecule i containing k different functional groups

$$\log \gamma_i^g = \sum_k v_{ki}(\log \Gamma_k - \log \Gamma_k^*) \tag{6.3}$$

where v_{ki} is the number of groups of type k in component i, Γ_k the activity coefficient of group k in the solution environment referred to by an arbitrary standard state, and Γ_k^* the activity coefficient of group k in the standard state environment. This activity coefficient is referred to the same state as Γ_k. The standard state chosen for a group is that of the pure molecular species i under consideration. This agrees with normal practice and ensures that the molecular activity coefficient is unity for the pure compound. As a consequence, the same functional group must be referred to different standard states depending on the molecular species under consideration. For example, if aliphatic alcohols are considered to be made up of methylene and hydroxyl groups, then the standard state of the hydroxyl group in methanol is its state in a solution containing 50% hydroxyl group and 50% methylene groups. The standard state of the hydroxyl group in propanol is its state in a solution containing 25% hydroxyl groups and 75% methylene groups.

The definition of the standard state is completed by specifying the temperature and pressure. Because the concern here is with condensed systems around atmospheric pressure, the pressure effect on free energy will be negligible and will not be considered further.

(4) Finally, the individual group contributions (Γ_k) in any environment containing groups of given kinds are assumed to be only a function of group concentrations, temperature, and pressure

$$\Gamma_k = \Gamma_k(X_1^g, X_2^g, \ldots, X_k^g, \ldots, T, P) \qquad (6.4)$$

where

$$X_k^g = \text{group fraction of group } k = \sum_j X_j v_{jk} \Big/ \sum_j X_j v_j$$

Based on these four assumptions and experimental data for binary n-hexane + methanol mixtures, Wilson and Deal calculated $\log \Gamma$ for the hydroxyl and methylene groups at different group concentrations. Using a graphical representation of $\log \Gamma_k$ vs. X_k^g for the n-hexane + methanol system, the authors then proceeded to estimate the activity coefficients of the n-heptane + ethanol system to within 15%, the maximum deviations occurring at the limits of infinite dilution.

Although Wilson and Deal specified a particular mathematical form for the combinatorial contributions (namely the Flory–Huggins model), they did not specify a mathematical relationship between the individual group contributions Γ_k and group concentrations X_k^g. Attempts to find the "best" compositional form for these group contributions, along with the most appropriate combinatorial form, led to the development of several predictive methods. As will now be seen, the Analytical Solution of Groups model (ASOG), the Analytical Group Solution model (AGSM), and the UNIQUAC Functional Group Activity Coefficients model (UNIFAC) are all based on Eq. (6.1), with a different mathematical form used for $\log \gamma_i^s$ and $\log \Gamma_k$.

A. Analytical Solution of Groups Model (ASOG)

The ASOG model, proposed by Derr and Deal (1969), follows naturally from the ideas of Wilson and Deal in that the activity coefficient is divided into two separate contributions

$$\log \gamma_i = \log \gamma_i^s + \log \gamma_i^g$$

The combinatorial part of the activity coefficient is described by a Flory–Huggins-type expression

$$\log \gamma_i^s = \log\left(v_i \bigg/ \sum_{j=1}^N v_j X_j\right) + 0.4343\left[1 - v_i \bigg/ \left(\sum_{j=1}^N v_j X_j\right)\right]$$

and the group contribution described by Eq. (6.3), and the Wilson equation

$$\log \gamma_i^g = \sum_k v_{ki}(\log \Gamma_k - \log \Gamma_k^{(i)}) \tag{6.5}$$

$$\ln \Gamma_k = -\ln\left(\sum_\ell X_\ell^g a_{k\ell}\right) + 1 - \sum_1 X_\ell^g a_{\ell k} \bigg/ \sum_m X_m^g a_{\ell m} \tag{6.6}$$

where the summations $\sum_k$, $\sum_\ell$, and $\sum_m$ are over all possible groups in the solution. In this expression v_j refers to the number of atoms (other than hydrogen atoms) in molecule j, v_{ki} is the number of atoms (other than hydrogen atoms) of group k in molecule i, and Γ_k is the group activity coefficient of group k, with the superscript (i) indicating the standard state is referred to pure component i as discussed earlier.[†]

The group Wilson parameters, $a_{k\ell}$ and $a_{\ell k}$, are determined from a least-squares analysis of the binary data. Whereas one might expect the various $a_{k\ell}$ parameters to be temperature independent, analysis of a large number of binary systems indicated these numerical values are inversely proportional to the absolute temperature

$$\ln a_{k\ell} = m_{k\ell} + n_{k\ell}/T \tag{6.7}$$

in the temperature regions commonly studied (20–110°C). This is consistent with the exponential form of the Wilson Λ_{ij} parameters as Eq. (6.7) can be written as

$$a_{k\ell} = \exp[m_{k\ell} + (n_{k\ell}/T)] \tag{6.8}$$

Table I lists the numerical values of $m_{k\ell}$ and $n_{k\ell}$ for several group interactions.

To illustrate the nomenclature and calculational procedures associated with the ASOG model, consider the following example.

[†] In the ASOG model it is assumed that the CH_3 group is equivalent to the CH_2, whereas for H_2O, and CH and C in alkanes, values of v_{ki} have been determined with reference to the observed and predicted activity coefficients. The values adopted as a result are $v_{H_2O} = 1.6$, $v_{CH} = 0.8$, and $v_C = 0.5$. Some examples of v_{ki} appear in Table II.

TABLE I
ASOG Group Parameters $m_{k\ell}$ and $n_{k\ell}$

k	CH$_2$	C=C	ArCH	CyCH	OH	CO	O
CH$_2$	$m_{k\ell}$ = 00 $n_{k\ell}$ = 00	1.33 −995	−0.74574 146.02	0.153 2.100	−41.3 7685	2.63 865	−0.095 32.5
C=C	−1.525 714.0	00 00	−0.575 340	— —	−0.160 −248	0.145 40.5	— —
ArCH	0.72974 −176.76	−1.60 247	00 00	0.329 156	2.27 −1110	0.927 186	— —
CyCH	−0.184 0.300	— —	0.53 −250	00 00	−12.0 −2230	3.28 −1045	— —
OH	4.713 −3060	10.58 −4550	−0.59 −940	5.64 −3220	00 00	−0.726 2.9	−0.67 −151
CO	−1.760 170	−1.64 110	−0.40 −215	−2.72 428	−0.330 1.30	00 00	−0.807 1.822
O	−0.51 166	— —	— —	— —	0.935 −152	1.08 1.1	00 00

TABLE II
Numerical Values of v_{ki} and v_i

Group	Name	$v_{ki}{}^a$	v_i
Alkanes CH_2	n-Hexane	CH_2:6	6
	2,3-Dimethylbutane	CH_2:5.6	6
	2,2,4-Trimethylpentane	CH_2:7.3	8
Alkenes C=C	1-Hexene	CH_2:4, C=C:2	6
	2-Methyl-1-butene	CH_2:2.8, C=C:2	5
Aromatics ArCH	Benzene	ArCH:6	6
	Toluene	CH_2:1, ArCH:6	7
	Ethylbenzene	CH_2:2, ArCH:6	8
	Styrene	C=C:2, ArCH:6	8
Cycloalkanes CyCH	Cyclohexane	CyCH:6	6
	Ethylcyclohexane	CH_2:2, CyCH:6	8
Alcohols OH	Ethanol	CH_2:2, OH:1	3
	2-Propanol	CH_2:2.8, OH:1	4
	t-Butanol	CH_2:3.5, OH:1	5
Ketones CO	Acetone	CH_2:2, CO:2	4
	2-Butanone	CH_2:3, CO:2	5
Ethers O	Diethylether	CH_2:4, 0:1	5
	Diisopropylether	CH_2:5.6, 0:1	7

[a] Values based on $v_{CH} = 0.8$ and $v_C = 0.5$.

EXAMPLE 1

Using the ASOG model, estimate the activity coefficients for n-heptane (1) + benzene (2) system at 298.15 K and $X_1 = 0.5$. From Table II, it is found n-heptane has 7 "CH_2" groups, thus $v_1 = 7$ and $v_{CH_2,1} = 7$, and benzene has 6 "ArCH" groups, thus $v_2 = 6$ and $v_{ArCH,2} = 6$. The required group interaction parameters are obtained from Table I[†] and Eq. (6.8)

$$a_{ArCH,CH_2} = \exp[0.72974 - 176.76/298.15] = 1.146$$

$$a_{CH_2,ArCH} = \exp[-0.74574 + 146.02/298.15] = 0.7742$$

[†] Table I was compiled in a "piecemeal" fashion by combining values reported in several different articles. The values are used to illustrate the ASOG method and are not intended to represent the best parameters. For those wishing to use the ASOG method, Tochigi et al. (1981) presented ASOG group parameters based on

$$a_{k\ell} = (v_\ell/v_k) \exp(-b_{k\ell}/T)$$

A. ASOG

For pure n-heptane

$$\ln \Gamma^{(1)}_{CH_2} = \ln \Gamma^{(1)}_{CH_2}(X^g_{CH_2} = 1) = 0$$

For pure benzene

$$\ln \Gamma^{(2)}_{ArCH} = \ln \Gamma^{(2)}_{ArCH}(X^g_{ArCH} = 1) = 0$$

For the binary mixture

$$X^g_{ArCH} = (0.5)(6)/[(0.5)(6) + (0.5)(7)] = 0.462$$

$$X^g_{CH_2} = (0.5)(7)/[(0.5)(6) + (0.5)(7)] = 0.538$$

$$\ln \Gamma_{ArCH} = -\ln[0.462 + (0.538)(1.146)] + 1 - 0.462/[0.462 \\ + (0.538)(1.146)] - (0.538)(0.7742)/[0.538 + (0.462)(0.7742)]$$

$$= 0.03100$$

The activity coefficient of benzene is now estimated via Eqs. (6.4–6.6)

$$\ln \gamma_2 = \ln\{6/[(0.5)(6) + (0.5)(7)]\} + 1 - 6/[(0.5)(6) + (0.5)(7)] \\ + 6(0.03100 - 0.000)$$

$$= 0.183$$

$$\gamma_2 = 1.20$$

This value compares favorably with the experimental value of $\gamma_2 = 1.241$ at $X_1 = 0.5564$ determined by Jain et al. (1973). Similarly, the activity coefficient of n-heptane in the binary mixture can be estimated as

$$\ln \Gamma_{CH_2} = -\ln[0.538 + (0.462)(0.7742)] + 1 - 0.538/[0.538 \\ + (0.462)(0.7742)] - (0.462)(1.146)/[0.462 + (0.538)(1.146)]$$

$$= 0.01862$$

$$\ln \gamma_1 = \ln\{7/[(0.5)(6) + (0.5)(7)]\} + 1 - 7/[(0.5)(6) + (0.5)(7)] \\ + 7(0.01862 - 0.00000)$$

$$= 0.1275$$

$$\gamma_1 = 1.14$$

Jain et al. reported an experimental value of $\gamma_1 = 1.144$ at $X_1 = 0.5564$.

As was seen in the previous example, the application of the ASOG model is relatively straightforward and requires only a prior knowledge of group interaction parameters. The important point here is that at a fixed temperature these parameters depend only on the nature of the groups and, by assumption, are independent of the nature of the molecule. Therefore, group parameters obtained from available experimental data for some mixtures

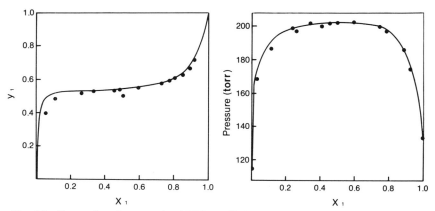

Fig. 6.1 Comparison between the ASOG predictions (———) and the experimental liquid–vapor equilibria data (●) for the ethanol (1) + isooctane (2) system. The compositions of the vapor phase y_1 were calculated from Eq. (3.95) using the ASOG model to predict the liquid phase activity coefficients. The experimental data were taken from Ratcliff and Chao (1969).

can be used to predict activity coefficients in other mixtures that contain not the same molecules but the same groups. For example, to predict activity coefficients in the binary system 2-butanone + nitrobenzene would require group interaction parameters for characterizing interactions between methyl, phenyl, keto, and nitro groups. These parameters can be evaluated from other binary mixtures containing these four functional groups, i.e., acetone + benzene, nitropropane + toluene, and acetone + nitropropane.

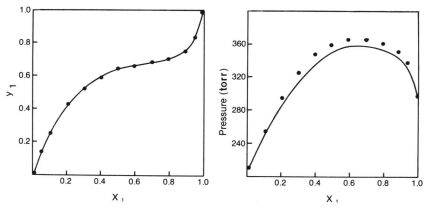

Fig. 6.2 Comparison between the ASOG predictions (———) and the experimental liquid–vapor equilibria data (●) for the chloroform (1) + methanol (2) system at 35°C. The experimental data were taken from Kireev and Sitnikov (1941).

A. ASOG

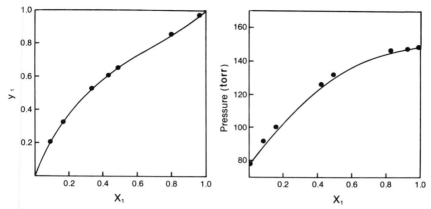

Fig. 6.3 Liquid-vapor equilibria for the benzene (1) + isooctane (2) system at 35°C. The solid lines (———) represent the ASOG predictions and the solid circles (●) represent experimental data taken from Weissman and Wood (1960).

Figures 6.1–6.3 further demonstrate the predictive ability of the ASOG model for ethanol + isooctane, methanol + chloroform, and benzene + isooctane mixtures. As can be seen from the various figures, the overall agreement between experimental and predicted values is good.

In using group contributional methods it must be remembered that the reduction of a large body of experimental data for just a few representative parameters reduces the predictive ability of the model for any one particular system or small set of systems, as the group parameters represent averages over the entire data set. This loss in predictive ability in more than compensated for by the fact that the group contributional model can be used to estimate the thermodynamic properties of systems not included in the original data reduction, provided the system contains only molecules having known group parameters.

With an expression for the activity coefficient of the components in a liquid mixture, the exact thermodynamic relationship

$$\Delta \bar{H}^{ex}/RT^2 = \sum_{i=1}^{N} X_i (\partial \ln \gamma_i / \partial T) \tag{6.9}$$

should enable one to estimate enthalpies of mixing. Nguyen and Ratcliff (1974), combining the ASOG formulation and Eq. (6.9), derived the following expression for excess enthalpies:

$$\Delta \bar{H}^{ex} = \sum_{i=1}^{N} X_i \sum_k v_{ki}(H_k - H_k^{(i)}) \tag{6.10}$$

with

$$H_k/RT^2 = -\partial \ln \Gamma_k/\partial T$$

$$= \sum_\ell \left(X_\ell^g b_{k\ell} \Big/ \sum_\ell X_\ell^g a_{k\ell} \right) + \sum_\ell \left(X_\ell^g b_{\ell k} \Big/ \sum_m X_m^g a_{\ell m} \right)$$

$$- \sum_\ell \left[X_\ell^g a_{\ell k} \left(\sum_m X_m^g b_{\ell m} \right) \Big/ \left(\sum_m X_m^g a_{\ell m} \right)^2 \right] \quad (6.11)$$

$$a_{k\ell} = \exp[m'_{k\ell} + n'_{k\ell}/T] \quad \text{and} \quad b_{k\ell} = -a_{k\ell} n'_{k\ell}/T^2$$

All summations in Eq. (6.11) are over groups.

This group method for predicting enthalpies of mixing is commonly referred to as the Analytical Group Solution Method (AGSM). It is able to predict excess enthalpies to within 10% for a wide variety of endothermic systems Lai et al. (1978) and also for highly nonideal exothermic systems Siman and Vera (1979). The enthalpy predictions, however, require a set of group parameters different from that used in this activity coefficient predictions.

B. The Analytical Group Solution Method (AGSM)

The AGSM derived by Ratcliff and co-workers[†] contains many of the features of the ASOG model, with the exception that the combinatorial part of the activity coefficient was first described using the Brönsted–Koefoed congruence principle

$$\ln \gamma_i = 2.303 B \left(v_i - \sum_{j=1}^N X_j v_j \right) + \sum_k v_{ki}(\ln \Gamma_k - \ln \Gamma_k^{(i)}) \quad (6.12)$$

with

$$\ln \Gamma_k = -\ln \left(\sum_\ell X_\ell^g a_{k\ell} \right) + 1 - \sum_\ell X_\ell^g a_{\ell k} \Big/ \sum_m X_m^g a_{\ell m} \quad (6.13)$$

The coefficient of B is a function of temperature only and has a numerical value of 4.6 J/mol at 25°C.

Although the initial tests of Eq. (6.12) were impressive, subsequent studies by Ronc and Ratcliff (1975) indicated that the Brönsted–Koefoed equation could not properly represent the size contributions in systems having polar

---

[†] See Lai et al. (1978), Maripuri and Ratcliff (1971a,b), Nguyen and Ratcliff (1971a,b), (1974), (1975), Ronc and Ratcliff (1975), and Sayegh and Ratcliff (1976).

C. UNIFAC

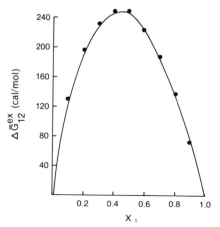

Fig. 6.4 Comparison between the experimentally determined excess free energies and the AGSM predictions for *n*-pentanol (1) + 2-methyl-pentane (2) mixtures at 25°C. The experimental values, denoted by solid circles, were taken from Sayegh and Ratcliff (1976).

components. To overcome this deficiency, the authors suggested that the combinatorial contributions be described by a Flory–Huggins-type expression in terms of group numbers (Eq. (6.2)) multiplied by a smooth function of size ratio. For a mixture containing two molecular species, the combinatorial part is written as

$$\ln \gamma_i^s = (2.9239 - 5.4777p + 12.8016p^2)[\ln v_i/(X_1 v_1 + X_2 v_2) + 1 - v_i/(X_1 v_1 + X_2 v_2)] \quad (6.14)$$

where

$$p = \frac{v_1}{v_2} = \frac{\text{number of groups in shorter molecule}}{\text{number of groups in longer molecule}}$$

Figure 4 compares the AGSM predictions to the experimentally determined excess Gibbs free energies for binary mixtures containing 1-pentanol + 2-methylpentane. The RMS deviation for this system is 2.8% indicating that the group contributional model can provide reasonable predictions of binary properties.

C. The UNIQUAC Functional Group Activity Coefficients Model (UNIFAC)

The UNIFAC group contribution model, developed by Fredenslund *et al.* (1975) is a reliable and fast method for predicting liquid phase activity coefficients of nonelectrolyte mixtures. As its name implies, the UNIFAC model incorporates the group-solution concepts into the basic UNIQUAC description of solution nonideality.

In a multicomponent mixture, the UNIFAC equation for the activity coefficient of molecular species i is

$$\ln \gamma_i = \ln(\phi_i/X_i) + \tfrac{1}{2}zq_i\ln(\theta_i/\phi_i) + \ell_i - (\phi_i/X_i)\sum_j X_j\ell_j$$
$$+ \sum_k \nu_{ki}(\ln \Gamma_k - \ln \Gamma_k^{(i)}) \qquad (6.15)$$

where the group activity coefficient is found via

$$\ln \Gamma_k = Q_k\left[1 - \ln\left(\sum_m \theta_m^g \psi_{mk}\right) - \sum_m \frac{\theta_m^g \psi_{mk}}{\sum_n \theta_n^g \psi_{nm}}\right] \qquad (6.16)$$

and

$$\psi_{mn} = \exp(-a_{mn}/T) \qquad (6.17)$$

Summations in Eq. (6.16) extend over all possible groups.

Volume and surface fractions of the molecular components (ϕ_i and θ_i) are calculated from

$$\phi_i = X_i r_i \bigg/ \left(\sum_j X_j r_j\right) \qquad (6.18)$$

$$\theta_i = X_{qi} \bigg/ \left(\sum_j X_j q_j\right) \qquad (6.19)$$

the liquid phase compositions and the pure component properties r_i and q_i

$$r_i = \sum_k \nu_{ki} R_k \quad \text{and} \quad q_i = \sum_k \nu_{ki} Q_k$$

where ν_{ki} is always an integer and refers to the number of groups of type k in molecule i. Group parameters R_k and Q_k are obtained from the van der Waals group volumes and surface areas V_{wk} and A_{wk}, given by Bondi (1968)

$$R_k = V_{wk}/15.17 \quad \text{and} \quad Q_k = A_{wk}/(2.5 \times 10^9)$$

The normalization factors 15.17 and 2.5×10^9 represent the volume and external surface area of a CH_2 unit in polyethylene. Table III lists numerical values of R_k and Q_k for several different molecular groups.

Area fractions of the individual groups (θ_m^g and θ_n^g) can be calculated in a similar manner.

$$\theta_m^g = X_m^g Q_m \bigg/ \left(\sum_n X_n^g Q_n\right)$$

with X_m^g being the mole fraction composition of group m in the mixture.

C. UNIFAC

TABLE III
UNIFAC Group Volume and Surface Area Parameters[a]

Main Group	Subgroup	R_k	Q_k	Example
CH_2	CH_3	09011	0.848	n-Hexane: $2\,CH_3$, $4\,CH_2$
	CH_2	0.6744	0.540	
	CH	0.4469	0.228	2-Methylpropane: $3\,CH_3$, $1\,CH$
	C	0.2195	0.000	Isooctane: $5\,CH_3$, $1\,C$, $1\,CH$, $1\,CH_2$
$C=C$	$CH_2=CH$	1.3454	1.176	1-Hexene: $1\,CH_3$, $3\,CH_2$, $1\,CH_2=CH$
	$CH=CH$	1.1167	0.867	2-Hexene: $2\,CH_3$, $2\,CH_2$, $1\,CH=CH$
	$CH_2=C$	1.1173	0.988	2-Methyl-1-butene: $2\,CH_3$, $1\,CH_2$, $1\,CH_2=C$
	$CH=C$	0.8886	0.676	2-Methyl-2-butene: $3\,CH_3$, $1\,CH=C$
	$C=C$	0.6605	0.485	2,3-Dimethyl-2-butene: $4\,CH_3$, $1\,C=C$
ACH	ACH	0.5313	0.400	Benzene: $6\,ACH$
	AC	0.3652	0.120	Styrene: $5\,ACH$, $1\,AC$, $1\,CH_2=CH$
$ACCH_2$	$ACCH_3$	1.2663	0.968	Toluene: $1\,ACCH_3$, $5\,ACH$
	$ACCH_2$	1.0396	0.660	Ethylbenzene: $5\,ACH$, $1\,ACCH_2$, $1\,CH_3$
	$ACCH$	0.8121	0.348	Cumene: $5\,ACH$, $2\,CH_3$, $1\,ACCH$
OH	OH	1.0000	1.200	2-Propanol: $2\,CH_3$, $1\,CH$, $1\,OH$
CH_3OH	CH_3OH	1.4311	1.432	Methanol: $1\,CH_3OH$
CH_2CO	CH_3CO	1.6724	1.488	2-Butanone: $1\,CH_3CO$, $1\,CH_3$, $1\,CH_2$
	CH_2CO	1.4457	1.180	3-Pentanone: $1\,CH_2CO$, $2\,CH_3$, $1\,CH_2$
CH_2O	CH_3O	1.1450	1.088	Dimethyl ether: $1\,CH_3$, $1\,CH_3O$
	CH_2O	0.9183	0.780	Diethyl ether: $2\,CH_3$, $1\,CH_2$, $1\,CH_2O$
	CHO	0.6908	0.468	Diisopropylether: $4\,CH_3$, $1\,CH$, $1\,CHO$
CCN	CH_3CN	1.8701	1.724	Acetonitrile: $1\,CH_3CN$
	CH_2CN	1.6434	1.416	Propionitrile: $1\,CH_3$, CH_2CN
CNH_2	CH_3NH_2	1.5959	1.544	Methylamine: $1\,CH_3NH_2$
	CH_2NH_2	1.3692	1.236	Propylamine: $1\,CH_3$, $1\,CH_2$, $1\,CH_2NH_2$
	$CHNH_2$	1.1417	0.924	Isopropylamine: $2\,CH_3$, $1\,CHNH_2$
CNH	CH_3NH	1.4337	1.244	Dimethylamine: $1\,CH_3$, $1\,CH_3NH$
	CH_2NH	1.2070	0.936	Diethylamine: $2\,CH_3$, $1\,CH_2$, $1\,CH_2NH$
	$CHNH$	0.9795	0.624	Diisopropylamine: $4\,CH_3$, $1\,CH$, $1\,CHNH$
$(C)_3N$	CH_3N	1.1865	0.940	Trimethylamine: $2\,CH_3$, $1\,CH_3N$
	CH_2N	0.9597	0.632	Triethylamine: $3\,CH_3$, $2\,CH_2$, $1\,CH_2N$
CCl_3	$CHCl_3$	2.8700	2.410	Chloroform: $1\,CHCl_3$
	CCl_3	2.6401	2.184	1,1,1-Trichloroethane: $1\,CH_3$, $1\,CCl_3$
CCl_4	CCl_4	3.3900	2.910	Carbon tetrachloride: $1\,CCl_4$

[a] A more complete listing of the UNIFAC group parameters can be found in Gmehling et al. (1982). Reprinted in part with permission, copyright 1982 American Chemical Society.

TABLE
Main Groups of UNIFAC

	CH_2	$C=C$	ACH	$ACCH_2$	OH	CH_3OH
CH_2	0.0	−200.0	61.13	76.50	986.5	697.2
$C=C$	2520.0	0.0	340.7	4102.0	693.9	1509.0
ACH	−11.12	−94.74	0.0	167.0	636.1	637.3
$ACCH_2$	−69.70	−269.7	−146.8	0.0	803.2	603.2
OH	156.4	8694.0	89.60	25.82	0.0	−137.1
CH_3OH	16.51	−52.39	−50.00	−44.50	249.1	0.0
CH_2CO	26.76	−82.92	140.1	365.8	164.5	108.7
CH_2O	83.36	76.44	52.13	65.69	237.7	339.7
CCN	24.82	34.78	−22.97	−138.4	185.4	157.8
CNH_2	−30.48	79.40	−44.85	—	−164.0	−481.7
CNH	65.33	−41.32	−22.31	223.0	−150.0	−500.4
$(C)_3N$	−83.98	−188.0	−223.9	109.9	28.60	−406.8
CCl_3	36.70	−185.1	288.5	33.61	742.1	649.1
CCl_4	−78.45	−293.7	−4.700	134.7	856.3	860.1

[a] A more complete listing of the UNIFAC parameters can be found in Gmehling et al. (1982). Reprinted

The various group interaction parameters a_{mn} are evaluated from experimental phase-equilibrium data. At the present time, interaction parameters have been determined for 340 different group pairs. Undoubtedly, this number will continue to grow with time, as the UNIFAC model is one of the more popular group contribution methods developed in the 1970s.

The following example illustrates the UNIFAC method.

EXAMPLE 2

Estimate the activity coefficients in an equimolar mixture of 1-hexanol (1) + n-hexane (2) at 298.15 K.

From Table III it can be seen 1-hexanol has 1 CH_3, 5 CH_2 and 1 OH group and n-hexane has 2 CH_3 and 4 CH_2 groups.

Pure component properties:

1-hexanol $r_1 = (1)(0.9011) + (5)(0.6744) + (1)(1.000) = 5.2731$
$q_1 = (1)(0.848) + (5)(0.540) + (1)(1.200) = 4.748$
$\ell_1 = (5)(5.2731 - 4.748) - (5.2731 - 1.0000) = -1.6476$

n-hexane $r_2 = (2)(0.9011) + (4)(0.6744) = 4.4998$
$q_2 = (2)(0.848) + (4)(0.540) = 3.856$
$\ell_2 = (5)(4.4998 - 3.856) - (4.4998 - 1.000) = -0.2808$

Binary group interactions are calculated from Table IV

C. UNIFAC

IV
Interaction Parameters[a]

CH_2CO	CH_2O	CCN	CNH_2	CHN	$(C)_3N$	CCl_3	CCl_4
476.4	251.5	597.0	391.5	255.7	206.8	24.90	104.3
524.5	289.3	405.9	396.0	273.6	658.8	4584.0	5831.0
25.77	32.14	212.5	161.7	122.8	90.49	−231.9	3.000
−52.10	213.1	6096.0	—	−49.29	23.50	−12.14	−141.2
84.00	28.06	6.712	83.02	42.70	−323.0	−98.12	143.1
23.39	−180.6	36.23	359.3	266.0	53.90	−139.4	−67.80
0.0	5.202	481.7	—	—	—	−354.6	−39.20
52.38	0.0	—	—	141.7	—	−154.3	47.67
−287.5	—	0.0	—	—	—	−15.62	−54.86
—	—	—	0.0	63.72	−41.11	—	−99.81
—	−49.30	—	108.8	0.0	−189.2	—	71.23
—	—	—	38.89	865.9	0.0	−352.9	−8.283
552.1	−20.93	74.04	—	—	−293.7	0.0	−30.10
372.0	113.9	492.0	261.1	91.13	−126.0	51.90	0.0

in part with permission, copyright American Chemical Society.

$$\Psi_{CH_2,OH} = \exp(-986.5/298.15) = 0.03656$$

$$\Psi_{OH,CH_2} = \exp(-156.4/298.15) = 0.5918$$

For pure n-hexane $\ln \Gamma_{CH_2}^{(2)} = \ln \Gamma_{CH_2}^{(2)}(X_{CH_2}^g = 1.0) = 0.0$

For pure 1-hexanol $\ln \Gamma_{OH}^{(1)} = \ln \Gamma_{OH}^{(1)}(X_{OH}^g = 1/7, X_{CH_2}^g = 6/7)$

$\theta_{OH}^{(1)} = (1)(1.2000)/[(1)(1.2000) + (1)(0.848) + (5)(0.540)] = 0.2527$

$\theta_{CH_2}^{(1)} = 0.7473$

$$\ln \Gamma_{OH}^{(1)} = 1.200 \left[1 - \ln[0.2527 + (0.7473)(0.03656)] \right.$$

$$\left. - \frac{0.2527}{0.2527 + (0.7473)(0.03656)} - \frac{(0.7473)(0.5918)}{0.7473 + (0.2527)(0.5918)} \right]$$

$$= 1.0528$$

$$\ln \Gamma_{CH_2}^{(1)} = Q_k \left[1 - \ln\{0.7473 + (0.2527)(0.5918)\} \right.$$

$$\left. - \frac{0.7473}{0.7473 + (0.2527)(0.5918)} - \frac{(0.2527)(0.03656)}{0.2527 + (0.7473)(0.03656)} \right]$$

$$= 0.2426 Q_k$$

For the binary mixture

$$\phi_1 = (0.5)(5.2731)/[(0.5)(5.2731) + (0.5)(4.4998)] = 0.5396$$

$$\theta_1 = (0.5)(4.748)/[(0.5)(4.748) + (0.5)(4.4998)] = 0.5518$$

$$\theta^g_{OH} = (0.5)(1.200)/[(0.5)(4.748) + (0.5)(3.856)] = 0.1395$$

$$\theta^g_{CH_2} = (0.5)(3.548 + 4.4998)/[(0.5)(4.748) + (0.5)(3.856)] = 0.8605$$

$$\ln \Gamma_{OH} = 1.200 \left[1 - \ln\{0.1395 + (0.8605)(0.03656)\} \right.$$
$$\left. - \frac{0.1395}{0.1395 + (0.8605)(0.03656)} - \frac{(0.8605)(0.5918)}{0.8605 + (0.1395)(0.5918)} \right]$$
$$= 1.6925$$

$$\ln \Gamma_{CH_2} = Q_k \left[1 - \ln\{0.8605 + (0.1395)(0.5918)\} \right.$$
$$\left. - \frac{0.8605}{0.8605 + (0.1395)(0.5918)} - \frac{(0.1395)(0.03656)}{0.1395 + (0.8605)(0.03656)} \right]$$
$$= 0.1163 Q_k$$

The activity coefficients are

$$\ln \gamma_1 = \ln(0.5396/0.5) + (5)(4.748) \ln(0.5518/0.5396) - 1.6476$$
$$- (0.5396/0.5)[(0.5)(-1.6476) + (0.5)(-0.2808)]$$
$$+ (1)(1.6925 - 1.0528) + (3.548)(0.1163 - 0.2426) = 0.1915$$

$$\gamma_1 = 1.211$$

$$\ln \gamma_2 = \ln(0.4604/0.5) + (5)(3.856) \ln(0.4482/0.4604) - 0.2808$$
$$- (0.4604/0.5)[(0.5)(-1.6476) + (0.5)(-0.2808)]$$
$$+ (3.856)(0.1163 - 0.0000) = 4.556$$

$$\gamma_2 = 1.577$$

Wieczorek and Stecki (1978) reported experimental values of $\gamma_1 = 1.336$ and $\gamma_2 = 1.640$ at $X_1 = 0.49452$.

Table V shows good agreement for predicted and calculated activity coefficients for the benzene + cyclohexane + n-hexane system at 298.15 K. Whereas this system is generally believed to be simple in terms of intermolecular interactions, readers should not be misled into believing that the UNIFAC method describes only simple systems. Fredenslund *et al.* (1977b) have presented numerous comparisons between the UNIFAC predictions and experimental liquid–vapor equilibrium data for such systems as chloroform +

D. Summary

TABLE V
Comparison between the UNIFAC Predictions and Experimental Vapor–Liquid Equilibria Data for the n-Hexane (1) + Cyclohexane (2) + Benzene (3) System at 298.15 K[a]

X_1	X_2	Experimental values			UNIFAC predictions		
		γ_1	γ_2	γ_3	γ_1	γ_2	γ_3
0.251	0.342	1.13	1.09	1.21	1.10	1.06	1.22
0.416	0.272	1.07	1.07	1.30	1.05	1.03	1.26
0.589	0.193	1.03	1.07	1.39	1.03	1.06	1.39
0.325	0.552	1.05	1.02	1.46	1.03	1.02	1.41
0.213	0.378	1.14	1.09	1.21	1.07	1.10	1.18
0.139	0.388	1.18	1.12	1.17	1.13	1.07	1.17
0.088	0.516	1.15	1.12	1.15	1.12	1.07	1.19
0.219	0.493	1.08	1.05	1.31	1.08	1.02	1.40
0.342	0.398	1.05	1.06	1.36	1.04	1.02	1.32
0.251	0.256	1.17	1.13	1.18	1.18	1.08	1.14
0.178	0.136	1.36	1.26	1.07	1.30	1.15	1.07

[a] The experimental data were taken from Li et al. (1974).

methanol + ethyl acetate, and methanol + ethanol + water. Based on this larger set of comparisons, one can conclude that the UNIFAC method provides reasonable predictions of solution behavior, particularly in light of its many simplifying assumptions.

D. Summary

All group contribution methods are necessarily approximations because any group within a given molecule is not *truly* independent of the other groups within that molecule. But it is precisely this independence that is the essential basis of every group contribution method. One can allow for interdependence of groups within a molecule by carefully defining what atoms constitute a particular group. Increasing distinction of groups, however, also increases the number of group interactions that must be characterized. Ultimately, if one carries group distinction to the extreme limit, it is possible to recover the individual molecules. In that event, the advantages associated with the group contribution method are lost. Judgement and experience is used to define the functional groups so as to achieve a compromise between predictive accuracy and engineering utility.

The primary aim of the ASOG, AGSM, and UNIFAC methods is to provide an accurate and reliable method for predicting vapor–liquid equilibrium of nonelectrolyte systems. The tables of group contribution parameters, therefore, are based entirely on vapor–liquid equilibrium data whenever possible. Application of the various models to other thermodynamic properties such as enthalpies of mixing, will require a different set (or sets) of parameters.

Problems

6.1 Using the ASOG model, predict the total pressure and vapor-phase compositions above binary benzene + di-n-butylether mixtures at 308.15 K. For calculational purposes, use the liquid mole fraction compositions listed in Table VI.

6.2 Estimate enthalpies of mixing for n-hexane (1) + chlorobutane (2) mixtures at 25°C at the compositions $X_2 = 0.25, 0.55$, and 0.75. The AGSM group parameters are given by

$$a_{CH_2,Cl} = \exp[1.921 - 82/T]$$

$$a_{Cl,CH_2} = \exp[1.193 - 261.8/T]$$

The experimentally determined values are 397.1, 516.4, and 366.5 J/mol from reference Lai et al. (1978).

TABLE VI
Experimental Vapor–Liquid Equilibria
Data for Benzene (1) + di-n-Butylether
Mixtures at 308.15 K[a]

X_1	Y_1	P (kPa)
0.0000	0.0000	1.541
0.0769	0.5155	2.936
0.1439	0.6829	4.164
0.2144	0.7781	5.462
0.2803	0.8340	6.694
0.4214	0.9039	9.318
0.5639	0.9432	11.944
0.6350	0.9569	13.241
0.7789	0.9778	15.827
0.9038	0.9912	18.035
1.0000	1.0000	19.753

[a] Experimental data reproduced with permission from Ott et al. (1981). Copyright Academic Press, New York.

TABLE VII
Activity Coefficients and Excess Gibbs Free Energies of 1-Dodecanol (1) + n-Hexane (2) Mixtures at 328.211 K[a]

X_1	γ_1	γ_2	$\Delta \bar{G}_{12}^{ex}$ (J/mol)
0.09998	2.7832	1.0567	414.8
0.19408	1.8263	1.1332	594.0
0.36786	1.2900	1.2899	694.7
0.46527	1.1680	1.3838	671.1
0.56916	1.0908	1.4880	602.3
0.67877	1.0430	1.6019	491.1
0.76204	1.0212	1.6910	384.8
0.87306	1.0053	1.8128	218.7

[a] Experimental data reproduced with permission from Wieczorek, (1978). Copyright Academic Press, New York.

6.3 Derive an expression for the excess enthalpy of a binary liquid mixture based on the UNIFAC model. Assume all of the UNIFAC parameters (R_k, Q_k and $a_{k\ell}$) are independent of temperature.

6.4 The UNIFAC model incorporates the group-solution concepts into the UNIQUAC model. Under what conditions are the two solution models identical?

6.5 Binary mixtures containing an alcohol and a nonpolar hydrocarbon generally exhibit relatively large deviations from ideality. Using the UNIFAC group contribution method, predict the activity coefficients and the excess Gibbs free energy for the 1-dodecanol + n-hexane system at 328.21 K. For calculational purposes, use the mole fraction compositions listed in Table VII.

6.6 Group contribution methods provide a simple, though fairly accurate method for estimating activity coefficients. Using the UNIFAC method, estimate the infinite dilution activity coefficients γ_A^∞ of 2-propanone, 2-butanone, 2-pentanone, and 4-methyl-2-pentanone in n-octane at 55°C. The experimental values, as determined by (1982) are $\gamma_A^\infty = 4.91$, 3.79, 3.35, and 2.78, respectively.

7
Simple Associated Solutions

The theories described in the two previous chapters attempt to explain all solution nonideality in terms of physical intermolecular forces. The Regular Solution model, the Wilson model and the Nonrandom Two Liquid (NRTL) model relate the activity coefficients to physical properties that reflect the relative size of the individual molecules and the physical forces (primarily London dispersion forces) operating between them. As such, these models cannot properly describe the molecular interactions in solutions containing hydrogen bonding, charge transfer complexes or other types of strong specific orientations involving two adjacent molecules. The limited success that physical models experience in describing mathematically the solution nonideality of associated solutions appears to be fortuitous, and little significance is placed on the various binary parameters. As mentioned in Chapter 5, the binary coefficients of the Wilson and NRTL models seldom satisfy the theoretical normalization conditions when they are determined from experimental data.

An alternative approach to the study of solution properties is based on a completely different premise, that molecules in the liquid phase interact with each other to form "new" chemical species and that the observed solution nonideality is a consequence of the chemical reactions. Dolezalek (1908) originally proposed this chemical theory of solution behavior at about the same time van Laar was publishing his ideas on liquid mixtures. Needless to say, these two researchers clashed several times in the literature, each time severely criticizing the other's views. [For a discussion of the arguments between Dolezalek and van Laar, see Hildebrand and Scott (1950b).]

At one time Dolezalek (1918) explained deviations in argon + nitrogen mixtures by postulating the existence of argon dimers. Because of this and similar misinterpretations of solution behavior, Dolezalek's important contributions to thermodynamics were not realized until many years later. The Ideal Associated Solution model traces its origins back to the early work of Dolezalek.

A. Thermodynamic Properties in Associated Solutions

Before discussing the Ideal Associated Solution model in detail, one must first derive a fundamental thermodynamic principle governing associated solutions. In certain liquid mixtures the formation of molecular complexes between like and/or unlike molecules occurs. This can be represented by chemical reactions of the general form

$$iA + jB \rightleftharpoons A_iB_j \quad i,j \geq 0 \text{ and } i+j \geq 2$$

For simplicity, one should initially restrict one's attention to an associated solution containing only 1:1 stoichiometric complexes between A and B

$$A + B \rightleftharpoons AB$$

The number of moles of the various chemical species actually in the solution ($\hat{n}_A$, $\hat{n}_B$, and $\hat{n}_{AB}$)

$$n_A = \hat{n}_A + \hat{n}_{AB} \qquad d\hat{n}_A = dn_A + d\hat{n}_{AB} \tag{7.1}$$

$$n_B = \hat{n}_B + \hat{n}_{AB} \qquad dn_B = d\hat{n}_B + d\hat{n}_{AB} \tag{7.2}$$

are related to the formal number of moles of the two components (n_A and n_B). The symbol (^) is used to designate the properties of the associated solution. The total free energy of this *ternary* solution can be described in terms of the chemical potentials of the three species

$$G = \hat{n}_A \hat{\mu}_A + \hat{n}_B \hat{\mu}_B + \hat{n}_{AB} \hat{\mu}_{AB} \tag{7.3}$$

with the total differential dG at constant T and P, given by

$$dG_{T,P} = \hat{\mu}_A d\hat{n}_A + \hat{\mu}_B d\hat{n}_B + \hat{\mu}_{AB} d\hat{n}_{AB} + \hat{n}_A d\hat{\mu}_A + \hat{n}_B d\hat{\mu}_B + \hat{n}_{AB} d\hat{\mu}_{AB} \tag{7.4}$$

The Gibbs–Duhem equation for multicomponent systems

$$\sum_i n_i d\mu_i = 0$$

enables the last three terms on the right-hand side of Eq. (7.4) to be eliminated, therefore

$$dG_{T,P} = \hat{\mu}_A d\hat{n}_A + \hat{\mu}_B d\hat{n}_B + \hat{\mu}_{AB} d\hat{n}_{AB} \tag{7.5}$$

Using the condition of chemical equilibrium

$$\hat{\mu}_{AB} = \hat{\mu}_A + \hat{\mu}_B$$

the combination of Eqs. (7.1), (7.2), and (7.5) gives

$$dG_{T,P} = \hat{\mu}_A(d\hat{n}_A + d\hat{n}_{AB}) + \hat{\mu}_B(d\hat{n}_B + d\hat{n}_{AB}) = \hat{\mu}_A dn_A + \hat{\mu}_B dn_B \tag{7.6}$$

Alternatively, the solution could have been described as a binary system, in which case

$$dG_{T,P} = \mu_A \, dn_A + \mu_B \, dn_B \tag{7.7}$$

Because these two equations must be identical for all values of dn_A and dn_B, the macroscopic chemical potentials μ_A and μ_B must be equal to the chemical potentials of the *uncomplexed* molecules

$$\mu_A = \hat{\mu}_A \quad \text{and} \quad \mu_B = \hat{\mu}_B \tag{7.8}$$

As will soon be discovered, Eq. (7.8) is independent of any assumptions regarding the mode of association and depends only on the assumption that the complexes are in thermodynamic equilibrium with one another.

Now consider a system in which component A self-associates ($i \geq 1$), and component B is completely inert ($j = 0$)

$$iA_1 \rightleftharpoons A_i$$

with the condition of thermodynamic equilibrium requiring

$$\hat{\mu}_{A_i} = i\hat{\mu}_{A_1} \tag{7.9}$$

The mass balance relationships for components A and B are

$$n_A = \sum_i i\hat{n}_{A_i} \qquad dn_A = \sum_i i\, d\hat{n}_{A_i} \tag{7.10}$$

$$n_B = \hat{n}_B \qquad dn_B = d\hat{n}_B \tag{7.11}$$

The total differential $dG_{T,P}$ for this self-associated solution can be written

$$dG_{T,P} = \sum_i \hat{\mu}_{A_i} \, d\hat{n}_{A_i} + \hat{\mu}_B \, d\hat{n}_B \tag{7.12}$$

in terms of the chemical potentials of the individual i-mers ($\hat{\mu}_{A_i}$) and the chemical potential of the uncomplexed B molecule ($\hat{\mu}_B$). Suitable mathematical manipulation of Eqs. (7.9–7.12) yields

$$dG_{T,P} = \hat{\mu}_{A_1} \, dn_A + \hat{\mu}_B \, dn_B \tag{7.13}$$

Comparing Eq. (7.13) to Eq. (7.7), one again finds that

$$\mu_A = \hat{\mu}_{A_1} \quad \text{and} \quad \mu_B = \hat{\mu}_B \tag{7.14}$$

the macroscopic chemical potentials must be equal to the chemical potentials of the uncomplexed (monomeric) molecules. Interested readers are encouraged to derive Eq. (7.14) for the more general case involving A_iB_j complexes. Several more examples of this important principle are in the problems at the end of the chapter.

B. The Ideal Associated Solution Model: Systems Having a Single AB-type Complex

The basic idea of this model involves the treatment of a nonideal binary solution as if it were an ideal ternary (or higher order) solution, the nonideality arising from the complexes existing in the solution. Hence, negative deviations from Raoult's law or exothermic heats of mixing are attributed to the formation of new chemical species, and positive deviations from Raoult's law or exothermic heats of mixing are attributed to the formation of new chemical species, and positive deviations from Raoult's law or endothermic heats of mixing are attributed to the self-association of one (or more) of the components.

For systems having only a 1:1 molecular complex

$$A + B \rightleftharpoons AB$$

the mole fraction equilibrium constant can be expressed

$$K_{AB} = \hat{X}_{AB}/(\hat{X}_A \hat{X}_B) = r(1-r)/[(X_A - r)(X_B - r)] \qquad (7.15)$$

$$r = \hat{n}_{AB}/(n_A + n_B) \quad \text{and} \quad X_A = n_A/(n_A + n_B)$$

in terms of the number of moles of the complex ($\hat{n}_{AB}$) and the formal amounts of components A and B (n_A, n_B). The thermodynamic principle developed in the last section enables one to express the Gibbs free energy of the *apparent* binary mixture as

$$\begin{aligned} G^{mix} &= n_A \mu_A + n_B \mu_B = n_A \hat{\mu}_A + n_B \hat{\mu}_B \\ &= n_A \mu_A^\bullet + n_A RT \ln \hat{X}_A + n_B \mu_B^\bullet + n_B RT \ln \hat{X}_B \end{aligned} \qquad (7.16)$$

The two activity coefficients γ_A and γ_B are not necessary because the Ideal Associated Solution model assumes an ideal ternary (multicomponent) solution when properly described. The excess free energy of mixing per stoichiometric mole of solution, $\Delta \bar{G}_{AB}^{ex}$, is obtained by subtracting the free energy of the ideal binary mixture

$$\begin{aligned} \Delta \bar{G}_{AB}^{ex} &= (n_A + n_B)^{-1}[G^{mix} - G^{id}] \\ &= (n_A + n_B)^{-1}[n_A \mu_A^\bullet + n_A RT \ln \hat{X}_A + n_B \mu_B^\bullet + n_B RT \ln \hat{X}_B \\ &\quad - n_A \mu_A^\bullet - n_A RT \ln X_A - n_B \mu_B^\bullet - n_B RT \ln X_B] \\ &= X_A RT \ln(\hat{X}_A/X_A) + X_B RT \ln(\hat{X}_B/X_B) \end{aligned} \qquad (7.17)$$

Notice that $\Delta \bar{G}_{AB}^{ex}$ is calculated relative to the formal number of moles in the liquid mixture $n_A + n_B$. This is the general convention for reporting thermodynamic excess properties of multicomponent solutions, regardless of whether they are associated or unassociated solutions. Reporting experimental data relative to the *true* number of moles in the solution ($\hat{n}_A + \hat{n}_B + \cdots$) is confusing, and the numerical values would then depend on how many of each chemical entity is *presumed* to be present. The formal number of moles is an unambiguous quantity and is not subject to various interpretations. Remember that solution properties are meaningless if they cannot be transmitted from the experimentalist to the practicing engineers.

Combining Eqs. (7.15) and (7.17) permits $\Delta \bar{G}_{AB}^{ex}$ to be written in a more convenient form, as

$$\Delta \bar{G}_{AB}^{ex} = X_A RT \ln[(X_A - r)/X_A(1 - r)] + X_B RT \ln[(X_B - r)/X_B(1 - r)] \quad (7.18)$$

Comparing Eq. (7.18) with the definition of the excess free energy

$$\Delta \bar{G}_{AB}^{ex} = X_A RT \ln \gamma_A + X_B RT \ln \gamma_B \quad (7.19)$$

it is seen that the usual binary activity coefficients are

$$\gamma_A = [(X_A - r)/X_A(1 - r)] \quad \text{and} \quad \gamma_B = [(X_B - r)/X_B(1 - r)] \quad (7.20)$$

The equilibrium constant for the AB complex can be determined from either the binary $\Delta \bar{G}_{AB}^{ex}$ data through Eq. (7.18), or from the binary activity coefficients via Eq. (7.20).

The excess enthalpy $\Delta \bar{H}_{AB}^{ex}$ is found through the standard thermodynamic relationship

$$\partial(\Delta \bar{G}_{AB}^{ex}/T)/\partial(1/T) = \Delta \bar{H}_{AB}^{ex}$$

and is given by

$$\Delta \bar{H}_{AB}^{ex} = \left(\frac{\partial \Delta \bar{G}_{AB}^{ex}/T}{\partial r}\right)\left(\frac{\partial r}{\partial K_{AB}}\right)\left(\frac{\partial K_{AB}}{\partial 1/T}\right)$$

$$= -\left[\frac{X_A X_B R}{(1-r)(X_A - r)} + \frac{X_A X_B R}{(1-r)(X_B - r)}\right]$$

$$\times \left[\frac{(X_A - r)^2 (X_B - r)^2}{X_A X_B (1 - 2r)}\right]\left[\frac{-K_{AB} h_{AB}}{R}\right]$$

$$= r h_{AB} \quad (7.21)$$

where h_{AB} is the molar enthalpy of complex formation.

To illustrate how the equilibrium parameters for the AB-complex are obtained from experimental data, consider the molar excess enthalpies of chloroform + triethylamine mixtures as determined by Hepler and Fenby (1973) at 298.15 K. As shown in Table I, the large negative excess enthalpy of mixing ($\Delta \bar{H}_{AB}^{ex} \approx -4$ kJ/mol at $X_A = 0.5$) provides a good indication of complexation in the liquid phase. Substituting $r = \Delta \bar{H}_{AB}^{ex}/X_A X_B$ into Eq. (7.15)

B. Single AB-type Complex

TABLE I
Molar Excess Enthalpies of Chloroform (1) + Triethylamine (2) System[a]

X_2	$\Delta \bar{H}^{ex}_{12}$ (kJ/mol)
0.1037	−1.17
0.2624	−2.86
0.2763	−2.92
0.4179	−3.90
0.4950	−4.03
0.6099	−3.94
0.6917	−3.28
0.8117	−2.12
0.8474	−1.77

[a] Table I is reproduced from Hepler and Fenby (1973). Copyright Academic Press, New York.

gives

$$X_A X_B / \Delta \bar{H}^{ex}_{AB} = -[(K_{AB} + 1)/K_{AB} h^2_{AB}] \Delta \bar{H}^{ex}_{AB} + [(K_{AB} + 1)/(K_{AB} h_{AB})] \quad (7.22)$$

a linear relationship between $X_A X_B / \Delta \bar{H}^{ex}_{AB}$ and $\Delta \bar{H}^{ex}_{AB}$, with the equilibrium constant and molar enthalpy of the complex calculated from the slope and intercept

$$\text{slope} = -(K_{AB} + 1)/(K_{AB} h^2_{AB}) \quad (7.23)$$
$$\text{intercept} = (K_{AB} + 1)/(K_{AB} h_{AB}) \quad (7.24)$$

Figure 7.1 shows that a plot of $X_A X_B / \Delta \bar{H}^{ex}_{AB}$ vs $\Delta \bar{H}^{ex}_{AB}$ is linear and in accordance with Eq. (7.22). Treating their binary data in this manner, Hepler

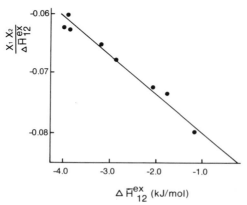

Fig. 7.1 Graphical determination of K_{AB} and h_{AB} for the triethylamine–chloroform complex.

and Fenby calculated values of $K_{AB} = 4.7$ and $h_{AB} = -14.2$ kJ/mol for the triethylamine–chloroform complex.

C. The Ideal Associated Solution Model: Systems Having Both AB-type and AB$_2$-type Complexes

The Ideal Associated Solution model can be extended to mixtures having more than a single molecular complex. McGlashan and Rastogi's (1958) treatment of the p-dioxane (A) + chloroform (B) system is a classical example of interpreting solution nonideality with both AB and AB$_2$ association complexes. For systems having 1:1 and 1:2 molecular complexes

$$A + B \xrightleftharpoons{K_{AB}} AB$$

$$A + 2B \rightleftharpoons AB_2$$

the two mole fraction equilibrium constants are expressed as

$$K_{AB} = \hat{X}_{AB}/(\hat{X}_A \hat{X}_B) \tag{7.25}$$

$$K_{AB_2} = \hat{X}_{AB_2}/(\hat{X}_A \hat{X}_B^2) \tag{7.26}$$

Mass balance considerations relate the stoichiometric mole fraction X_B to the *true* mole fraction of chemical species B by

$$X_B = \frac{(1 + K_{AB})\hat{X}_B + K_{AB_2}\hat{X}_B^2(2 - \hat{X}_B)}{1 + K_{AB}\hat{X}_B(2 - \hat{X}_B) + K_{AB_2}\hat{X}_B^2(3 - 2\hat{X}_B)} \tag{7.27}$$

The excess Gibbs free energy of mixing

$$\begin{aligned}\Delta \bar{G}_{AB}^{ex} &= X_A RT \ln(\hat{X}_A/X_A) + X_B RT \ln(\hat{X}_B/X_B) \\ &= (1 - X_B)RT \ln\left[(1 - \hat{X}_B)/(1 + K_{AB}\hat{X}_B + K_{AB_2}\hat{X}_B^2)(1 - X_B)\right] \\ &\quad + X_B RT \ln(\hat{X}_B/X_B) \end{aligned} \tag{7.28}$$

is obtained by substituting Eqs. (7.25–7.27) into Eq. (7.17). Notice the above expression contains two equilibrium constants that can be evaluated from binary vapor pressure data. Comparison of Eqs. (7.28) and (7.19) reveals that the ordinary activity coefficients are

$$\gamma_A = (1 - \hat{X}_B)/[(1 + K_{AB}\hat{X}_B + K_{AB_2}\hat{X}_B^2)(1 - X_B)] \tag{7.29}$$

$$\gamma_B = \hat{X}_B/X_B \tag{7.30}$$

C. Both AB-type and AB$_2$-type Complexes

with $\hat{X}_B$ calculated from X_B through Eq. (7.27). (It practice it is easier to compute X_B for a given value of $\hat{X}_B$, rather than to solve the cubic equation for $\hat{X}_B$ at a particular value of X_B.)

Through suitable mathematical manipulation of Eqs. (7.29) and (7.30), the formation constants of the two chemical complexes can be related to the activity coefficients

$$(1 - \gamma_A X_A - X_B \gamma_B)/(\gamma_A X_A \gamma_B X_B) = K_{AB} + K_{AB_2} X_B \gamma_B \quad (7.31)$$

or the corresponding activities

$$(1 - a_A - a_B)/a_A a_B = K_{AB} + K_{AB_2} + a_B \quad (7.32)$$

A plot of $(1 - a_A - a_B)/a_A a_B$ vs a_B should be linear, with the slope corresponding to K_{AB_2} and the intercept to K_{AB}.

The excess enthalpy of mixing $\Delta \bar{H}_{AB}^{ex}$ is related to $\Delta \bar{G}_{AB}^{ex}$ by the formula

$$\Delta \bar{H}_{AB}^{ex} = -RT^2(\partial(\Delta \bar{G}_{AB}^{ex}/RT)/\partial T)$$

Carrying out this differentiation with $\Delta \bar{G}_{AB}^{ex}$ given by Eq. (7.28) gives

$$\Delta \bar{H}_{AB}^{ex} = [(1 - X_B)\hat{X}_B/(1 + K_{AB}\hat{X}_B + K_{AB_2}\hat{X}_B^2)][K_{AB}h_{AB} + K_{AB_2}h_{AB_2}\hat{X}_B] \quad (7.33)$$

with

$$h_{AB} = RT^2(\partial \ln K_{AB}/\partial T) \quad (7.34a)$$

$$h_{AB_2} = RT^2(\partial \ln K_{AB_2}/\partial T) \quad (7.34b)$$

As mentioned earlier, McGlashan and Rastogi's (1958) treatment of p-dioxane and chloroform mixtures is a classic example of a system having two types of molecular complexes. The data given in Table II are taken from McGlashan and Rastogi. Notice the activity coefficients of both components are less than one, indicating negative deviations from ideality. The Ideal Associated Solution model attributes negative deviations from Raoult's law to the formation of new chemical species between unlike molecules. Figure 7.2 depicts the graphical determination of the two equilibrium constants from the binary data. The positive slope suggests the presence of the 1:2 association complex. (A zero slope would have indicated the presence of just a 1:1 complex.)

Although it is theoretically possible to calculate the equilibrium constants and hydrogen bond enthalpies of the 1:1 and 1:2 complexes from enthalpy data, one generally calculates the two equilibrium constants from vapor pressure measurements and the hydrogen bond enthalpies from calorimetric

TABLE II
Properties of Binary Mixtures of
p-Dioxane (A) + Chloroform (B) at 50°C[a]

X_A	$a_A (= \gamma_A X_A)$	$a_B (= \gamma_B X_B)$
0.0000	1.0000	0.0000
0.0932	0.9085	0.0453
0.1248	0.8732	0.0625
0.1757	0.8147	0.0924
0.2000	0.7899	0.1065
0.2626	0.7199	0.1439
0.3615	0.5908	0.2250
0.4750	0.4462	0.3289
0.5555	0.3400	0.4255
0.6718	0.2142	0.5605
0.8780	0.0473	0.8613
0.9398	0.0200	0.9391
1.0000	0.0000	1.0000

[a] Reproduced from McGlashan and Rastogi (1958), with permission of the copyright owners.

data. Matsui et al. (1973) presented a unique method for determining numerical values of h_{AB} and h_{AB_2} from

$$h_{AB} = (\Delta \bar{H}_A^{ex})_{X_B=1}^{\infty}(1 + K_{AB})/K_{AB} \tag{7.35}$$

$$h_{AB_2} = [(\Delta \bar{H}_A^{ex})_{X_B=1}^{\infty}(1 + K_{AB} + K_{AB_2})/K_{AB_2}] - [(\Delta \bar{H}_A^{ex})_{X_A=1}^{\infty}(K_{AB} + 1)/K_{AB}] \tag{7.36}$$

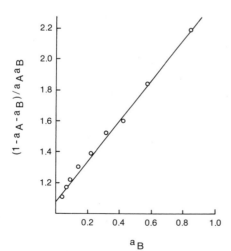

Fig. 7.2 Graphical determination of the equilibrium constants for the 1:1 and 1:2 p-dioxane–chloroform complexes.

C. Both AB-type and AB$_2$-type Complexes

enthalpy of solution data in the pure solvents.

Sondern and Perkampus's (1982a,b) thermodynamic study of liquid benzene (A) + SbCl$_3$ (B) mixtures is another example of calculating association constants from vapor pressure measurements. The calculational method becomes more involved, however, because the authors could only determine the activity coefficient of benzene γ_A

$$\ln \gamma_A = [-6.673 + (2467.4/T)]X_B^2 + [17.012 - (68310.7/T)]X_B^3 \\ + [-7.020 + 2866.5/T)]X_B^4 \qquad (7.37)$$

as a function of liquid phase composition.

The values of the unmeasured $\ln \gamma_A$ function can be computed by applying the Gibbs–Duhem equation. In the first step the polynomial, Eq. (7.37), is differentiated with respect to X_B at fixed T ($T = 350$ K) and P

$$(d \ln \gamma_A/dX_B) = 2(0.377)X_B - 3(2.504)X_B^2 + 4(1.170)X_B^3 \qquad (7.38)$$

Substituting this expression into the Gibbs–Duhem equation gives

$$-d \ln \gamma_B = [(1 - X_B)/X_B] d \ln \gamma_A \\ = [(1 - X_B)/X_B][0.754 X_B - 7.512 X_B^2 + 4.680 X_B^3] d X_B \qquad (7.39)$$

Now integrate the right-hand side from pure SbCl$_3$ ($X_B = 1$) to the mole fraction of interest X_B. Choosing these limits, the following is used

$$\lim_{X_B \to 1} \ln \gamma_B = 0$$

that defines the lower limit of integration on the left-hand side. Integration and rearrangement yields

$$\ln \gamma_B = -0.485 - 0.754 X_B + 4.133 X_B^2 - 4.064 X_B^3 + 1.170 X_B^4$$

Having established an expression relating γ_B to the liquid phase composition, the two association constants K_{AB} and K_{AB_2} can be determined graphically. Figure 7.3 depicts the graphical determination of the two equilibrium constants $K_{AB} \approx 0.7$ and $K_{AB_2} \approx 0.8$.

Complexes involving SbCl$_3$ and aromatic hydrocarbons are known to exist in the solid state. Menshutkin (1911) established SbCl$_3$–C$_x$H$_y$ and 2SbCl$_3$–C$_x$H$_y$ stoichiometries in numerous binary systems of SbCl$_3$, and aromatic hydrocarbons from differential thermal analysis measurements. The existence of only these two stoichiometries can be explained by consulting the published X-ray structure data (Hulme and Mullem, 1976; Holme and Szymanski, 1969; and Lipka and Mootz, 1978). All complexes investigated

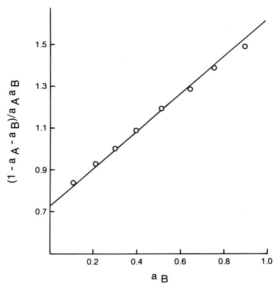

Fig. 7.3 Graphical determination of the equilibrium constants for the 1:1 and 1:2 benzene–$SbCl_3$ complex.

as of 1981 show the same structural pattern. Generally they are built up from layers of $SbCl_3$ alternating with layers of aromatic molecules.

D. Summary

As was just seen, the Ideal Associated Solution model can describe triethylamine + chloroform, p-dioxane + chloroform, and benzene + $SbCl_3$ systems by assuming the formation of heterogeneous molecular complexes. In many instances it is tempting to postulate the existence of higher-order complexes. Theoretically, there is no limit to the number of different chemical species that can be postulated. Practically, it must be realized that any solution theory can describe a given system if there are a sufficient number of adjustable parameters. These parameters (K_i and h_i) should always be viewed cautiously unless they are supported by independent physico-chemical measurements. This is one of the reasons that researchers compare equilibrium constants determine via spectroscopy, vapor pressure measurements, calorimetry, etc.

Criticisms of the Ideal Associated Solution model involves the basic assumption that all solution nonideality is caused by the formation of new

D. Summary

chemical species. From the past three chapters, it was learned that mixtures of saturated hydrocarbons exhibit deviations from ideality, and these deviations are describable with simple solution models based entirely on nonspecific interactions. Should one expect all the physical contributions to suddenly vanish, just because the solution contains two polar molecules rather than two nonpolar molecules? The answer is obviously no.

Arnett *et al.* (1967, 1970) with their "pure base" calorimetric method for determining enthalpies of hydrogen bond formation, tried to separate physical and chemical effects. But the sensitivity of the numerical results to the selection of model compound and inert solvent (Duer and Bertrand, 1970) raises important doubts regarding the overall effectiveness of the pure base method. Salija, *et al.* (1977) used a somewhat similar rationale in their comparison of enthalpies of transfer of alkenes and the corresponding alkanes from the vapor state to methanol, dimethylformamide, benzene, and cyclohexane. The more exothermic values for the alkenes in methanol and dimethylformamide were attributed to dipole-induced dipole interactions between the solvent and the polarizable π-bond.

Still others have preferred to evaluate the thermodynamic parameters associated with complex formation by studying the behavior of a solute in binary solvent mixtures containing both a complexing and inert solvent. Higuchi *et al.* (see Fung and Higuchi, 1971 and Anderson *et al.* 1980) developed a stoichiometric complexation model that enables the estimation of equilibrium constant based on the variation of solubility with complexing solvent concentration. Inherent in this model is the basic assumption that the solubility of the uncomplexed solute is independent of binary solvent composition. Essentially this assumption amounts to saying the physical interactions between any two molecules are identical, irrespective of their chemical nature. By now it should be realized that the thermodynamic properties of benzene + cyclohexane mixtures are different from the thermodynamic properties of cyclohexane + n-heptane mixtures.

The chemical and physical descriptions of solutions represent extreme, one-sided statements of what thermodynamicists believe to be the actual situation. Generally, both physical and chemical forces should be taken into account. A comprehensive theory of liquid solutions should provide for a smooth transition from one limit of an entirely physical description to the other limit of a completely chemical description. It is, of course, difficult to formulate theories that take into account both the physical and chemical effects, as the mathematics become complex and the number of adjustable parameters rapidly increases. Nevertheless, a few classical attempts have been made and in the next chapter three such approaches pertaining to alcohol–hydrocarbon mixtures will be discussed in detail.

Problems

7.1 Treat a mixture of two carboxylic acids (components A and B) as a pentanary solution

$$2A \rightleftharpoons A_2$$
$$2B \rightleftharpoons B_2$$
$$A_2 + B_2 \rightleftharpoons 2AB$$

containing monomers (A, B), dimers (A_2, B_2), and "mixed" dimers. Show that the macroscopic chemical potentials must be equal to the chemical potentials of the monomeric molecules

$$\mu_A = \hat{\mu}_{A_1} \quad \text{and} \quad \mu_B = \hat{\mu}_{B_1}$$

7.2 Using the equilibrium constant $K_{AB} = 4.7$ determined by Hepler and Fenby, estimate the values of $\Delta \bar{G}_{AB}^{ex}$ for the triethylamine + chloroform system at the mole fraction compositions listed in Table III. Compare your predicted values to the smoothed experimental values of Handa and Jones (1975) as at 298 K, also given in Table III.

7.3 Ohta et al. (1980) determined the molar excess Gibbs free energies of 2-butanone (A) + chloroform (B) mixtures from isothermal vapor–liquid equilibrium measurements. Using the values listed in Table IV

TABLE III
Molar Excess Gibbs Free Energies For Chloroform (1) + Triethylamine (2) Mixtures

X_1	$\Delta \bar{G}_{12}^{ex}$ (kJ/mol)
0.1	−0.30
0.2	−0.56
0.3	−0.75
0.4	−0.87
0.5	−0.91
0.6	−0.86
0.7	−0.74
0.8	−0.54
0.9	−0.29

TABLE IV
Vapor–Liquid Equilibrium at 318.15 K
2-Butanone (A) + Chloroform (B)[a]

X_A	γ_A	γ_B	$\Delta \bar{G}_{AB}^{ex}$ (J/mol)
0.072	0.424	0.979	−216.0
0.147	0.452	0.963	−393.6
0.224	0.513	0.927	−550.9
0.276	0.569	0.905	−602.2
0.347	0.639	0.862	−667.6
0.422	0.739	0.782	−713.6
0.463	0.756	0.765	−721.7
0.576	0.860	0.664	−689.1
0.670	0.907	0.604	−613.4
0.752	0.948	0.545	−504.7
0.837	0.962	0.513	−373.9
0.913	0.989	0.478	−198.1
0.944	0.990	0.450	−143.2

[a] Reproduced from Ohta et al. (1980), with permission of the copyright owners.

Problems   147

TABLE V
Vapor–Liquid Equilibrium at 328.15 K
2-Butanone (A) + Chloroform (B)[a]

X_A	γ_A	γ_B	$\Delta \bar{G}_{AB}^{ex}$ (J/mol)
0.079	0.440	0.990	−202.6
0.148	0.471	0.974	−365.7
0.245	0.588	0.931	−504.1
0.300	0.633	0.903	−569.8
0.347	0.704	0.857	−606.6
0.386	0.709	0.845	−643.5
0.444	0.775	0.797	−653.6
0.563	0.859	0.711	−640.4
0.662	0.912	0.647	−568.1
0.752	0.953	0.580	−468.0
0.873	0.985	0.527	−258.7
0.948	0.998	0.512	−100.5

[a] Reproduced from Ohta *et al.* (1980), with permission of the copyright owners.

and V evaluate K_{AB} and K_{AB_2} at both temperatures. Estimate h_{AB} and h_{AB_2} from the temperature dependence of the equilibrium constants. Ohta *et al.* reported

$K_{AB} = 1.40$ and $K_{AB_2} = 0.54$ at   318.15 K

$K_{AB} = 1.25$ and $K_{AB_2} = 0.45$ at   328.15 K

$h_{AB} = -10.0$ kJ/mol   and $h_{AB_2} = -15.9$ kJ/mol

7.4 Estimate the excess enthalpies of 2-butanone + chloroform mixtures at 308.15 K. For calculational simplicity, select 10 mole fractions covering the range $X_B = 0.00$–1.00. Space the compositions far enough apart so $\Delta \bar{H}_{AB}^{ex}$ can be graphed as a function of X_B. The necessary equilibrium parameters were determined in problem 7.3. (Remember to extrapolate the two equilibrium constants to 308.15 K.) Compare your predicted values to the experimental excess enthalpies of Ohta *et al.* listed in Table VI.

7.5 Beath and Williamson determined the enthalpies of mixing for chloroform + diethyl ether mixtures at 25°C. Using their data given in Table VII, and the Ideal Associated Solution model, estimate K_{AB} and h_{AB}. Assume that only a 1:1 complex is formed.

7.6 By differentiating the excess molar Gibbs free energy with respect to pressure, show that the excess molar volume of an A + B + AB + AB$_2$

TABLE VI
Excess Enthalpies of Binary
2-Butanone (A) + Chloroform
(B) Mixtures at 308.15 K[a]

X_A	$\Delta \bar{H}_{AB}^{ex}$ (J/mol)
0.134	−1042.3
0.205	−1451.9
0.270	−1742.8
0.321	−1934.1
0.388	−2076.9
0.425	−2115.1
0.509	−2097.5
0.625	−1852.3
0.712	−1540.0
0.794	−1164.1
0.868	−774.3
0.927	−442.0

[a] Reprinted from Ohta (1980) with permission from the copyright owners.

TABLE VII
Excess Enthalpies of Mixing at
25°C for Diethyl Ether (A) +
Chloroform (B)[a]

X_B	$\Delta \bar{H}_{AB}^{ex}$ (J/mol)
0.1487	−1201
0.3054	−2108
0.4090	−2513
0.4709	−2609
0.5378	−2634
0.5984	−2547
0.6944	−2129
0.8726	−1029

[a] Reproduced with permission from Beath and Williamson (1969). Copyright Academic Press, New York.

Ideal Associated Solution is given by

$$\Delta \bar{V}_{AB}^{ex} = \frac{\hat{X}_A \hat{X}_B (K_{AB} v_{AB} + K_{AB_2} v_{AB_2} \hat{X}_B)}{1 + \hat{X}_A \hat{X}_B (K_{AB} + 2\hat{X}_{AB} K_{AB_2})}$$

where

$$v_{AB} = -RT \left(\frac{\partial \ln K_{AB}}{\partial P} \right)_{T, X_A} \quad \text{and} \quad v_{AB_2} = -RT \left(\frac{\partial \ln K_{AB_2}}{\partial P} \right)_{T, X_A}$$

7.7 Apelblat, Tamir, and Wagner measured the vapor–liquid equilibrium properties for binary mixtures containing acetone and chloroform at 25°C and 35.17°C. Using their data given in Tables VIII and IX, calculate K_{AB} and K_{AB_2} at both temperatures. The numerical values obtained by the authors were

$$\ln K_{AB} = (1240/T) - 4.021$$

and

$$\ln K_{AB_2} = (2421/T) - 8.237$$

7.8 Liquid carboxylic acids exist almost completely as dimers. When two liquid carboxylic acids are mixed, mixed dimers are formed

$$A_2 + B_2 \rightleftharpoons 2 AB \quad K_{eq} = (X_{AB})^2/(X_{A_2} X_{B_2})$$

TABLE VIII
Vapor–Liquid Equilibrium Data in the Acetone (A) + Chloroform (B) System at 25°C[a]

X_A	γ_A	γ_B
0.000	0.517	1.000
0.094	0.546	0.960
0.216	0.559	0.922
0.293	0.602	0.893
0.353	0.653	0.851
0.405	0.701	0.813
0.480	0.831	0.744
0.593	0.889	0.702
0.666	0.928	0.687
0.707	0.942	0.669
0.822	0.966	0.654
0.895	0.986	0.580
0.951	0.996	0.553
1.000	1.000	0.543

[a]Reproduced from Apelblat et al. (1980), with the permission of the copyright owner.

TABLE IX
Vapor–Liquid Equilibrium Data in the Acetone (A) + Chloroform (B) Systems at 35.17°C[a]

X_A	γ_A	γ_B
0.000	0.406	1.000
0.128	0.491	0.985
0.160	0.504	0.960
0.250	0.557	0.947
0.378	0.728	0.854
0.456	0.805	0.788
0.578	0.890	0.702
0.706	0.946	0.638
0.749	0.960	0.594
0.755	0.965	0.592
0.827	0.973	0.560
1.000	1.000	0.497

[a]Reproduced from Apelblat et al. (1980), with the permission of the copyright owner.

Using the Ideal Associated Solution concept and assuming that negligible amounts of monomers exist in either pure or mixed solutions, derive the equation for ΔG^{mix} if K_{eq} has the purely statistical value of 4.

7.9 For the Chloroform (A) + Triethylamine (B) system, show that the enthalpy of hydrogen bond formation can be calculated from either the partial molar enthalpy of solution of B (at infinite dilution) in A

$$h_{AB} = (\Delta \bar{H}_B^{\text{ex}})_{X_A = 1}^\infty (1 + K_{AB})/K_{AB}$$

or the partial molar enthalpy of solution of A (at infinite dilution) in B

$$h_{AB} = (\Delta \bar{H}_A^{\text{ex}})_{X_B = 1}^\infty (1 + K_{AB})/K_{AB}$$

Using the following information

$$(\Delta \bar{H}_A^{\text{ex}})_{X_B = 1}^\infty = -12.68 \text{ kJ/mol} \quad \text{and} \quad (\Delta \bar{H}_B^{\text{ex}})_{X_A = 1}^\infty = -11.55 \text{ kJ/mol}$$

$$K_{AB} = 4.7$$

Calculate the value of h_{AB}.

8

Estimation of Thermodynamic Excess Properties of Ternary Alcohol-Hydrocarbon Systems from Binary Data

The marked influence of hydrogen bonding on the solution and spectroscopic properties of alcohols has made them especially amenable to investigative studies. The breaking of hydrogen bonds that occurs on dilution of alcohols in nonpolar (inert) solvents results in large endothermic enthalpies of mixing, changes in the partial molar volumes, viscosities, ultrasonic absorption, infra-red spectra, NMR chemical shifts, vapor pressures, and other properties,[†] all of which have been used to study the self-association characteristics of alcohol molecules. Despite the vast amount of experimental data presently available in the literature, a continuing controversy persists over the number and size of associated species in solutions.

Much of the earlier research, as summarized by Pimentil and McClellan (1960), treated alcohol polymerization as a stepwise process resulting in a continuum of species. Thus, at low concentrations, dimers would be the predominate polymeric species; with the larger *polymer chains* becoming more significant with increasing alcohol concentration. Van Ness et al. (1967) compared infrared data to heat of mixing data for ethanol + n-heptane and ethanol + toluene mixtures and concluded that the results were best explained by a model containing monomers, cyclic dimers, and linear

---

[†] Note the following references for the property listed: Endothermic enthalpies of mixing, Anderson et al. (1975), Aveyard and Mitchell (1967), Ramalho and Ruel (1968), Salvani et al. (1965), Smith and Brown (1973), Treszczanowicz and Treszczanowicz (1981), Treszczanowicz et al. (1981), and Van Ness et al. (1967). Changes in partial molar volumes, Treszczanowicz and Benson (1978), (1980). Viscosities, Salamon et al. (1975a,b). Ultrasonic absorption, Emara and Atkinson (1974), Garland et al. (1971), *Lana and Zana* (1970), Musa and Eisner (1959), Rassing and Jensen (1970), Sodovyev et al. (1968), and Zana and Lang (1975). Infra-red spectra, Dos Santos et al. (1965), Fletcher (1969), (1972), Fletcher and Heller (1968), and Hammaker et al. (1968). NMR chemical shifts Dixon (1970), Servanton–Gadouleau et al. (1975), and Tucker and Becker (1973). Vapor pressures, Anderson et al. (1978, 1979), French and Stokes (1981), Rytting et al. (1978), Stokes and Adamson (1977), and Tucker et al. (1969).

A. Kretschmer–Wiebe Association Model

polymers having 20 or more monomer units per chain. Tucker *et al.* (1969) noted that the simplest model ($1-3-\infty$) that can adequately describe the vapor pressure data for ethanol + n-hexadecane systems contains two equilibrium constants, one constant for trimer formation and the second for the sequential addition of the monomer.

Still other studies have indicated that a single polymer of definite size may dominate. Fletcher and Heller (1968) explained the infra-red data of 1-octanol in n-decane (from dilute solutions to the pure alcohol) in terms of a monomer–tetramer self-association model. Dixon (1970) also found that the monomer–tetramer model gave the best correlations for his proton magnetic resonance data on the hydroxyl shift for methanol in cyclohexane. Anderson *et al.* (1978) explained the vapor pressures of several n-alkanols in isooctane with a monomer–pentamer model.

Rather than confuse the molecular interpretation of alcohol–hydrocarbon systems further, attention will be focused primarily on the development of thermodynamic models. Examples to be used are alcohol–hydrocarbon systems with no concern as to the physical significance of the equilibrium parameters (constants). In many respects this adds an empirical flavor to the theoretical models.

The majority of predictive expressions presented in the past three chapters provide reasonably accurate predictions for multicomponent systems containing only nonspecific interactions. These expressions provide only fair predictions for the more complex associated solutions, such as alcohol–hydrocarbon mixtures. Because alcohol solutions comprise a substantial portion of liquid mixtures of practical importance, it becomes imperative to consider predictive methods that will be applicable to self-associated molecules.

The predictive expressions presented in this chapter are based on articles by Renon and Prausnitz (1967), Nitta and Katayama (1973), Nagata (1973a, b, c, d, 1977, 1978), Nagata *et al.* (1977), and Nagata and Ogasawara (1982).

A. Kretschmer–Wiebe Association Model

The excess Gibbs free energy $\Delta \bar{G}^{ex}_{ABC}$ for a ternary solution containing an alcohol (component A) and two inert hydrocarbons (components B and C) is described by the sum of two separate contributions, one contribution representing chemical interactions and the other representing physical interactions

$$\Delta \bar{G}^{ex}_{ABC} = (\Delta \bar{G}^{ex}_{ABC})_{ph} + (\Delta \bar{G}^{ex}_{ABC})_{ch} \tag{8.1}$$

The chemical contribution is based on the Kretschmer–Wiebe (1954) self-association model that basically assumes the alcohol forms continuous linear hydrogen-bonded polymers $A_1, A_2, A_3, \ldots, A_i, \ldots$ by successive chemical reactions

$$A_1 + A_1 \rightleftharpoons A_2$$
$$A_2 + A_1 \rightleftharpoons A_3$$
$$\vdots$$
$$A_i + A_1 \rightleftharpoons A_{i+1}$$
$$\vdots$$

described by a single equilibrium constant of the form (Kretschmer and Wiebe, 1954)

$$K_A = \hat{C}_{A_{i+1}}/\hat{C}_{A_i}\hat{C}_{A_1}\bar{V}_A = [\hat{\phi}_{A_{i+1}}/\hat{\phi}_{A_i}\hat{\phi}_{A_1}][i/(i+1)] \quad (8.2)$$

where $\hat{C}_{A_i}$ and $\hat{\phi}_{A_i}$ refer to the molar concentration and volume fraction of the i-mer respectively, calculated using the molar volume of the monomer multiplied by i. The inclusion of cyclic polymers into the model increases the complexity of the derived expressions, well beyond the intended scope of this book.

From Eq. (8.2), the concentration of an individual alcohol i-mer can be expressed as[†]

$$\hat{C}_{A_i} = \hat{n}_{A_i}/V = \hat{C}_{A_{i-1}}\hat{C}_{A_1}K_A\bar{V}_A = \hat{C}_{A_1}(K_A\hat{C}_{A_1})^{i-1} \quad (8.3)$$

The stoichiometric volume fraction of the alcohol (ϕ_A) is the sum of the volume fractions of each individual alcohol species

$$\phi_A = \sum_{i=1}^{\infty} \hat{\phi}_{A_i} = (1/K_A)\sum_{i=1}^{\infty} i(K_A\hat{\phi}_{A_1})^i = \hat{\phi}_{A_1}(1 - K_A\hat{\phi}_{A_1})^{-2} \quad (8.4)$$

It should be noted that $|K_A\phi_{A_1}| < 1$ for the infinite series to converge. The volume fraction of the alcohol monomer in the ternary solution is given

$$\hat{\phi}_{A_1} = [(2K_A\phi_A + 1) - (1 + 4K_A\phi_A)^{1/2}]/2K_A^2\phi_A \quad (8.5)$$

The chemical part of the Gibbs free energy is based on the Flory–Huggins athermal model and may be written as

$$(g_{ABC}^m)_{ch} = RT\left[\sum_{i=1}^{\infty} \hat{n}_{A_i} \ln \hat{\phi}_{A_i} + \hat{n}_B \ln \hat{\phi}_B + \hat{n}_C \ln \hat{\phi}_C\right] \quad (8.6)$$

The respective chemical potentials are obtained through differentiation with

---

[†] For notational simplicity the solid circle superscripts (●) are being left off the molar volumes of the pure components. This simplified notation will be retained throughout the remaining chapters.

A. Kretschmer–Wiebe Association Model

respect to the number of moles of each chemical species

$$(\hat{\mu}_{A_i} - \mu^{\bullet}_{A_i})/RT = \ln \hat{\phi}_{A_i} + 1 - \bar{V}_{A_i}/\hat{V}_{sol} \tag{8.7a}$$

$$(\hat{\mu}_B - \mu^{\bullet}_B)/RT = \ln \hat{\phi}_B + 1 - \bar{V}_B/\hat{V}_{sol} \tag{8.7b}$$

and

$$(\hat{\mu}_C - \mu^{\bullet}_C)/RT = \ln \hat{\phi}_C + 1 - \bar{V}_C/\hat{V}_{sol} \tag{8.7c}$$

The superscript (•) indicates that the property refers to a *pure* liquid state and $\hat{V}_{sol}$ is the *true* molar volume of the solution

$$1/\hat{V}_{sol} = \sum_{i=1}^{\infty} \hat{\phi}_{A_i}/\bar{V}_{A_i} + \hat{\phi}_B/\bar{V}_B + \hat{\phi}_C/\bar{V}_C$$

$$= (1/K_A\bar{V}_A) \sum_{i=1}^{\infty} (K_A\hat{\phi}_{A_1})^i + \hat{\phi}_B/\bar{V}_B + \hat{\phi}_C/\bar{V}_C$$

$$= \hat{\phi}_{A_1}/[\bar{V}_A(1 - K_A\hat{\phi}_{A_1})] + \phi_B/\bar{V}_B + \phi_C/\bar{V}_C \tag{8.8}$$

and for the pure alcohol

$$1/\hat{V}^*_{sol} = \hat{\phi}^*_{A_1}/[\bar{V}_A(1 - K_A\hat{\phi}^*_{A_1})] \tag{8.9}$$

It was shown in Chapter 7 that the chemical potential of the stoichiometric component is equal to the chemical potential of the monomer, that is

$$\mu_A = \hat{\mu}_{A_1} \tag{7.14'}$$

Combining Eqs. (8.7a) and (7.14') gives

$$\mu_A = \mu^{\bullet}_{A_1} + RT[\ln\hat{\phi}_{A_1} + 1 - \bar{V}_A/\hat{V}_{sol}] \tag{8.10}$$

For the pure alcohol, Eq. (8.10) becomes

$$\mu^*_A = \mu^{\bullet}_{A_1} + RT[\ln \hat{\phi}^*_{A_1} + 1 - \bar{V}_A/\hat{V}^*_{sol}] \tag{8.11}$$

Subtraction of Eq. (8.11) from Eq. (8.10) yields

$$\mu_A = \mu^*_A + RT[\ln(\hat{\phi}_{A_1}/\hat{\phi}^*_{A_1}) + (\bar{V}_A/\hat{V}^*_{sol}) - (\bar{V}_A/\hat{V}_{sol})] \tag{8.12}$$

The expression for $(\Delta \bar{G}^{ex}_{ABC})_{ch}$ is derived by combining Eqs. (8.7a–c) and Eq. (8.12)

$$(\Delta \bar{G}^{ex}_{ABC})_{ch} = RT[X_A \ln(\hat{\phi}_{A_1}/\hat{\phi}^*_{A_1}X_A) + X_B \ln(\phi_B/X_B) + X_C \ln(\phi_C/X_C)$$
$$- (X_A\bar{V}_A + X_B\bar{V}_B + X_C\bar{V}_C)/\hat{V}_{sol}$$
$$+ X_B + X_C + (X_A\bar{V}_A/\hat{V}^*_{sol})] \tag{8.13}$$

Through slight mathematical manipulations, Eq. (8.13) can be written in a more convenient form of

$$(\Delta \bar{G}^{ex}_{ABC})_{ch} = RT[X_A \ln(\hat{\phi}_{A_1}/\hat{\phi}^*_{A_1}X_A) + X_B \ln(\phi_B/X_B)$$
$$+ X_C \ln(\phi_C/X_C) + X_A K_A(\hat{\phi}_{A_1} - \hat{\phi}^*_{A_1})] \tag{8.14}$$

The simplest equation that might be expected to adequately describe the physical contribution to the excess Gibbs free energy $(\Delta G_{ABC}^{ex})_{ph}$ of a ternary alcohol plus two inert hydrocarbon system is

$$(\Delta G_{ABC}^{ex})_{ph} = \left(\hat{n}_B \bar{V}_B + \hat{n}_C \bar{V}_C + \sum_{i=1}^{\infty} \hat{n}_{A_i} \bar{V}_{A_i}\right)^{-1} \left[\hat{n}_B \bar{V}_B \hat{n}_C \bar{V}_C A_{BC} \right.$$
$$\left. + \sum_{i=1}^{\infty} \hat{n}_{A_i} \bar{V}_{A_i} \hat{n} \bar{V}_{B_i} A_{A_iB} + \sum_{i=1}^{\infty} \hat{n}_{A_i} \bar{V}_{A_i} \hat{n}_C \bar{V}_C A_{A_iC}\right] \quad (8.15)$$

in which the A_{ij} terms represent binary interaction parameters. Obviously, Eq. (8.15) contains too many parameters for useful applications, but reasonable assumptions enable the number of parameters to be greatly reduced. The solubility parameter approach of Scatchard–Hildebrand mentioned in Chapter 5 provides a basis for estimation of interaction parameters involving alcohol complexes

$$A_{A_iK} = (\delta_{A_i} - \delta_K)^2, \quad K = B, C \quad (8.16)$$

Assuming that the solubility parameter of an alcohol polymer $(\delta_{A_{i+1}})$ is a weighted molar volume average of the solubility parameters of the i-mer and monomer

$$\delta_{A_{i+1}} = (\bar{V}_{A_i} \delta_{A_i} + \bar{V}_{A_1} \delta_{A_1})/(\bar{V}_{A_i} + \bar{V}_{A_1}) = (i\delta_{A_i} + \delta_{A_1})/(i+1) = \delta_{A_i} \quad (8.17)$$

gives

$$A_{A_iB} = A_{A_1B} \quad \text{and} \quad A_{A_iC} = A_{A_1C} \quad (8.18)$$

Combination of Eqs. (8.15–8.18) enables $(\Delta G_{ABC}^{ex})_{ph}$ to be expressed

$$(\Delta G_{ABC}^{ex})_{ph} = (n_A \bar{V}_A + n_B \bar{V}_B + n_C \bar{V}_C)^{-1}$$
$$[n_A \bar{V}_A n_B \bar{V}_B A_{A_1B} + n_A \bar{V}_A n_C \bar{V}_C A_{A_1C} + n_B \bar{V}_B n_C \bar{V}_C A_{BC}] \quad (8.19)$$

in terms of three binary interaction parameters. Subsitution of Eqs. (8.14) and (8.19) into Eq. (8.1) yields the following expression for the total Gibbs free energy of the ternary system (per stoichiometric mole of solution)

$$(\Delta \bar{G}_{ABC}^{ex}) = RT[X_A \ln(\hat{\phi}_{A_1}/\hat{\phi}_{A_1}^* X_A) + X_B \ln(\phi_B/X_B) + X_C \ln(\phi_C/X_C)$$
$$+ X_A K_A(\hat{\phi}_{A_1} - \hat{\phi}_{A_1}^*)] + (X_A \bar{V}_A + X_B \bar{V}_B + X_C \bar{V}_C)$$
$$\times [\phi_A \phi_B A_{A_1B} + \phi_A \phi_C A_{A_1C} + \phi_B \phi_C A_{BC}] \quad (8.20)$$

Inspection of Eq. (8.20) reveals that for model systems obeying this expression the properties of the two contributive alcohol–hydrocarbon systems would obey

$$(\Delta \bar{G}_{AK}^{ex})' = RT[X_A' \ln(\hat{\phi}_{A_1}'/\hat{\phi}_{A_1}^* X_A') + X_B \ln(\phi_B'/X_K') + X_A' K_A(\hat{\phi}_{A_1}' - \hat{\phi}_{A_1}^*)]$$
$$+ \phi_A' \phi_K' A_{A_1K}(X_A' \bar{V}_A + X_K' \bar{V}_K), \quad K = B, C \quad (8.21)$$

A. Kretschmer–Wiebe Association Model

and the properties of the contributive binary hydrocarbon mixture would obey

$$(\Delta \bar{G}_{BC}^{ex})' = RT[X'_B \ln(\phi'_B/X'_B) + X'_C \ln(\phi'_C/X'_C)] \\ + \phi'_B\phi'_C A_{BC}(X'_B \bar{V}_B + X'_C \bar{V}_C) \quad (8.22)$$

The symbol (') is intended to denote that the compositions now correspond to a binary system (i.e., $X'_A + X'_B = 1$) rather than the ternary system. The various binary interaction terms A_{A_1B}, A_{A_1C}, and A_{BC} are determined from the parametrization of the binary data according to Eqs. (8.21) and (8.23). Unlike many of the predictive equations presented in Chapter 4, Eq. (8.20) requires that the binary properties be represented in a specific form. For the purpose of calculations in the exercises at the end of this chapter, the binary data will be expressed in the appropriate manner. In gathering data from the literature, this will rarely be true.

With an expression for the excess Gibbs free energy, the excess enthalpy of mixing can be obtained through the standard thermodynamic relationship

$$\Delta \bar{H}^{ex} = \frac{\partial(\Delta \bar{G}^{ex}/T)}{\partial(1/T)}$$

Rather than proceed in this manner, the $\Delta \bar{H}_{ABC}^{ex}$ can be described as the sum of two separate contributions, one contribution representing chemical interactions and the second representing physical interactions

$$\Delta \bar{H}_{ABC}^{ex} = (\Delta \bar{H}_{ABC}^{ex})_{ch} + (\Delta \bar{H}_{ABC}^{ex})_{ph} \quad (8.23)$$

The chemical part of the total excess enthalpy represents a change in the number of each individual chemical entity (i-mers) between the ternary (or binary) solution and the pure alcohol

$$(\Delta H^{ex})_{ch} = h_A \sum_{i=1}^{\infty} (i-1)\hat{n}_{A_i} - h_A \sum_{i=1}^{\infty} (i-1)\hat{n}_{A_i}^* \quad (8.24)$$

where h_A is the enthalpy of formation of an alcohol–alcohol hydrogen bond (assumed to be independent off the degree of imerization) related to the association constant through the van't Hoff equation

$$[d \ln K_A/d(1/T)] = -h_A/R \quad (8.25)$$

Combining Eqs. (8.3) and (8.24), one writes

$$(\Delta H^{ex})_{ch} = h_A \hat{n}_{A_1} \sum_{i=1}^{\infty} (i-1)(K_A \hat{\phi}_{A_1})^{i-1} - h_A \hat{n}_{A_1}^* \sum_{i=1}^{\infty} (i-1)(K_A \hat{\phi}_{A_1}^*)^{i-1}$$
$$= h_A K_A n_A (\hat{\phi}_{A_1} - \hat{\phi}_{A_1}^*) \quad (8.26)$$

8 Ternary Alcohol–Hydrocarbon Systems

The physical contribution to the excess enthalpy $(\Delta \bar{H}^{ex}_{ABC})_{ph}$

$$(\Delta \bar{H}^{ex}_{ABC})_{ph} = (X_A \bar{V}_A + X_B \bar{V}_B + X_C \bar{V}_C)[\phi_A \phi_B B_{A_1B} + \phi_A \phi_C B_{A_1C} + \phi_B \phi_C B_{BC}] \tag{8.27}$$

is expressed in terms of three new binary interaction parameters. Because the derivation of Eq. (8.27) is identical to that for $(\Delta \bar{G}^{ex}_{ABC})_{ph}$ it will not be presented here.

The total excess enthalpy (per stoichiometric mole of solution) is then given by

$$\begin{aligned}\Delta \bar{H}^{ex}_{ABC} &= h_A X_A K_A (\hat{\phi}_{A_1} - \hat{\phi}^*_{A_1}) \\ &+ (X_A \bar{V}_A + X_B \bar{V}_B + X_C \bar{V}_C)[\phi_A \phi_B B_{A_1B} + \phi_A \phi_C B_{A_1C} + \phi_B \phi_C B_{BC}]\end{aligned} \tag{8.28}$$

The interaction parameters B_{A_1B}, B_{A_1C}, and B_{BC} are evaluated from the binary enthalpy data through the appropriate reduction of Eq. (8.28).

As stated previously, Eq. (8.28) could have been developed through the differentiation of Eq. (8.20). For educational purposes, the differentiation of the chemical contribution $(\Delta \bar{G}^{ex}_{ABC})_{ch}$ will be presented in detail. Remember that the athermal Flory–Huggins model assumes that the molar volumes of the individual *pure* components ($\bar{V}_A$, $\bar{V}_B$, and $\bar{V}_C$) are independent of temperature.

Application of the chain rule permits the $(\Delta \bar{H}^{ex}_{ABC})_{ch}$ expression to be derived in a simple manner, as

$$\begin{aligned}(\Delta \bar{H}^{ex}_{ABC})_{ch} &= R\left(\frac{\partial \Delta \bar{G}^{ex}_{ch}/RT}{\partial 1/T}\right) \\ &= R\left(\frac{\partial \Delta \bar{G}^{ex}_{ch}/RT}{\partial K_A}\right)\left(\frac{\partial K_A}{\partial 1/T}\right) = -h_A K_A \left(\frac{\partial \Delta \bar{G}^{ex}_{ch}/RT}{\partial K_A}\right)\end{aligned} \tag{8.29}$$

Differentiation of $(\Delta \bar{G}^{ex}_{ABC})_{ch}/RT$ with respect to K_A gives

$$\begin{aligned}(\Delta \bar{H}^{ex}_{ABC})_{ch} &= -h_A K_A \left\{X_A \left[\frac{\partial \ln(\hat{\phi}_{A_1}/\hat{\phi}^*_{A_1})}{\partial K_A}\right] + X_A(\hat{\phi}_{A_1} - \hat{\phi}^*_{A_1})\right. \\ &\quad \left. + K_A X_A \left(\frac{\partial \hat{\phi}_{A_1}}{\partial K_A} - \frac{\partial \hat{\phi}^*_{A_1}}{\partial K_A}\right)\right\} \\ &= -K_A h_A X_A \left[\left(\frac{\partial \hat{\phi}_{A_1}}{\partial K_A}\right)\left(\frac{1 + K_A \hat{\phi}_{A_1}}{\hat{\phi}_{A_1}}\right)\right. \\ &\quad \left. - \left(\frac{\partial \hat{\phi}^*_{A_1}}{\partial K_A}\right)\left(\frac{1 + K_A \hat{\phi}^*_{A_1}}{\hat{\phi}^*_{A_1}}\right) + (\hat{\phi}_{A_1} - \hat{\phi}^*_{A_1})\right]\end{aligned} \tag{8.30}$$

A. Kretschmer–Wiebe Association Model

Using Eq. (8.4) (with $\phi_A = 1.0$), one can evaluate $\partial \hat{\phi}_{A_1}^* / \partial K_A$ as

$$\partial \hat{\phi}_{A_1}^* / \partial K_A = -2\hat{\phi}_{A_1}^{*2}/(1 + K_A \hat{\phi}_{A_1}^*) \tag{8.31}$$

Similarly

$$\left(\frac{\partial \hat{\phi}_{A_1}}{\partial K_A} \right) = -\frac{2\hat{\phi}_{A_1}^2}{1 + K_A \hat{\phi}_{A_1}} \tag{8.32}$$

Substitution of these two partial derivatives into Eq. (8.30) gives

$$(\Delta \bar{H}_{ABC}^{ex})_{ch} = -h_A X_A K_A \left[\left(\frac{-2\hat{\phi}_{A_1}^2}{1 + K_A \hat{\phi}_{A_1}} \right) \left(\frac{1 + K_A \hat{\phi}_{A_1}}{\hat{\phi}_{A_1}} \right) \right.$$
$$\left. + \left(\frac{2\hat{\phi}_{A_1}^{*2}}{1 + K_A \hat{\phi}_{A_1}^*} \right) \left(\frac{1 + K_A \hat{\phi}_{A_1}^*}{\hat{\phi}_{A_1}^*} \right) + (\hat{\phi}_{A_1} - \hat{\phi}_{A_1}^*) \right]$$
$$= K_A X_A h_A (\hat{\phi}_{A_1} - \hat{\phi}_{A_1}^*)$$

This expression is identical to Eq. (8.26) and demonstrates that the enthalpy expression is internally consistent with the corresponding Kretschmer–Wiebe expression for the excess Gibbs free energy. In some instances it is easier to derive the expression for $\Delta \bar{H}^{ex}$ by differentiating the free energy expression directly. In other cases it is easier to derive the $\Delta \bar{H}^{ex}$ expression by considering the chemical part of the excess enthalpy as a change in the number of moles of each individual chemical entity between the ternary (or binary) solution and the pure *unmixed* components. The method we chosen is up to each person.

The Kretschmer–Wiebe association model can not be used to describe excess volumes, as the ideal molar volume approximation $\Delta V^{ex} = 0$ is inherent in the basic model. The ideal molar volume approximation is also incorporated into the Mecke–Kempter and Attenuated Equilibrium Constant (AEC) models that will be discussed in the next two sections of this chapter. It is important to remember the basic limitations of each solution model so that its mathematical expressions are not extended beyond their intended applications. Far too often researchers arrive at erroneous conclusions because they try to interpret a solution's nonideality with an inappropriate mathematical description.

The application of the Kretschmer–Wiebe association model to ternary alcohol plus two inert hydrocarbon systems requires only an a prior knowledge of the binary properties and one equilibrium constant. At a fixed temperature, K_A depends only on the alcohol and is independent of the hydrocarbon co-solvent. Figures 8.1 and 8.2 show calculated chemical contributions to the binary enthalpies for values of K_A ranging from 50 to 300. For convenience, the molar enthalpy of an alcohol–alcohol hydrogen bond

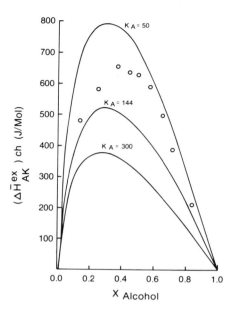

Fig. 8.1 Calculated values of the chemical contributions to the enthalpies as a function of alcohol concentration. The three curves represent different values of the equilibrium constant. Open circles (○) denote the experimental enthalpies of 1-hexanol + n-heptane mixtures determined by Nunez et al., (1976).

is assumed to be -25 kJ/mol. The enthalpy curves calculated via Eq. (8.26) exhibit the experimentally observed skewness with respect to mole fraction composition, as indicated by the open circles. The molar volumes were selected so that values of $(\Delta \bar{H}^{ex}_{AK})_{ch}$ would correspond to either 1-hexanol +

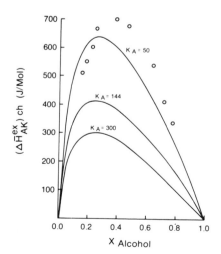

Fig. 8.2 Calculated values of the chemical contributions to the enthalpies as a function of alcohol concentration. The three curves represent different values of the equilibrium constant. Open circles (○) denote the experimental data of the 1-hexanol + cyclohexane system as determined by Nunez et al., (1976).

A. Kretschmer–Wiebe Association Model

TABLE I
Kretschmer–Wiebe
Association Constants
of Alcohols at 50°C[a]

Alcohol	K_A
Methanol	450
Ethanol	190
1-Propanol	90
1-Butanol	80
1-Pentanol	72
1-Hexanol	66
1-Octanol	58
1-Decanol	54

[a] Table I is reproduced with permission from Nagata, (1975). Copyright 1975 American Chemical Society.

n-heptane or 1-hexanol + cyclohexane mixtures. Differences between the experimental enthalpies and values calculated via Eq. (8.26) are attributed to the physical interactions $(\Delta \bar{H}_{AK}^{ex})_{ph}$, thus enabling the evaluation of the binary interaction parameter $B_{A_1 K}$.

Table I lists Kretschmer–Wiebe equilibrium constants of several alcohols, as reported by Nagata (1975). No physical significance is placed on the various equilibrium parameters because continuous linear association models are an oversimplified description of alcohol–hydrocarbon mixtures. For example, the Kretschmer–Wiebe association model assumes the excess volumes are equal to zero; an assumption that is not verified by direct experimental evidence. The basic model ignores the existence of cyclic hydrogen bonded species. Yet the interpretation of some solution properties indicates the presence of cyclic dimers, trimers, tetramers, and/or pentamers. Despite these inherent limitations, the Kretschmer–Wiebe association model enables the estimation of many multicomponent thermodynamic excess properties from binary data.

The Scatchard–Hildebrand equation is just one of the expressions that can be used to describe physical interactions. Several of the expressions developed in Chapter 5 were also suggested. Nagata (1973a,b, 1978) proposed the following two formulas

$$(\Delta \bar{G}_{ABC}^{ex})_{ph} = -\sum_{i=1}^{3} X_i \ln \left(\sum_j \phi_j \tau_{ji} \right) \tag{8.33}$$

and

$$(\Delta \bar{H}^{\text{ex}}_{\text{ABC}})_{\text{ph}} = \sum_{i=1}^{3} X_i \left[\sum_{j=1}^{3} \phi_j \left(\frac{\partial \tau_{ji}}{\partial (1/T)} \right) \middle/ \sum_{j=1}^{3} \phi_j \tau_{ji} \right] \quad (8.34)$$

where

$$\tau_{ij} = \exp[-(u_{ij} - u_{jj})/RT]$$

for $(\Delta \bar{G}^{\text{ex}}_{\text{ABC}})_{\text{ph}}$ and $(\Delta \bar{H}^{\text{ex}}_{\text{ABC}})_{\text{ph}}$ of ternary alcohol plus two hydrocarbon mixtures. From the standpoint of practicality, the use of two coefficients per binary solution provides a better mathematical representation of the experimental data. In addition, Eqs. (8.33) and (8.35) presumably offer several advantages when the Kretschmer–Wiebe association model is extended to alcoholic solutions containing an active solvent (Nagata, 1977). Needless to say, the evaluation of τ_{ij} and τ_{ji} from the binary is relatively complex and may require reiterative computer calculations.

To eliminate the time consuming parametrization of binary data, Acree and Rytting (1982c) developed a general equation for estimation of excess enthalpies

$$\bar{H}^{\text{ex}}_{\text{ABC}} = X_A h_A K_A (\hat{\phi}_{A_1} - \hat{\phi}^*_{A_1}) + (X_A + X_B)(\phi_A + \phi_B)(\Delta \bar{H}^{\text{ex}}_{\text{AB}})'_{\text{ph}}$$
$$+ (X_A + X_C)(\phi_A + \phi_C)(\Delta \bar{H}^{\text{ex}}_{\text{AC}})'_{\text{ph}} + (X_B + X_C)(\phi_B + \phi_C)(\Delta \bar{H}^{\text{ex}}_{\text{BC}})'_{\text{ph}}$$
$$(8.35)$$

which combines numerical values of $(\Delta \bar{H}^{\text{ex}}_{ij})_{\text{ph}}$ rather than binary coefficients. In this expression $(\Delta \bar{H}^{\text{ex}}_{ij})'_{\text{ph}}$ refers to the numerical value of the physical contribution in the i–j binary solution, calculated at the mole fraction composition (X'_i, X'_j), such that

$$X'_i = 1 - X'_j = X'_i/(X'_i + X'_j)$$

Essentially Acree and Rytting treated the physical interactions of the ternary solution with a Kohler-type expression.

The results of calculations with Eq. (8.35) are compared to experimental enthalpies for several ternary alcohol-hydrocarbon systems in Table II. In general, Eq. (8.35) provides reasonably accurate predictions ranging from quite good for ternary alcohol-saturated hydrocarbon systems to only fair for ternary alcohol-aromatic hydrocarbon systems. But it should be remembered that the predictive method was developed for systems containing an alcohol and two inert hydrocarbons. Modifications of Eq. (8.35) to include association between the alcohol and another component introduces additional equilibrium constant(s) and would require a prior knowledge of each

B. Mecke–Kempter Association Model

TABLE II

Comparison between Predictions of Eq. (8.35) and Experimental Enthalpies for Several Ternary Alcohol–Hydrocarbon Solutions

Systems	Deviations* (%)	Data reference[a]
$1\text{-}C_5H_{11}OH + C_6H_{14} + C_6H_{12}$	4.1	1
$1\text{-}C_6H_{13}OH + C_7H_{16} + C_6H_{12}$	4.7	2
$2\text{-}C_3H_7OH + C_6H_6 + C_6H_{12}$	4.4	3
$2\text{-}C_3H_7OH + C_6H_6 + C_6H_{11}CH_3$	3.8	4
$2\text{-}C_3H_7OH + C_6H_5CH_3 + C_6H_{12}$	2.6	5
$C_2H_5OH + C_6H_6 + C_6H_{14}$	9.6	6
$C_2H_5OH + p\text{-}C_6H_4(CH_3)_2 + C_6H_{12}$	4.4	7
$1\text{-}C_3H_7OH + p\text{-}C_6H_4(CH_3)_2 + C_6H_{12}$	4.7	7
$2\text{-}C_3H_7OH + p\text{-}C_6H_4(CH_3)_2 + C_6H_{12}$	2.6	7
$1\text{-}C_3H_7OH + C_6H_{14} + C_6H_{12}$	2.0	8
$1\text{-}C_3H_7OH + C_7H_{16} + C_6H_{12}$	3.3	8
$1\text{-}C_3H_7OH + C_8H_{18} + C_6H_{12}$	3.1	8

* Deviations (%) = $(100/N) \sum_{i=1}^{N} |[(\Delta \bar{H}_{ABC}^{ex})^{obs} - (\Delta \bar{H}_{ABC}^{ex})^{Eq.\,(8.35)}]/(\Delta \bar{H}_{ABC}^{ex})^{obs}|$

[a] References (1) Posa et al. (1972); (2) Nunez et al. (1976); (3) Nagata et al. (1978); (4) Nagata et al. (1977); (5) Nagata et al. (1980); (6) Jones and Lu (1966); (7) Nagata and Ogasawara (1982); (8) Vesely et al. (1982).

equilibrium constant and enthalpy of formation before predictions could be made. Readers interested in the extension of the Kretschmer–Wiebe association model to ternary alcohol-active hydrocarbon systems are encouraged to read several articles by Nagata (1973a,b, 1978).

B. Mecke–Kempter Association Model

The Kretschmer–Wiebe association model is just one of the solution models currently used to describe liquid mixtures containing both alcohol and hydrocarbon molecules. The Mecke–Kempter association model is equally as popular. The basic model assumes that the alcohol forms continuous linear hydrogen-bonded polymers $A_1, A_2, \ldots, A_i$, by successive chemical reactions

$$A_i + A_1 \rightleftharpoons A_{i+1}$$

described by a single isodesmic equilibrium constant of the form (Kempter and Mecke, 1940; Treszczanowicz, 1973)

$$K_A = (\hat{\phi}_{A_{i+1}}/\hat{\phi}_{A_i}\hat{\phi}_{A_1}) \tag{8.36}$$

with the volume fraction of the i-mer calculated using the molar volume of the monomer multiplied by i.

The overall volume fraction of the alcohol (ϕ_A) is the sum of the volume fractions of each individual alcohol species

$$\phi_A = \sum_{i=1}^{\infty} \hat{\phi}_{A_i} = \sum_{i=1}^{\infty} \hat{\phi}_{A_1}(K_A \hat{\phi}_{A_1})^{i-1} = \hat{\phi}_{A_1}/(1 - K_A \hat{\phi}_{A_1}) \qquad (8.37)$$

It should be noted that $|K_A \hat{\phi}_{A_1}| < 1$ for the infinite series to converge.

The excess Gibbs free energy $\Delta \bar{G}_{ABC}^{ex}$ for a ternary solution containing an alcohol and two inert hydrocarbons is again separated into two contributions

$$\Delta \bar{G}_{ABC}^{ex} = (\Delta \bar{G}_{ABC}^{ex})_{ch} + (\Delta \bar{G}_{ABC}^{ex})_{ph}$$

The physical contribution is described by Eq. (8.19). (Although the equation for the nonspecific interactions is identical for the two different association models, the binary interaction parameters will be considerably different.)

The chemical part of the Gibbs free energy is based on the Flory–Huggins athermal model and may be written as

$$(g_{ABC}^m)_{ch} = RT \left[\sum_{i=1}^{\infty} \hat{n}_{A_i} \ln \hat{\phi}_{A_i} + \hat{n}_B \ln \hat{\phi}_B + \hat{n}_C \ln \hat{\phi}_C \right]$$

The respective chemical potentials are obtained through differentiation with respect to the number of moles of each chemical species

$$(\hat{\mu}_{A_i} - \hat{\mu}_{A_i}^\bullet)/RT = \ln \hat{\phi}_{A_i} + 1 - (\bar{V}_{A_i}/\hat{V}_{sol})$$
$$(\hat{\mu}_B - \mu_B^\bullet)/RT = \ln \hat{\phi}_B + 1 - (\bar{V}_B/\hat{V}_{sol})$$
$$(\hat{\mu}_C - \mu_C^\bullet)/RT = \ln \hat{\phi}_C + 1 - (\bar{V}_C/\hat{V}_{sol})$$

where $\hat{V}_{sol}$ is the *true* molar volume of the ternary solution

$$1/\hat{V}_{sol} = \sum_{i=1}^{\infty} \hat{\phi}_{A_i}/\bar{V}_{A_i} + \hat{\phi}_B/\bar{V}_B + \hat{\phi}_C/\bar{V}_C$$

$$= \sum_{i=1}^{\infty} \hat{\phi}_{A_1}(K_A \hat{\phi}_{A_1})^{i-1}/i\bar{V}_{A_1} + \phi_B/\bar{V}_B + \phi_C/\bar{V}_C$$

$$= (1/K_A \bar{V}_A) \sum_{i=1}^{\infty} (1/i)[K_A \phi_A/(1 + K_A \phi_A)]^i + \phi_B/\bar{V}_B + \phi_C/\bar{V}_C$$

$$= \ln(1 + K_A \phi_A)/K_A \bar{V}_A + \phi_B/\bar{V}_B + \phi_C/\bar{V}_C \qquad (8.38)$$

and for the pure alcohol

$$1/\hat{V}_{sol}^* = [\ln(1 + K_A)/\bar{V}_A K_A] \qquad (8.39)$$

B. Mecke–Kempter Association Model

Remembering that the chemical potential of the stoichiometric component is equal to the chemical potential of the monomer, one can combine Eqs. (8.7 a–c) and (8.12) to yield the following expression for $(\Delta \bar{G}^{ex}_{ABC})_{ch}$

$$(\Delta \bar{G}^{ex}_{ABC})_{ch} = RT[X_A \ln(\hat{\phi}_{A_1}/\hat{\phi}^*_{A_1} X_A) + X_B \ln(\hat{\phi}_B/X_B) + X_C \ln(\hat{\phi}_C/X_C)$$
$$- (X_A \bar{V}_A + X_B \bar{V}_B + X_C \bar{V}_C)/\hat{V}_{sol} + X_B$$
$$+ X_C + (X_A \bar{V}_A/\hat{V}^*_{sol})] \tag{8.13}$$

Substituting Eqs. (8.38) and (8.39) into Eq. (8.13) permits rearrangement of the excess Gibbs free energy into a more convenient form

$$(\Delta \bar{G}^{ex}_{ABC})_{ch} = RT\{X_A \ln(\hat{\phi}_{A_1}/\hat{\phi}^*_{A_1} X_A) + X_B \ln(\phi_B/X_B) + X_C \ln(\phi_C/X_C)$$
$$- [(X_A \bar{V}_A + X_B \bar{V}_B + X_C \bar{V}_C) \ln(1 + K_A \phi_A)/V_A K_A]$$
$$+ [X_A \ln(1 + K_A)/K_A]\} \tag{8.40}$$

The total excess free energy $\Delta \bar{G}^{ex}_{ABC}$ (per stoichiometric mole of solution) is obtained by adding together the expressions for the chemical and physical contributions

$$\Delta \bar{G}^{ex}_{ABC} = RT\left[X_A \ln \frac{\hat{\phi}_{A_1}}{\hat{\phi}^*_{A_1} X_A} + X_B \ln \frac{\phi_B}{X_B} + X_C \ln \frac{\phi_C}{X_C} \right.$$
$$\left. - \frac{(X_A \bar{V}_A + X_B \bar{V}_B + X_C \bar{V}_C) \ln(1 + K_A \phi_A)}{K_A \bar{V}_A} + \frac{X_A \ln(1 + K_A)}{K_A} \right]$$
$$+ (X_A \bar{V}_A + X_B \bar{V}_B + X_C \bar{V}_C)[\phi_A \phi_B A_{A_1B} + \phi_A \phi_C A_{A_1C} + \phi_B \phi_C A_{BC}] \tag{8.41}$$

The three interaction parameters A_{A_1B}, A_{A_1C}, and A_{BC} are evaluated from the binary free energies through the appropriate reduction of Eq. (8.41). Because the $\Delta \bar{G}^{ex}_{ch}$ expressions for the Mecke–Kempter model are different than the $\Delta \bar{G}^{ex}_{ch}$ expression for the Kretschmer–Wiebe model, the interaction parameters for the binary alcohol–hydrocarbon mixtures must also be different for the two solution models. The binary interaction parameter for the hydrocarbon–hydrocarbon mixture will be the same for both association models.

As stated previously, the expression for $\Delta \bar{H}^{ex}$ can be derived by differentiating the free energy expression directly. Instead of proceeding in this manner, one can choose to describe the $\Delta \bar{H}^{ex}_{ABC}$ as the sum of two contributions

$$\Delta \bar{H}^{ex}_{ABC} = (\Delta \bar{H}^{ex}_{ABC})_{ch} + (\Delta \bar{H}^{ex}_{ABC})_{ph}$$

The chemical part of the total excess enthalpy represents a change in number of each individual alcohol species (*i*-mers) between the ternary (or binary)

8 Ternary Alcohol–Hydrocarbon Systems

solution and the *pure* alcohol

$$(\Delta H_{ABC}^{ex})_{ch} = h_A \sum_{i=1}^{\infty} (i-1)\hat{n}_{A_i} - h_A \sum_{i=1}^{\infty} (i-1)\hat{n}_{A_i}^*$$

$$= h_A \sum_{i=1}^{\infty} \frac{i-1}{i} (K_A \hat{\phi}_{A_1})^i - h_A \sum_{i=1}^{\infty} \frac{i-1}{i} (K_A \hat{\phi}_{A_1}^*)^i$$

$$= \frac{h_A \hat{n}_{A_1}}{K_A \hat{\phi}_{A_1}} \sum_{i=1}^{\infty} \frac{i-1}{i} \left(\frac{K_A \phi_A}{1 + K_A \phi_A}\right)^i - \frac{h_A n_{A_1}}{K_A \hat{\phi}_{A_1}^*} \sum_{i=1}^{\infty} \frac{i-1}{i} \left(\frac{K_A}{1 + K_A}\right)^i$$

(8.42)

Evaluation of the infinite sums gives

$$(\Delta \bar{H}_{ABC}^{ex})_{ch} = -h_A X_A [\ln(1 + K_A \phi_A)/K_A \phi_A - \ln(1 + K_A)/K_A] \quad (8.43)$$

The chemical contribution to the excess enthalpy is identical to expressions presented previously (Nagata, 1973a, b, 1978; Treszczanowski, 1973). Substitution of Eqs. (8.27) and (8.43) into Eq. (8.23) yields the following expression for the excess enthalpy of the alcohol plus two hydrocarbon system:

$$\Delta \bar{H}_{ABC}^{ex} = -h_A X_A [\ln(1 + K_A \phi_A)/K_A \phi_A - \ln(1 + K_A)/K_A]$$
$$+ (X_A \bar{V}_A + X_B \bar{V}_B + X_C \bar{V}_C)(\phi_A \phi_B A_{A_1 B} + \phi_A \phi_C A_{A_1 C} + \phi_B \phi_C A_{BC})$$

(8.44)

Figure 8.3 shows the behavior of Eq. (8.43) for various values of K_A ranging from 100 to 300. (For graphical purposes, calculations of $(\Delta \bar{H}_{AK}^{ex})_{ch}$ were performed assuming a binary solution.) Notice that the calculated enthalpy curves are highly skewed at mole fraction compositions corresponding to

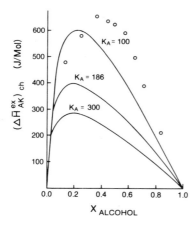

Fig. 8.3 Calculated values of the chemical contributions to the enthalpies as a function of alcohol concentration. The three curves represent different values of the Mecke–Kempter association constant. Open circles (○) denote the experimental data of the 1-hexanol + n-heptane system.

C. Attenuated Equilibrium Constant (AEC) Model

TABLE III
Mecke–Kempter Association
Constants of Alcohols
at 50°C[a]

Alcohol	K_A
Methanol	450
Ethanol	190
1-Propanol	110
1-Butanol	95
1-Pentanol	88
1-Hexanol	85
1-Octanol	72
1-Decanol	66

[a] Reproduced with permission from Nagata and Kazuma, (1977). Copyright 1977 the American Chemical Society.

low alcohol concentrations and closely resemble the shape of the experimental enthalpy curve of 1-hexanol + n-heptane mixtures, as indicated by the open circles. The Mecke–Kempter equilibrium constants are listed in Table III for several pure alcohols.

C. Attenuated Equilibrium Constant (AEC) Model

Up to this point discussions were limited to alcohol–hydrocarbon mixtures. Continuous linear association models, though, have been applied to liquid mixtures containing other self-associating molecules. Tobalky and Thatch (1962) showed that the self-association of 2-n-butylbenzimidazole and of benzotriazole in benzene could be described by a two-parameter sequential equilibrium constant model.[†] Robeson et al. (1981) concluded that the self-association characteristics of methyl lithocholate in carbon tetrachloride

[†] A one-parameter sequential equilibrium constant model assumes that the stepwise addition of monomer $A_i + A_1 \rightleftharpoons A_{i+1}$ can be represented by a single equilibrium constant. A two-parameter model ($1 - n - \infty$)

$$nA_1 \rightleftharpoons A_n, \quad K_n$$
$$A_i + A_1 \rightleftharpoons A_{i+1}, \quad K_A, \quad i \geq n$$

distinguishes between the equilibrium constant for the smallest self-associated species (n-mer) and the equilibrium constant for the remaining polymers.

solutions could be described by a two-parameter attenuated equilibrium constant model. Garland and Christian (1975) proposed a two-parameter attenuated equilibrium constant model to explain the *stacking* of nucleoside bases in aqueous solutions.

As was stated at the beginning of this chapter, the primary emphasis is on the development of thermodynamic models. Although the Attenuated Equilibrium Constant (AEC) model may not be truly applicable to alcohol–hydrocarbon mixtures, the AEC model is as thermodynamically valid as the other two solution models and describes the properties of several solutions containing self-associating molecules. It is for this reason that the AEC model is being included with the Kretschmer–Wiebe and Mecke–Kempter association models.

In the AEC model, it is assumed for any association step

$$A_1 + A_{i-1} \rightleftharpoons A_i$$

that the equilibrium constant describing the chemical reaction

$$K_i = (\hat{C}_{A_i}/\hat{C}_{A_{i-1}}\hat{C}_{A_1}\bar{V}_A) = K_A/i \tag{8.45}$$

is attenuated by an integer i ($i = 2, 3, \cdots$). The molar concentrations are calculated using the ideal molar volume approximation. The formal concentration of the associating species C_A is expressed by

$$C_A = \hat{C}_{A_1} + 2\hat{C}_{A_2} + 3\hat{C}_{A_3} + \cdots + i\hat{C}_{A_i} + \cdots$$

$$= \sum_{i=1}^{\infty} \frac{i(K_A\bar{V}_A)^{i-1}\hat{C}_{A_1}^i}{i!} = \hat{C}_{A_1} \sum_{i=1}^{\infty} \frac{(K_A\bar{V}_A\hat{C}_{A_1})^{i-1}}{(i-1)!} \tag{8.46}$$

The infinite series converges for all values of K_A, $\bar{V}_A$, $\hat{C}_{A_1}$, giving

$$C_A = \hat{C}_{A_1} \exp(K_A\hat{\phi}_{A_1})$$
$$\phi_A = \hat{\phi}_{A_1} \exp(K_A\hat{\phi}_{A_1}) \tag{8.47}$$

The chemical part of the Gibbs free energy is based on the athermal Flory–Huggins model and may be written as

$$(g_{ABC}^m)_{ch} = RT\left[\sum_{i=1}^{\infty} \hat{n}_{A_i} \ln \hat{\phi}_{A_i} + \hat{n}_B \ln \hat{\phi}_B + \hat{n}_C \ln \hat{\phi}_C\right]$$

with the summation extending over all polymeric species in solution. The chemical potentials are obtained through the appropriate differentiation

$$(\hat{\mu}_{A_i} - \mu_{A_i}^\bullet)/RT = \ln \hat{\phi}_{A_i} + 1 - \bar{V}_{A_i}/\hat{V}_{sol}$$
$$(\hat{\mu}_B - \mu_B^\bullet)/RT = \ln \hat{\phi}_B + 1 - \bar{V}_B/\hat{V}_{sol}$$
$$(\hat{\mu}_C - \mu_C^\bullet)/RT = \ln \hat{\phi}_C + 1 - \bar{V}_C/\hat{V}_{sol}$$

C. Attenuated Equilibrium Constant (AEC) Model

where $\hat{V}_{sol}$ is the *true* molar volume of the solution

$$\frac{1}{\hat{V}_{sol}} = \sum_{i=1}^{\infty} \frac{\hat{\phi}_{A_i}}{\bar{V}_{A_i}} + \frac{\phi_B}{\bar{V}_B} + \frac{\phi_C}{\bar{V}_C}$$

$$= \frac{1}{K_A \bar{V}_A} \sum_{i=1}^{\infty} \frac{(K_A \hat{\phi}_{A_1})^i}{i!} + \frac{\phi_B}{\bar{V}_B} + \frac{\phi_C}{\bar{V}_C}$$

$$= \frac{1}{K_A \bar{V}_A} \left[\sum_{i=0}^{\infty} \frac{(K_A \hat{\phi}_{A_1})^i}{i!} - 1 \right] + \frac{\phi_B}{\bar{V}_B} + \frac{\phi_C}{\bar{V}_C}$$

$$= \frac{\exp(K_A \hat{\phi}_{A_1}) - 1}{K_A \bar{V}_A} + \frac{\phi_B}{\bar{V}_B} + \frac{\phi_C}{\bar{V}_C} \tag{8.48}$$

and for the pure alcohol

$$\frac{1}{\hat{V}_{sol}^*} = \frac{\exp(K_A \hat{\phi}_{A_1}^*) - 1}{K_A \bar{V}_A} \tag{8.49}$$

Remembering that the chemical potential of the stoichiometric component is equal to the chemical potential of the monomeric species, Eqs. (8.7a–c) and (8.12) can be combined to yield the following expression for $(\Delta \bar{G}_{ABC}^{ex})_{ch}$

$$(\Delta \bar{G}_{ABC}^{ex})_{ch} = RT \left[X_A \ln(\hat{\phi}_{A_1}/\hat{\phi}_{A_1}^* X_A) + X_B \ln(\phi_B/X_B) + X_C \ln(\phi_C/X_C) \right.$$
$$\left. - \frac{X_A \bar{V}_A + X_B \bar{V}_B + X_C \bar{V}_C}{\hat{V}_{sol}} + X_B + X_C + \frac{X_A \bar{V}_A}{\hat{V}_{sol}^*} \right] \tag{8.13}$$

Substitution of Eqs. (8.48) and (8.49) into Eq. (8.13) permits rearrangement of the excess Gibbs free energy into a more convenient form

$$(\Delta \bar{G}_{ABC}^{ex})_{ch} = RT\{X_A \ln(\hat{\phi}_{A_1}/\hat{\phi}_{A_1}^* X_A) + X_B \ln(\phi_B/X_B) + X_C \ln(\phi_C/X_C)$$
$$- (X_A \bar{V}_A + X_B \bar{V}_B + X_C \bar{V}_C)[\exp(K_A \hat{\phi}_{A_1}) - 1]/K_A \bar{V}_A$$
$$+ X_A[\exp(K_A \hat{\phi}_{A_1}^*) - 1]/K_A\} \tag{8.50}$$

The total excess free energy $\Delta \bar{G}_{ABC}^{ex}$ (per stoichiometric mole of solution) is obtained by adding together the expressions for the chemical and physical contributions

$$\Delta \bar{G}_{ABC}^{ex} = RT\{X_A \ln(\hat{\phi}_{A_1}/\hat{\phi}_{A_1}^* X_A) + X_B \ln(\phi_B/X_B) + X_C \ln(\phi_C/X_C)$$
$$- (X_A \bar{V}_A + X_B \bar{V}_B + X_C \bar{V}_C)[\exp(K_A \hat{\phi}_{A_1}) - 1]/K_A \bar{V}_A$$
$$+ X_A[\exp(K_A \hat{\phi}_{A_1}^*) - 1]/K_A\}$$
$$+ (X_A \bar{V}_A + X_B \bar{V}_B + X_C \bar{V}_C)(\phi_A \phi_B A_{A_1B} + \phi_A \phi_C A_{A_1C} + \phi_B \phi_C A_{BC})$$
$$\tag{8.51}$$

The three binary interaction parameters A_{A_1B}, A_{A_1C}, and A_{BC} are evaluated from the binary data through the appropriate reduction of Eq. (8.51).

The expression for the excess enthalpy can be obtained through direct differentiation of $\Delta \bar{G}^{ex}_{ABC}$

$$\Delta \bar{H}^{ex}_{ABC} = -h_A K_A (\partial \Delta \bar{G}^{ex}_{ch}/RT/\partial K_A) + R(\partial \Delta \bar{G}^{ex}_{ph}/RT/\partial 1/T) \quad (8.52)$$

Remember that the anthermal Flory–Huggins model assumes that the molar volumes of the pure components are independent of temperature; i.e., $\bar{V}_i \neq \bar{V}_i(T)$. Performing the indicated differentation gives

$$\Delta \bar{H}^{ex}_{ABC} = -h_A K_A \left\{ X_A \left(\frac{\partial \ln(\hat{\phi}_{A_1}/\hat{\phi}^*_{A_1})}{\partial K_A} \right) - \frac{X_A [\exp(K_A \hat{\phi}^*_{A_1}) - 1]}{K_A} \right.$$
$$+ \frac{(X_A \bar{V}_A + X_B \bar{V}_B + X_C \bar{V}_C)[\exp(K_A \hat{\phi}_{A_1}) - 1]}{\bar{V}_A K_A^2}$$
$$- \frac{(X_A \bar{V}_A + X_B \bar{V}_B + X_C \bar{V}_C)}{\bar{V}_A K_A} \left(\frac{\partial \exp K_A \hat{\phi}_{A_1}}{\partial K_A} \right)$$
$$\left. + \frac{K_A}{\bar{V}_A} \left(\frac{\partial \exp K_A \hat{\phi}^*_{A_1}}{\partial K_A} \right) \right\}$$
$$+ (X_A \bar{V}_A + X_B \bar{V}_B + X_C \bar{V}_C) \left\{ \phi_A \phi_B \left[A_{A_1B} + (1/T) \left(\frac{\partial A_{A_1B}}{\partial 1/T} \right) \right] \right.$$
$$+ \phi_A \phi_C \left[A_{A_1C} + (1/T) \left(\frac{\partial A_{A_1C}}{\partial 1/T} \right) \right]$$
$$\left. + \phi_B \phi_C \left[A_{BC} + (1/T) \left(\frac{\partial A_{BC}}{\partial 1/T} \right) \right] \right\}$$
(8.53)

Mathematical manipulations enable Eq. (8.53) to be further simplified

$$\Delta \bar{H}^{ex}_{ABC} = -h_A \{(X_A \bar{V}_A + X_B \bar{V}_B + X_C \bar{V}_C)[\exp(K_A \hat{\phi}_{A_1}) - 1]/\bar{V}_A K_A$$
$$- X_A[\exp(K_A \hat{\phi}^*_{A_1}) - 1]/K_A\}$$
$$+ (X_A \bar{V}_A + X_B \bar{V}_B + X_C \bar{V}_C)[\phi_A \phi_B B_{A_1B} + \phi_A \phi_C B_{A_1C} + \phi_B \phi_C B_{BC}]$$
(8.54)

where

$$B_{ij} = A_{ij} + (1/T)(\partial A_{ij}/\partial 1/T)$$

Figures 8.4 and 8.5 show calculated chemical contributions to binary enthalpies for values of K_A ranging from 175 to 450. For convenience, the molar enthalpy of a hydrogen bond is again assumed to be -25 kJ/mol. The

C. Attenuated Equilibrium Constant (AEC) Model

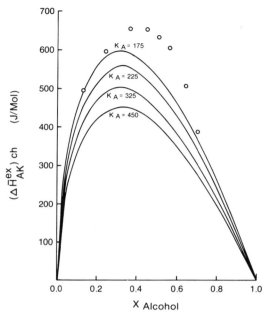

Fig. 8.4 Calculated values of the chemical contributions to the excess enthalpies as a function of alcohol concentration. The four curves represent different values of the equilibrium constant. The open circles (○) denote the experimental enthalpies of 1-hexanol + n-heptane mixture determined by Nunez et al., (1976).

molar volumes of the components were selected so that values of $(\Delta \bar{H}_{AK}^{ex})_{ch}$ would correspond to mixtures of 1-hexanol + n-heptane and 1-hexanol + cyclohexane. Notice that the chemical contributions to the excess enthalpies are highly skewed at mole fraction compositions corresponding to low alcohol concentrations. Superimposed on Figs. 8.4 and 8.5 are the experimental enthalpies for the two binary systems.

Differences between the experimental enthalpies and $(\Delta \bar{H}_{AK}^{ex})_{ch}$ are attributed to the physical interactions. Comparison of Figs. 8.4 and 8.5 for a given value of K_A reveals that calculated physical contributions for the 1-hexanol + cyclohexane system are larger than those of the 1-hexanol + n-heptane system. This observation is consistent with the fact that the excess enthalpies of n-hexane + cyclohexane mixtures are slightly larger than the excess enthalpies of n-hexane + n-heptane mixtures, the latter being the more ideal solution. Researchers often estimate the nonspecific interactions of alcohol–hydrocarbon mixtures with the measured properties of a homolog–hydrocarbon mixture; the homolog being the hydrocarbon molecule that would be formed if the hydroxyl moiety is simply replaced by a hydrogen.

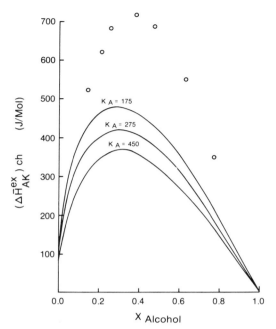

Fig. 8.5 Calculated values of the chemical contributions to the excess enthalpies as a function of alcohol concentration. The three curves represent different values of the equilibrium constant. The open circles (○) denote experimental enthalpies of 1-hexanol + cyclohexane mixtures determined by Nunez et al., (1976).

As an indication of the potential applicability of the AEC model to alcohol–hydrocarbon mixtures, $\Delta \bar{H}^{ex}_{ABC}$ has been estimated for the 1-hexanol + n-heptane + cyclohexane system using

$$\Delta \bar{H}^{ex}_{ABC} = -h_A \left\{ \frac{(X_A \bar{V}_A + X_B \bar{V}_B + X_C \bar{V}_C)[\exp(K_A \hat{\phi}_{A_1}) - 1]}{\bar{V}_A K_A} \right.$$

$$\left. - \frac{X_A[\exp(K_A \hat{\phi}^*_{A_1}) - 1]}{K_A} \right\} + (X_A + X_B)(\phi_A + \phi_B)(\Delta \bar{H}^{ex}_{AB})^*_{ph}$$

$$+ (X_A + X_C)(\phi_A + \phi_C)(\Delta \bar{H}^{ex}_{AC})^*_{ph} + (X_B + X_C)(\phi_B + \phi_C)(\Delta \bar{H}^{ex}_{BC})^*_{ph}$$

(8.55)

the $(\Delta \bar{H}^{ex}_{ABC})_{ch}$ expression of the AEC model coupled with a Kohler-type representation of $(\Delta \bar{H}^{ex}_{ABC})_{ph}$. Equation (8.55) with $K_A = 325$ provides reasonably accurate predictions for the ternary excess enthalpies, with an average absolute deviation of 4.7%. In comparison, the experimental enthalpies differ from the predictions of the Kretschmer–Wiebe and Mecke–Kempter models by 4.7 and 4.4%, respectively. The numerical value of the equilibrium constant $K_A^{25°C} = 325$ was based on the excess enthalpies for six binary 1-hexanol + n-alkane systems. This preliminary value will undoubtedly change as the AEC model is more thoroughly tested with additional enthalpy and free energy data. Regardless of the outcome of testing on alcohol–hydrocarbon systems, the AEC model describes solution properties for several other self-associating molecules.

D. Two-Constant Kretschmer–Wiebe Association Model

Up to this point, it was assumed that the association characteristics of alcohol mixtures could be described by a single equilibrium constant, either in terms of volume fractions or molarities. The next logical extension of previous discussions would be to look at a two-equilibrium constant description of alcohol solutions where the equilibrium constant for the formation of the dimeric species K_2

$$A_1 + A_1 \rightleftharpoons A_2$$

$$K_2 = \hat{\phi}_{A_2}/2\hat{\phi}_{A_1}^2 \tag{8.56}$$

is different than the equilibrium constant for each succeeding chemical reaction

$$A_i + A_1 \rightleftharpoons A_{i+1} \quad (i \geq 2)$$

$$K_A = (\hat{\phi}_{A_{i+1}}/\hat{\phi}_{A_i}\hat{\phi}_{A_1})[i/(i+1)] \tag{8.57}$$

Some support for this type of self-association can be found in theoretical arguments based on statistical thermodynamics.

The stoichiometric volume fraction of the alcohol (ϕ_A) is the sum of the volume fractions of each individual alcohol species

$$\phi_A = \hat{\phi}_{A_1} + \sum_{i=2}^{\infty} \hat{\phi}_{A_i} = \hat{\phi}_{A_1} + (K_2/K_A^2) \sum_{i=2}^{\infty} i(K_A\hat{\phi}_{A_1})^i$$

$$= \hat{\phi}_{A_1} - K_2\hat{\phi}_{A_1}/K_A + (K_2/K_A^2) \sum_{i=1}^{\infty} i(K_A\hat{\phi}_{A_1})^i$$

$$= \hat{\phi}_{A_1} - K_2\hat{\phi}_{A_1}/K_A + (K_2/K_A)[\hat{\phi}_{A_1}/(1 - K_A\hat{\phi}_{A_1})^2] \tag{8.58}$$

Unfortunately, there is no simple way to express $\hat{\phi}_{A_1}$ in terms of K_A, K_2, and ϕ_A. The best way to solve Eq. (8.58) would be to use a reiterative method. Although this might seem like an unsurmountable problem if the calculations are being performed by hand, computers will rapidly converge on the correct value after just a few iterations.

The chemical part of the excess Gibbs free energy is derived from the athermal Flory–Huggins model. Because the details of this derivation are identical to those presented earlier for the one-constant Kretschmer–Wiebe model, only the final result is stated

$$(\Delta \bar{G}^{ex}_{ABC})_{ch}/RT = X_A \ln(\hat{\phi}_{A_1}/\hat{\phi}^*_{A_1} X_A) + X_B \ln(\phi_B/X_B) + X_C \ln(\phi_C/X_C)$$
$$- [X_A \bar{V}_A + X_B \bar{V}_B + X_C \bar{V}_C/\bar{V}_A]$$
$$\times [(K_A - K_2)/K_A + K_2/K_A(1 - K_A\hat{\phi}_{A_1})] \hat{\phi}_{A_1}$$
$$+ X_A \hat{\phi}^*_{A_1}[(K_A - K_2)/K_A + K_2/K_A(1 - K_A\hat{\phi}^*_{A_1})] \quad (8.59)$$

The expression for the chemical contributions to the excess enthalpy can be obtained through differentiation of $(\Delta \bar{G}^{ex}_{ABC})_{ch}$ according to the chain rule

$$(\Delta \bar{H}^{ex}_{ABC})_{ch} = R(\partial \Delta \bar{G}^{ex}_{ch}/RT/\partial 1/T)$$
$$= -h_A K_A(\partial \Delta \bar{G}^{ex}_{ch}/RT/\partial K_A) - h_2 K_2(\partial \Delta \bar{G}^{ex}_{ch}/RT/\partial K_2) \quad (8.60)$$

Performing the actual differentiation, one finds that $(\Delta \bar{H}^{ex}_{ABC})_{ch}$ can be expressed as

$$(\Delta \bar{H}^{ex}_{ABC})_{ch} = X_1 h_2 K_2[(\hat{\phi}_{A_1}/\phi_A) - \hat{\phi}^*_{A_1}]$$
$$+ X_2 h_A K_A\{(\hat{\phi}_{A_1} - \hat{\phi}^*_{A_1}) - [(\hat{\phi}^2_{A_1}/\phi_A) - \hat{\phi}^{*2}_{A_1}]\} \quad (8.61)$$

TABLE IV
Predictive Ability of the One-Constant and Two-Constant Kretschmer–Wiebe Association Model[a]

Ternary system	Absolute arithmetic mean deviation (J/mol)	
	One-constant model[b]	Two-constant model[b]
Ethanol + p-xylene + cyclohexane	35.3	25.8
1-Propanol + p-xylene + cyclohexane	22.3	12.9
2-Propanol + p-xylene + cyclohexane	31.3	13.6
Ethanol + toluene + cyclohexane	23.8	18.1
2-Propanol + toluene + cyclohexane	22.5	9.8
2-Propanol + benzene + methylcylohexane	17.4	14.4

[a] Reproduced from Nagata and Ogasawara (1982), with permission of the copyright owners.

[b] The basic model was extended to include association between the alcohol and unsaturated hydrocarbons. Physical contributions to the excess enthalpy were described by the UNIQUAC model.

in terms of the enthalpies of hydrogen bond formation for dimerization h_2, and linear polymerization $(i \geq 2)$ h_A.

The two-constant Kretschmer–Wiebe model $(K_A \neq K_2)$ just described correctly reduces to the one-constant model whenever $K_A = K_2$. The incorporation of the additional equilibrium constant and enthalpy of formation leads to a better set of descriptive equations as one might expect, based on the fact that the number of parameters doubled.

As far as the predictive ability of Eq. (8.61) is concerned, Nagata and Ogasawara (1982) compared the predictions of both the one-constant and two-constant Kretschmer–Wiebe models to experimental enthalpies for six different systems. The results of this comparison are given in Table IV. Based on these six systems, one can conclude that the two-constant model provides reasonable predictions for ternary alcohol plus two hydrocarbon systems, which in many instances are superior to the predictions of the corresponding one-constant model.

E. Summary

It is left up to the readers to decide from the evidence presented in the previous four chapters and from their own experience how much reliance to place upon the various predictive methods in estimating and interpreting the properties of nonelectrolyte solutions. The fact that so many predictive expressions were developed suggests that no single method adequately describes all types of solutions commonly encountered. Predictive methods should never be regarded as a satisfactory substitute for reliable measurements. It is only through experimental measurements that one learns the limitations of today's solution models. Just as more precise P–V–T measurements showed the inadequacies of the ideal gas law prompted the development of the more sophisticated equations of state, it is a personal belief that additional experimental data for both binary and multicomponent systems will encourage the development of better (and in some cases, more realistic) descriptions of liquid mixtures.

Problems

8.1 Calculate the monomeric volume fraction concentrations $\hat{\phi}_{A_1}$ of the Kretschmer–Wiebe association model corresponding to $\phi_A = 0.1, 0.2, 0.3, \ldots, 1.0$. For calculational purposes use $K_A = 100$. Using this value of K_A, calculate the values of $\hat{\phi}_{A_1}$ for the Mecke–Kempter and AEC model at the volume fraction compositions just listed.

8.2 Differentiate Eq. (8.40) with respect to pressure. What restrictions do the ideal molar volume approximation place on $(\partial K_A/\partial P)$?

8.3 Using the Mecke–Kempter association model and $B_{A_1K} = 12.02$ J/cm^3, calculate the excess enthalpies at 25°C for the 1-propanol + cyclohexane system at $X_{1\text{-propanol}} = 0.2496, 0.5858$, and 0.7743. The experimentally observed values are 597.9, 523.5, and 334.3 J/mol, respectively. (I Nagata and K. Kazuma, 1977).

8.4 Using the Kretschmer–Wiebe association model and $B_{A_1K} = 0.791$ J/cm^3, calculate the excess enthalpies for the 1-butanol (1) + n-hexane (2) system at the binary compositions listed in Table V. Does the Kretschmer–Wiebe association model satisfactorily describe the experimental data? Be sure to convert the equilibrium constant K_A to 25°C.

TABLE V
Excess Enthalpies for
Binary Mixtures of
1-Butanol (1) + n-Hexane (2)
at 25°C[a]

X_1	$\Delta \bar{H}^{ex}_{12}$ (J/mol)
0.1634	500
0.1772	513
0.3151	572
0.3323	573
0.5817	452
0.5960	439
0.5971	439
0.8173	213

[a] Experimental data were taken from Brown et al., (1964). Reprinted with permission of the copyright owners.

8.5 Repeat the calculations from the previous problem, this time using the Mecke–Kempter association model and $B_{A_1K} = 6.171$ J/cm^3. Which model best represents the enthalpies of the 1-butanol + n-hexane system? The numerical values of B_{A_1K} given in the two problems were determined by a least-squares analysis of the binary data, and therefore represent the best possible coefficients.

8.6 Many thermodynamic models for alcohol–hydrocarbon mixtures were derived from Eq. (8.1) by assuming a different form for $(\Delta \bar{G}^{ex}_{AB})_{ph}$. For example, Libermann and Wilhelm (1975) combined the Kretschmer–

Wiebe association model and the Bruin equation (Bruin, 1970) to obtain

$$\frac{\Delta \bar{G}^{ex}_{AB}}{RT} = \frac{2X_A X_B \ln A_{AB}}{(X_A + X_B A_{AB})(X_B + X_A A_{AB})} + X_A \ln(\hat{\phi}_{A_1}/\hat{\phi}^*_{A_1} X_A)$$
$$+ X_B \ln(\phi_B/X_B) + X_A K_A (\hat{\phi}_{A_1} - \hat{\phi}^*_{A_1})$$

Derive the corresponding expression for the enthalpy of mixing. For simplicity, leave the answer in terms of $(\partial A_{AB}/\partial T)$.

8.7 For a binary alcohol–hydrocarbon mixture obeying the Kretschmer–Wiebe association model, Eq. (8.21), show that the infinite dilution activity coefficients are given by

$$\ln \gamma^\infty_A = \ln(\bar{V}_A/\bar{V}_B \hat{\phi}^*_{A_1}) + (1 - \bar{V}_A/\bar{V}_B) - K_A \hat{\phi}^*_{A_1} + \frac{\bar{V}_A A_{A_1 B}}{RT}$$

and

$$\ln \gamma^\infty_B = \ln(\bar{V}_B/\bar{V}_A) + (1 - \bar{V}_B/\bar{V}_A) + K_A(\bar{V}_B/\bar{V}_A)\hat{\phi}^*_{A_1} + \frac{\bar{V}_B A_{A_1 B}}{RT}$$

9
Vapor–Liquid Equilibrium and Azeotropic Systems

The past four chapters have focused primarily on the development of solution models for the mathematical representation of thermodynamic properties, and for the estimation of multicomponent properties from binary data. Although the ability to correlate and predict thermodynamic quantities is important, engineering applications require knowledge of vapor–liquid equilibrium data, such as the boiling point of a mixture as a function of liquid-phase composition. As we saw in Chapter 3, the pressure and vapor-phase compositions above liquid mixtures are related to the activity coefficients in the liquid phase. After having developed several models for the excess Gibbs free energy, it is only natural to incorporate predictive methods into the previous discussions of vapor–liquid equilibrium.

A. Isothermal Vapor–Liquid Equilibrium in Binary Systems

The equilibrium between a homogeneous binary mixture and its vapor is described by the approximate relationship

$$P(1 - y_2) = P_1^\bullet (1 - X_2)\gamma_1 \tag{9.1}$$

$$P y_2 = P_2^\bullet X_2 \gamma_2 \tag{9.2}$$

which correlates the three intensive parameters P, X_2, and y_2 at constant temperature. If one regards the Raoult's law activity coefficients as known parameters, one can successively eliminate y_2, X_2, and P from the pair of equations. The result is

$$P = P_1^\bullet (1 - X_2)\gamma_1 + P_2^\bullet X_2 \gamma_2 \tag{9.3a}$$

$$X_2 = (P - P_1^\bullet \gamma_1)/(P_2^\bullet \gamma_2 - P_1^\bullet \gamma_1) \tag{9.3b}$$

A. Isothermal in Binary Systems

and

$$P = P_1^\bullet P_2^\bullet \gamma_1 \gamma_2 / [P_1^\bullet \gamma_1 y_2 + P_2^\bullet \gamma_2(1 - y_2)] \tag{9.4a}$$

$$y_2 = (P_1^\bullet P_2^\bullet \gamma_1 \gamma_2 - PP_2^\bullet \gamma_2)/(PP_1^\bullet \gamma_1 - PP_2^\bullet \gamma_2) \tag{9.4b}$$

and

$$X_2 = P_1^\bullet \gamma_1 y_2 / [P_1^\bullet \gamma_1 y_2 + P_2^\bullet \gamma_2(1 - y_2)] \tag{9.5a}$$

$$y_2 = P_2^\bullet \gamma_2 X_2 / [P_1^\bullet \gamma_1(1 - X_2) + P_2^\bullet \gamma_2 X_2] \tag{9.5b}$$

Equation (9.3) describes the bubble-point or boiling-point curve, and Eq. (9.4) the dew-point or condensation curve. The last set of equations correlate the liquid- and vapor-phase compositions, all at constant temperature.

The activity coefficients are related to the excess molar Gibbs free energy by the expression

$$RT \ln \gamma_i = \Delta \bar{G}_i^{ex} = \left(\frac{\partial \Delta G^{ex}}{\partial n_i}\right)_{T, P, n_j \neq n_i} \tag{9.6}$$

For illustrative purposes, assume that the excess free energy of the binary mixture is adequately described by a one-parameter Redlich–Kister equation

$$\Delta \bar{G}_{12}^{ex} = X_1 X_2 A_{12} \tag{9.7}$$

with the activity coefficients obtained via the appropriate differentiation

$$RT \ln \gamma_1 = X_2^2 A_{12} \quad \text{and} \quad RT \ln \gamma_2 = (1 - X_2)^2 A_{12} \tag{9.8}$$

or

$$\gamma_1 = \exp\left[\frac{A_{12} X_2^2}{RT}\right] \quad \text{and} \quad \gamma_2 = \exp\left[\frac{A_{12}(1 - X_2)^2}{RT}\right] \tag{9.9}$$

These expressions for the activity coefficients, on substitution into Eq. (9.3), lead to the following equation, which gives the vapor pressure at the bubble point as a function of the liquid-phase composition at constant temperature

$$P = P_1^\bullet(1 - X_2) \exp\left[\frac{A_{12} X_2^2}{RT}\right] + P_2^\bullet X_2 \exp\left[\frac{A_{12}(1 - X_2)^2}{RT}\right] \tag{9.10}$$

In accordance with $P = P_1 + P_2$, the partial pressures are given by

$$P_1 = P(1 - y_2) = P_1^\bullet(1 - X_2) \exp[A_{12} X_2^2 / RT] \tag{9.11}$$

$$P_2 = P y_2 = P_2^\bullet X_2 \exp[A_{12}(1 - X_2)^2 / RT] \tag{9.12}$$

Substitution of Eq. (9.9) into Eq. (9.5b) yields the following equation for the vapor-phase composition:

$$y_2 = \frac{P_2^\bullet X_2 \exp[A_{12}(1-X_2)^2/RT]}{P_1^\bullet(1-X_2)\exp[A_{12}X_2^2/RT] + P_2^\bullet X_2 \exp[A_{12}(1-X_2)^2/RT]} \quad (9.13)$$

The relative volatility of a binary mixture, defined by $\alpha_{12} = (y_1/y_2)/(X_1/X_2)$ is a useful measure of the separation effected between the vapor and liquid phases. Chemical engineers use this quantity to estimate the minimum number of theoretical plates required to achieve a desired separation. The lack of experimental values for the relative volatilities of most ternary (and higher-order multicomponent) systems, coupled with the desire to predict these values from the available data, was one of the original reasons behind the development of predictive methods for thermodynamic excess properties.

Expressed in terms of the relative volatility, Eq. (9.13) takes the form

$$y_2 = X_2/[(1-X_2)\alpha_{12} + X_2] \quad (9.14)$$

where

$$\alpha_{12} = (P_1^\bullet/P_2^\bullet)\exp[A_{12}(2X_2 - 1)/RT] \quad (9.15)$$

If the bubble-point curve and the composition of the vapor phase as a function of liquid-phase composition are known, then the equilibrium diagram is fully specified. That is, the vapor pressures of the pure components and the numerical value of A_{12} determine the isothermal vapor–liquid equilibrium.

Equation (9.10) suggests a method for calculating activity coefficients from a knowledge of the total pressure as a function of liquid composition. Vapor pressures above binary liquids can be experimentally determined at several liquid-phase compositions, and a least-squares analysis of the data in accordance with Eq. (9.10) will give the best numerical value of the binary A_{12} parameter. Knowledge of the A_{12} parameter enables the calculation of the excess molar Gibbs free energy (Eq. 9.7) and the activity coefficients (Eq. 9.9). More sophisticated least squares-procedures would incorporate the virial coefficients of both pure liquids and any vapor-phase nonideality corrections into the expressions for the total pressure.

If the compositional dependence of the total pressure cannot be adequately described by a one-parameter Redlich–Kister equation, one could develop an analogous expression from the two-parameter mathematical representations of $\Delta \bar{G}_{12}^{ex}$ discussed previously (i.e., the Wilson equation, the NRTL equation and/or the two-parameter Redlich–Kister equation, etc.). Three-parameter representations of $\Delta \bar{G}_{12}^{ex}$ can be used if greater calculational accuracy is desired. Theoretically, there is no limit to the number of parameters that can be used. Practically though it must be realized that the number of experimental observations must increase drastically with an increasing

A. Isothermal in Binary Systems

number of parameters. It is for this reason that one rarely sees more than four or five Redlich–Kister parameters in $\Delta \bar{G}_{12}^{\text{ex}}$ representations. Furthermore, the accuracy of the experimental data generally does not warrant the use of a large number of parameters.

B. Isothermal Vapor–Liquid Equilibrium in Ternary Systems

The isothermal vapor–liquid equilibrium of a ternary system is specified, within the limits of the approximations introduced previously, by the set of equations

$$P(1 - y_2 - y_3) = P_1^\bullet(1 - X_2 - X_3)\gamma_1 \qquad (9.16\text{a})$$

$$Py_2 = P_2^\bullet X_2 \gamma_2 \qquad (9.16\text{b})$$

$$Py_3 = P_3^\bullet X_3 \gamma_3 \qquad (9.16\text{c})$$

Adding Eqs. (9.16a–c) gives the following equation for the bubble-point surface:

$$P = P_1^\bullet(1 - X_2 - X_3)\gamma_1 + P_2^\bullet X_2 \gamma_2 + P_3^\bullet X_3 \gamma_3 \qquad (9.17)$$

Eliminating X_2 and X_3 from Eq. (9.17) gives the equation for the dew-point surface

$$P = \frac{P_1^\bullet P_2^\bullet P_3^\bullet \gamma_1 \gamma_2 \gamma_3}{P_1^\bullet P_2^\bullet \gamma_1 \gamma_2 y_3 + P_1^\bullet P_3^\bullet \gamma_1 \gamma_3 y_2 + P_2^\bullet P_3^\bullet \gamma_2 \gamma_3 (1 - y_2 - y_3)} \qquad (9.18)$$

which is nothing more than the total pressure above the ternary system expressed in terms of vapor-phase compositions. The relationship between the vapor and liquid compositions is given by

$$y_2 = P_2^\bullet X_2 \gamma_2 / [P_1^\bullet(1 - X_2 - X_3)\gamma_1 + P_2^\bullet X_2 \gamma_2 + P_3^\bullet X_3 \gamma_3] \qquad (9.19)$$

$$y_3 = P_3^\bullet X_3 \gamma_3 / [P_1^\bullet(1 - X_2 - X_3)\gamma_1 + P_2^\bullet X_2 \gamma_2 + P_3^\bullet X_3 \gamma_3] \qquad (9.20)$$

whereas the dependence of the liquid-phase composition upon that of the vapor phase is expressed as

$$X_2 = y_2 P_1^\bullet P_2^\bullet \gamma_1 \gamma_3 / [(1 - y_2 - y_3) P_2^\bullet P_3^\bullet \gamma_2 \gamma_3 + y_2 P_1^\bullet P_3^\bullet \gamma_1 \gamma_3 + y_3 P_1^\bullet P_2^\bullet \gamma_1 \gamma_2] \qquad (9.21)$$

$$X_3 = y_3 P_1^\bullet P_2^\bullet \gamma_1 \gamma_2 / [(1 - y_2 - y_3) P_2^\bullet P_3^\bullet \gamma_2 \gamma_3 + y_2 P_1^\bullet P_3^\bullet \gamma_1 \gamma_3 + y_3 P_1^\bullet P_2^\bullet \gamma_1 \gamma_2] \qquad (9.22)$$

Because the activity coefficients γ_i are functions of composition, Eqs. (9.18), (9.21), and (9.22) are implicit in X_2 and X_3; whereas Eqs. (9.17), (9.19), and (9.20) give P, y_2 and y_3 explicitly. This latter set of equations can be used for the evaluation of both the total pressure and vapor-phase compositions, provided that the activity coefficients are known functions of composition.

For ideal mixtures, the activity coefficients are equal to unity and the bubble-point surface reduces to a plane passing through the points corresponding to the vapor pressures of the pure components, whereas the surfaces $P(y_2, y_3)$, $y_2(X_2, X_3)$, $y_3(X_2, X_3)$, $X_2(y_2, y_3)$, and $X_3(y_2, y_3)$ are hyperboloids.

For nonideal ternary mixtures, remember that the activity coefficients are mathematically related to the excess Gibbs free energy

$$\Delta \bar{G}^{ex}_{123} = RT[X_1 \ln X_1 + X_2 \ln X_2 + X_3 \ln X_3] \tag{9.23}$$

Initially, assume that $\Delta \bar{G}^{ex}_{123}$ can be reasonably approximated from the A_{ij} parameters of the three contributive binary systems

$$\Delta \bar{G}^{ex}_{123} = X_1 X_2 A_{12} + X_1 X_3 A_{13} + X_2 X_3 A_{23} \tag{9.24}$$

in which case the activity coefficients would obey

$$RT \ln \gamma_1 = A_{12} X_2^2 + A_{13} X_3^2 + (A_{12} + A_{13} - A_{23}) X_2 X_3 \tag{9.25a}$$

$$RT \ln \gamma_2 = A_{12} X_1^2 + A_{23} X_3^2 + (A_{12} + A_{23} - A_{13}) X_1 X_3 \tag{9.25b}$$

$$RT \ln \gamma_3 = A_{13} X_1^2 + A_{23} X_2^2 + (A_{13} + A_{23} - A_{12}) X_1 X_2 \tag{9.25c}$$

These relationships, upon substitution into Eq. (9.17), enable one to evaluate the vapor pressure of a regular ternary mixture (remember that $\Delta \bar{G}^{ex}_{123}$ of a Regular Solution is described by Eq. 9.24) as a function of composition at constant temperature. If it is desirable to evaluate the vapor-phase composition as a function of liquid-phase composition, Eq. (9.19) can be rewritten as follows:

$$y_2 = X_2 / [(1 - X_2 - X_3)\alpha_{12} + X_2 + X_3 \alpha_{32}] \tag{9.26a}$$

$$y_3 = X_3 / [(1 - X_2 - X_3)\alpha_{13} + X_2 \alpha_{23} + X_3] \tag{9.26b}$$

The relative volatilities are

$$\alpha_{12} = \frac{1}{\alpha_{21}} = \frac{P_1^\bullet}{P_2^\bullet} \exp\left[\frac{A_{12}(1 - 2X_1)}{RT} + \frac{(A_{13} - A_{23} - A_{12})X_3}{RT}\right] \tag{9.27a}$$

$$\alpha_{23} = \frac{1}{\alpha_{32}} = \frac{P_2^\bullet}{P_3^\bullet} \exp\left[\frac{A_{23}(1 - 2X_2)}{RT} + \frac{(A_{12} - A_{13} - A_{23})X_1}{RT}\right] \tag{9.27b}$$

C. Isobaric Equilibrium in Binary Systems

and

$$\alpha_{13} = \frac{1}{\alpha_{31}} = \frac{P_1^\bullet}{P_3^\bullet} \exp\left[\frac{A_{13}(1-2X_1)}{RT} + \frac{(A_{12}-A_{23}-A_{13})X_2}{RT}\right] \quad (9.27c)$$

The use of relative volatilities instead of activity coefficients is sometimes preferable because the exponential term contains only a linear function of mole fraction than a quadratic function.

If $\Delta \bar{G}_{123}^{ex}$ is approximated by the Wilson equation (see Chapter 5), then the relative volatilities would be expressed in terms of the six binary Wilson coefficients

$$\alpha_{12} = \frac{P_1^\bullet}{P_2^\bullet} \exp\left\{\ln\left[\frac{X_1\Lambda_{21} + X_2 + X_3\Lambda_{23}}{X_1 + X_2\Lambda_{12} + X_3\Lambda_{13}}\right] + \frac{X_1(\Lambda_{12}-1)}{X_1 + X_2\Lambda_{12} + X_3\Lambda_{13}} \right.$$
$$\left. + \frac{X_2(1-\Lambda_{21})}{X_1\Lambda_{21} + X_2 + X_3\Lambda_{23}} + \frac{X_3(\Lambda_{32}-\Lambda_{31})}{X_1\Lambda_{31} + X_2\Lambda_{32} + X_3}\right\} \quad (9.28a)$$

$$\alpha_{13} = \frac{P_1^\bullet}{P_3^\bullet} \exp\left\{\ln\left[\frac{X_1\Lambda_{31} + X_2\Lambda_{32} + X_3}{X_1 + X_2\Lambda_{12} + X_3\Lambda_{13}}\right] + \frac{X_1(\Lambda_{13}-1)}{X_1 + X_2\Lambda_{12} + X_3\Lambda_{13}} \right.$$
$$\left. + \frac{X_2(\Lambda_{23}-\Lambda_{21})}{X_1\Lambda_{21} + X_2 + X_3\Lambda_{23}} + \frac{X_3(1-\Lambda_{31})}{X_1\Lambda_{31} + X_2\Lambda_{32} + X_3}\right\} \quad (9.28b)$$

and

$$\alpha_{23} = \frac{P_2^\bullet}{P_3^\bullet} \exp\left\{\ln\left[\frac{X_1\Lambda_{31} + X_2\Lambda_{32} + X_3}{X_1\Lambda_{21} + X_2 + X_3\Lambda_{23}}\right] + \frac{X_1(\Lambda_{13}-\Lambda_{12})}{X_1 + X_2\Lambda_{12} + X_3\Lambda_{13}} \right.$$
$$\left. + \frac{X_2(\Lambda_{23}-1)}{X_1\Lambda_{21} + X_2 + X_3\Lambda_{23}} + \frac{X_3(1-\Lambda_{32})}{X_1\Lambda_{31} + X_2\Lambda_{32} + X_3}\right\} \quad (9.28c)$$

Naturally, the expressions for relative volatilities become more complex with increasing number of components and with increasing number of binary coefficients. Calculational procedures may at first seem insurmountable, but the modern computers can go through multicomponent calculations relatively quickly.

C. Isobaric Vapor–Liquid Equilibrium in Binary Systems

The equations developed so far apply to systems at constant temperature. A large majority of distillation processes, though, are performed under isobaric conditions. The vapor–liquid equilibrium at any constant pressure is completely specified by a knowledge of the liquid- and vapor-phase compositions and the equilibrium temperature. Equation (9.1) applies to isobaric conditions as well as isothermal conditions

$$y_i P = P_i^{\bullet}(T) X_i \gamma_i(T, X) \tag{9.29}$$

This time remember that the vapor pressures of the pure components $P_i^{\bullet} = P_i^{\bullet}(T)$ and the activity coefficients $\gamma_i = \gamma_i(T, X)$ are functions of temperature.[†] Proceeding as before, the total pressure above the binary mixture is written as

$$P = X_1 P_1^{\bullet}(T) \gamma_1(T, X) + X_2 P_2^{\bullet}(T) \gamma_2(T, X) \tag{9.30}$$

or dividing through by the pressure

$$1 = (P_1^{\bullet}(T)/P) X_1 \gamma_1(T, X) + (P_2^{\bullet}(T)/P) X_2 \gamma_2(T, X) \tag{9.31}$$

The Clausius–Clapeyron equation relates $P_i^{\bullet}(T)/P$ to the temperature in the manner

$$\ln(P_i^{\bullet}(T)/P) = \int_{T_i}^{T} (\Delta \bar{H}_i^{\text{vap}}/RT^2) dT \tag{9.32}$$

where T_i is the absolute boiling temperature of the i-th component under a reference pressure P, and $\Delta \bar{H}_i^{\text{vap}}$ is its molar latent enthalpy of vaporization. Assuming that $\Delta \bar{H}_i^{\text{vap}}$ is essentially constant over a limited temperature interval gives

$$\ln(P_i^{\bullet}(T)/P) = (\Delta \bar{H}_i^{\text{vap}}/R)(1/T_i - 1/T) \tag{9.33}$$

so that

$$P_i^{\bullet}/P = \exp[(\Delta \bar{H}_i^{\text{vap}}/R)(1/T_i - 1/T)] \tag{9.34}$$

$$P_i^{\bullet}/P = \exp[(\Delta \bar{S}_i^{\text{vap}}/RT)(T - T_i)] \tag{9.35}$$

where $\Delta \bar{S}_i^{\text{vap}}$ is the molar entropy of vaporization of pure component i.

The exponential relationship between $P_i^{\bullet}/P$ and $1/T$, if substituted into Eq. (9.30), shows that even for ideal mixtures (where $\gamma_i = 1$) the boiling temperature is not a linear function of liquid-phase compositions. The isothermal vapor pressure above an ideal mixture is a linear function of liquid-phase composition.

Equation (9.30) gives the boiling temperature isobar implicitly. The activity coefficients of the components, which are functions of liquid-phase composition and of temperature, must be specified in some way, such as by

---

[†] The added notation (T) and (T, X) will be dropped soon as it is too cumbersome to carry throughout the discussion. It is understood, however, that the functional dependence still exists.

C. Isobaric Equilibrium in Binary Systems

equations based upon Regular Solution theory. The resulting expressions, though, are complicated and implicit with regard to temperature, so that numerical calculations are far from simple and any detailed discussion is almost prohibitive. These difficulties can be overcome, however, by using an approach suggested by Malesinski (1961). The binary excess Gibbs free energy is written as:

$$\Delta \bar{G}_{12}^{ex} = RT[X_1 \ln(Py_1/P_1^\bullet X_1) + X_2 \ln(Py_2/P_2^\bullet X_2)] \tag{9.36}$$

or in a slightly rearranged form

$$\Delta \bar{G}_{12}^{ex} = RT[X_1 \ln(P/P_1^\bullet) + X_2 \ln(P/P_2^\bullet)] \\ + RT[X_1 \ln(y_1/X_1) + X_2 \ln(y_2/X_2)] \tag{9.37}$$

Incorporating the temperature dependence of $P_i^\bullet/P$, Eq. 9.37 can be rewritten as

$$\Delta \bar{G}_{12}^{ex} = X_1 \Delta \bar{S}_1^{vap}(T_1 - T) + X_2 \Delta \bar{S}_2^{vap}(T_2 - T) \\ + RT[X_1 \ln(y_1/X_1) + X_2 \ln(y_2/X_2)] \tag{9.38}$$

in terms of the molar entropies of vaporization. After suitable mathematical manipulations, it is found that the temperature can be expressed as

$$T = s_1 T_1 + s_2 T_2 - [\Delta \bar{G}_{12}^{ex}/(X_1 \Delta \bar{S}_1^{vap} + X_2 \Delta \bar{S}_2^{vap})] \\ + [RT/(X_1 \Delta \bar{S}_1^{vap} + X_2 \Delta \bar{S}_2^{vap})][X_1 \ln(y_1/X_1) + X_2 \ln(y_2/X_2)] \tag{9.39}$$

$$s_1 = 1 - s_2 = X_1 \Delta \bar{S}_1^{vap}/(X_1 \Delta \bar{S}_1^{vap} + X_2 \Delta \bar{S}_2^{vap})$$

a weighted mole fraction average of the boiling temperatures of the pure components, with the two additional terms representing the deviations from ideality and the ratios between the vapor- and liquid-phase compositions.

If both liquids obey Trouton's rule, $\Delta \bar{S}_1^{vap} = \Delta \bar{S}_2^{vap}$, then the entropy fractions become equal to the mole fractions. The third term in Eq. (9.39) represents the contribution arising from the nonideality of the liquid mixture. The larger the numerical value of $\Delta \bar{G}_{12}^{ex}$, the larger are the deviations from additivity expressed by the first two terms. Because $\Delta \bar{G}_{12}^{ex}$ is a parabolic-type function, the deviations from simple additivity are most marked near the equimolar composition. For binary mixtures having a positive excess free energy, $\Delta \bar{G}_{12}^{ex} > 0$, the isobar in a TX diagram lies below the straight line connecting the boiling temperatures of the pure liquids. Conversely, if $\Delta \bar{G}_{12}^{ex} < 0$, the isobar will lie above the straight line. This simple picture of the boiling temperature isobar may be modified, sometimes considerably, by the last term on the right-hand side of Eq. (9.39). It can be shown mathematically that the last term is always negative (or equal to zero in the case of the azeotropic point where $X_1^{az} = y_1^{az}$ and $X_2^{az} = y_2^{az}$. (The superscript az denotes the azeotrope.) The absolute value of the last term becomes smaller in magnitude as the composition of the vapor phase approaches that in the liquid phase. Consequently, for systems having a negative $\Delta \bar{G}_{12}^{ex}$, the third and fourth terms exert opposing influences on the shape of the isobar, which will lie above or below the additive line, depending on the sign of the difference between the last two terms over the relevant composition range. As one might expect, the absolute value of $|X_1 \ln(y_1/X_1) + X_2 \ln(y_2/X_2)|$ increases with increasing differences between the boiling temperatures of the two components, and may possibly conceal the effect of a negative $\Delta \bar{G}_{12}^{ex}$, so the binary mixture will boil at a temperature lower than the additive value.

D. Isobaric Vapor–Liquid Equilibrium in Ternary Systems

The boiling temperature of a ternary mixture can be expressed in a similar manner. Recalling that the excess Gibbs free energy is given by

$$\Delta \bar{G}_{123}^{ex} = RT[X_1 \ln(P/P_1^{\bullet}) + X_2 \ln(P/P_2^{\bullet}) + X_3 \ln(P/P_3^{\bullet})] + RT[X_1 \ln(y_1/X_1) + X_2 \ln(y_2/X_2) + X_3 \ln(y_3/X_3)] \quad (9.40)$$

one can incorporate the temperature dependence through the Clausius–Clapeyron equation. Thus, the equation of the boiling temperature isobar has the form

$$T = s_1 T_1 + s_2 T_2 + s_3 T_3 - \frac{\Delta \bar{G}_{123}^{ex}}{\sum_{i=1}^{3} X_i \Delta \bar{S}_i^{vap}} + \frac{RT \sum_{i=1}^{3} X_i \ln(y_i/X_i)}{\sum_{i=1}^{3} X_i \Delta \bar{S}_i^{vap}} \quad (9.41)$$

where

$$s_j = X_j \Delta \bar{S}_j^{vap} \bigg/ \left(\sum_{i=1}^{3} X_i \Delta \bar{S}_i^{vap} \right)$$

The equations for ternary mixtures can be easily extended to higher-order multicomponent systems. The equations also suggest how one might calculate the excess Gibbs free energy from a knowledge of the vapor–liquid equilibrium compositions and the partial pressures of the pure components as a function of temperature. The numerical value of $\Delta \bar{G}_{123}^{ex}$ refers to a particular temperature and pressure.

E. Azeotropes in Binary Mixtures Under Isobaric Conditions

Binary liquid systems with a vapor pressure extremum under isothermal conditions, or equivalently, with an extremum in the boiling temperature under isobaric conditions, are called azeotropic systems, and the mixture whose composition corresponds to the extremum point is called an azeotrope. Wade and Merriman (1911) introduced these terms into the scientific literature more than 70 years ago. Lecat (1926) subdivided azeotropes according to the characteristics of the extremum. Positive azeotropes were characterized by a minimum boiling temperature under isobaric conditions, i.e., a maximum in the vapor pressure under isothermal conditions. Negative azeotropes, on the other hand, had a maximum boiling temperature and a minimum vapor pressure. In multicomponent systems there is also the possibility that saddle azeotropes (sometimes called positive–negative azeotropes) may occur.

The problem of predicting the possible existence of azeotropes and the calculation of their compositions and boiling temperatures became more and more important as the growing number of experimental data demonstrated that azeotrope formation was not a rarity. With the aid of thermodynamics, the basic types of liquid–vapor equilibrium diagrams can be established. It has already been seen in previous sections of this chapter how the isothermal liquid–vapor equilibrium data, as well as the isobaric liquid–vapor equilibrium can be described in terms of the excess free energy. Attention will now be focused on a special type of liquid mixture, namely those mixtures having an azeotrope.

The general expressions developed thus far for the liquid–vapor equilibrium also apply to azeotropic systems. At constant pressure, remember that the boiling temperature of a binary mixture is given by Eq. (9.39). At the azeotrope, the mole fraction compositions in the liquid phase must equal the corresponding compositions in the vapor phase. (See Chapter 3 for the derivation.) This fact enables Eq. (9.39) to be greatly simplified as

$$T^{az} = s_1^{az}T_1 + s_2^{az}T_2 - (\Delta \bar{G}_{12}^{ex})^{az}/(X_1^{az} \Delta \bar{S}_1^{vap} + X_2^{az} \Delta \bar{S}_2^{vap}) \qquad (9.42)$$

The symbol az denotes that the temperature and compositions of the system correspond to the azeotrope.

To develop predictive expressions for the azeotropic composition and temperature it becomes necessary to specify the compositional dependence of $\Delta \bar{G}_{12}^{ex}$. In other words, to assume some mathematical relationship between $\Delta \bar{G}_{12}^{ex}$ and X_1. For calculational simplicity, assume that the excess free energy of the binary mixture can be adequately described by

$$\Delta \bar{G}_{12}^{ex} = X_1 X_2 A_{12}$$

Furthermore, it can also be stipulated that the A_{12} parameter does not vary with temperature, so that one can use numerical values for A_{12} evaluated at temperatures other than the azeotropic temperature. The development of predictive expressions and solution models often requires a compromise between thermodynamic rigor and calculational practicality. A predictive expression that requires an enormous amount of data input is of little use unless the required data are readily available.

Within the framework of the approximations discussed previously, the boiling temperature can be written as

$$T^{az} = s_1^{az} T_1 + s_2^{az} T_2 + X_1^{az} X_2^{az} A_{12}/(X_1^{az} \Delta \bar{S}_1^{vap} + X_2^{az} \Delta \bar{S}_2^{vap}) \quad (9.43)$$

Equation (9.43) contains two unknown quantities, the azeotropic temperature T^{az}, and the azeotropic composition X_1^{az}.

To eliminate one of the unknown quantities, one can utilize the fact that there is an extremum in the boiling temperature isobar, that is, $\partial T/\partial X_1 = 0$ at T^{az}. Performing the indicated differentiation gives

$$0 = \left(\frac{\partial T}{\partial X_1}\right)_P = \frac{T_1 \Delta \bar{S}_1^{vap} \Delta \bar{S}_2^{vap}}{(X_1 \Delta \bar{S}_1^{vap} + X_2 \Delta \bar{S}_2^{vap})^2} - \frac{T_2 \Delta \bar{S}_1^{vap} \Delta \bar{S}_2^{vap}}{(X_1 \Delta \bar{S}_1^{vap} + X_2 \Delta \bar{S}_2^{vap})^2}$$

$$+ \frac{A_{12}(X_2 - X_1)(X_1 \Delta \bar{S}_1^{vap} + X_2 \Delta \bar{S}_2^{vap}) - A_{12} X_1 X_2 (\Delta \bar{S}_1^{vap} - \Delta \bar{S}_2^{vap})}{(X_1 \Delta \bar{S}_1^{vap} + X_2 \Delta \bar{S}_2^{vap})^2}$$

$$(9.44)$$

Through suitable rearrangement of Eq. (9.44), the azeotropic composition X_2^{az} can be expressed

$$X_2^{az} = 0.5 + (T_1 - T_2) \Delta \bar{S}_2^{vap}/2A_{12} + [(X_2^{az})^2/2](1 - \Delta \bar{S}_2^{vap}/\Delta \bar{S}_1^{vap}) \quad (9.45)$$

in terms of the boiling point temperature and the entropies of vaporization of the pure components and the binary interaction parameter A_{12}. Eq. (9.45) is referred to as the Malesinski (1965) equation. The equal vaporization form of Eq. (9.45) is probably more familiar to readers than is the more complex form just presented.

After assuming an equation for the mathematical relationship for $\Delta \bar{G}_{12}^{ex}$ is a function of composition one could develop a predictive expression for the azeotropic composition in a slightly different manner. For a binary mixture obeying $\Delta \bar{G}_{12}^{ex} = X_1 X_2 A_{12}$, the vapor pressure of the azeotrope P^{az} at the azeotropic temperature (T^{az}) and composition (X_2^{az}) could be written as

$$P^{az} = P_1^{\bullet}(T^{az}) \exp[A_{12}(X_2^{az})^2/RT^{az}] \quad (9.46a)$$

$$P^{az} = P_2^{\bullet}(T^{az}) \exp[A_{12}(1 - X_2^{az})^2/RT^{az}] \quad (9.46b)$$

E. Azeotropes in Binary Mixtures

Taking advantage of the Clausius–Clapeyron equation and choosing the reference pressure as the azeotropic pressure, Eqs. (9.46a–b) can be rewritten

$$1 = \exp\{[\Delta \bar{S}_1^{\text{vap}}(T^{\text{az}} - T_1) + A_{12}(X_1^{\text{az}})^2]/RT^{\text{az}}\} \tag{9.47a}$$

$$1 = \exp\{[\Delta \bar{S}_2^{\text{vap}}(T^{\text{az}} - T_2) + A_{12}(1 - X_2^{\text{az}})^2]/RT^{\text{az}}\} \tag{9.47b}$$

It should be remembered that this set of equations contain the assumption that the enthalpy of vaporization of the components is constant in the temperature range $T_i \to T^{\text{az}}$.

From Eq. (9.47a) the following relationships are found:

$$T_1 - T^{\text{az}} = (A_{12}/\Delta \bar{S}_1^{\text{vap}})(X_2^{\text{az}})^2 \tag{9.48}$$

$$X_2^{\text{az}} = [(\Delta \bar{S}_1^{\text{vap}}/A_{12})(T_1 - T^{\text{az}})]^{1/2} \tag{9.49}$$

$$A_{12} = [\Delta \bar{S}_1^{\text{vap}}/(X_2^{\text{az}})^2](T_1 - T^{\text{az}}) \tag{9.50}$$

between the azeotropic parameters. Similarly, Eq. (9.47b) gives

$$T_2 - T^{\text{az}} = (A_{12}/\Delta \bar{S}_2^{\text{vap}})(1 - X_2^{\text{az}})^2 \tag{9.51}$$

$$1 - X_2^{\text{az}} = [\Delta \bar{S}_2^{\text{vap}}(T_2 - T^{\text{az}})/A_{12}]^{1/2} \tag{9.52}$$

$$A_{12} = [\Delta \bar{S}_2^{\text{vap}}/(1 - X_2^{\text{az}})^2](T_2 - T^{\text{az}}) \tag{9.53}$$

Subtracting Eqs. (9.48) and (9.51), rederives the Malesinski equation. Part of this derivation is given as

$$(T_1 - T^{\text{az}}) - (T_2 - T^{\text{az}}) = \frac{A_{12}}{\Delta \bar{S}_1^{\text{vap}}}(X_2^{\text{az}})^2 - \frac{A_{12}}{\Delta \bar{S}_2^{\text{vap}}}(1 - X_2^{\text{az}})^2$$

$$\frac{\Delta \bar{S}_1^{\text{vap}} \Delta \bar{S}_2^{\text{vap}}(T_1 - T_2)}{A_{12}} = \Delta \bar{S}_2^{\text{vap}}(X_2^{\text{az}})^2 - \Delta \bar{S}_1^{\text{vap}}(1 - X_2^{\text{az}})^2 \tag{9.54}$$

$$X_2^{\text{az}} = 0.5 + \frac{(T_1 - T_2)\Delta \bar{S}_2^{\text{vap}}}{2A_{12}} + \frac{(X_2^{\text{az}})^2}{2}\left(1 - \frac{\Delta \bar{S}_2^{\text{vap}}}{\Delta \bar{S}_1^{\text{vap}}}\right)$$

The Prigogine equation (Prigogine and Defay, 1954) for binary azeotropes is derived by dividing Eq. (9.52) into Eq. (9.49)

$$X_2^{\text{az}}/(1 - X_2^{\text{az}}) = (\Delta \bar{S}_1^{\text{vap}}/\Delta \bar{S}_2^{\text{vap}})^{1/2}[(T_1 - T^{\text{az}})/(T_2 - T^{\text{az}})]^{1/2} \tag{9.55}$$

Glancing at Eq. (9.55) shows that the prediction of the azeotropic composition requires a prior knowledge of the azeotropic temperature. One might naturally question the usefulness of such an equation. From an experimental viewpoint, it is extremely easy to determine the azeotropic temperature with a fairly high degree of precision. The broad temperature extremum, though, makes it difficult to experimentally measure the azeotropic compositions.

TABLE I

Comparison between the Observed Binary Azeotropic Compositions and the Values Calculated Using Either the Malesinski Equation (Eq. 9.45) or the Prigogine Equation (Eq. 9.55)[a]

	Mole fraction composition (X_2)		
Component(1) + component (2)	Eq. (9.45)	Eq. (9.55)	Experimental
Benzene + cyclohexane	0.472	0.474	0.460
Benzene + cyclohexene	0.330	0.385	0.342
Benzene + n-hexane	0.890	0.828	0.950
Benzene + n-heptane	0.0014	0.050	0.001
Benzene + 2,2-dimethylpentane	0.530	0.535	0.520
Pyridine + n-heptane	0.726	0.727	0.717
Pyridine + n-octane	0.375	0.376	0.352
Pyridine + n-nonane	0.082	0.084	0.064
Acetone + chloroform	0.606	0.615	0.658
Acetone + n-hexane	0.365	0.365	0.376
Acetone + carbon disulfide	0.609	0.609	0.608
Ethanol + n-hexane	0.585	0.580	0.668
Ethanol + chloroform	0.206	0.238	0.160
Aniline + pseudocumene	0.835	0.823	0.833
Aniline + p-cymene	0.632	0.658	0.618
Aniline + n-nonane	0.828	0.828	0.823
Aniline + n-decane	0.607	0.601	0.534
Aniline + n-undecane	0.389	0.390	0.293
Aniline + n-dodecane	0.276	0.239	0.178
Aniline + n-tridecane	0.113	0.115	0.070
Aniline + n-tetradecane	—	0.024	0.023
Acetic acid + pyridine	0.488	0.485	0.422
Acetic acid + n-hexane	0.905	0.905	0.908
Acetic acid + n-heptane	0.668	0.668	0.549
Acetic acid + n-octane	0.443	0.444	0.318
Acetic acid + n-nonane	0.270	0.271	0.174
Acetic acid + n-decane	0.129	0.132	0.074
Acetic acid + n-undecane	0.043	0.045	0.024
Propionic acid + n-heptane	0.920	0.909	0.973
Propionic acid + n-octane	0.687	0.686	0.703
Propionic acid + n-nonane	0.391	0.391	0.330
Propionic acid + n-decane	0.161	0.162	0.112
Carbon disulfide + n-pentane	0.828	0.829	0.895
Carbon disulfide + cyclopentane	0.392	0.392	0.348
Carbon disulfide + methyl formate	0.638	0.637	0.720

[a] Reprinted in part with permission from Kurtyka and Kurtyka, (1980). Copyright American Chemical Society.

F. Prediction of Ternary Azeotropes

Equation (9.55) provides a convenient method for estimating the azeotropic composition, which in many cases differs from the *true* (experimental) azeotropic composition by only a few relative percent.

As an example illustrating the application of Eq. (9.55), suppose one wants to predict the azeotropic composition of the pyridine (1) + 2,2,4 trimethylpentane (2) system from the azeotropic temperature $T^{az} = 95.75°C$. Knowing the boiling point temperatures of the pure components $T_1 = 115.4°C$ and $T_2 = 99.3°C$, one estimates X_2^{az}

$$X_2^{az}/(1 - X_2^{az}) = [(388.6 - 368.5)/(372.5 - 368.9)]^{1/2} = 2.363$$

$$X_2^{az} = 0.703$$

by assuming $\Delta \bar{S}_1^{vap} = \Delta \bar{S}_2^{vap}$. The calculated value differs from the experimental value of Lecat (1947), $X_2^{az} = 0.694$, by only a few percent.

Kurtyka and Kurtyka (1980) compared the predictive abilities of both the Malesinski equation and Prigogine equation to the experimental azeotropic compositions for 60 different systems. The results of this comparison are summarized in Table I. It is seen from Table I that the Malesinski equations ($\Delta \bar{S}_1^{vap} = \Delta \bar{S}_2^{vap}$) generally give better results than those obtained with the Prigogine equation ($\Delta \bar{S}_1^{vap} = \Delta \bar{S}_2^{vap}$). When differences in the vaporization entropies were taken into account, Kurtyka and Kurtyka found the Prigogine equation to be superior.[†]

F. Prediction of Ternary Azeotropes from Binary Data

One of the most important factors that determines the course of fractionation and distillation processes is whether or not azeotropes are formed in the mixtures to be separated. Several methods[‡] have been proposed for determining the azeotropic properties of three component systems, and most

---

[†] See Haase (1950), Hollo and Lengyel (1965), Horvath (1961), and Prigogine and Defay (1954).

[‡] The original statements of Kurtyka and Kurtyka are somewhat misleading in the case of positive azeotropes, as both the Malensinski and Prigogine equation reduce to (Hanna, 1982; Kurtyka, 1982)

$$X_2^{az} = d_1^{1/2}/(d_1^{1/2} + d_2^{1/2})$$

$$d_1 = T_1 - T^{az} \quad \text{and} \quad d_2 = T_2 - T^{az}$$

For negative azeotropes, however, the two equations do predict different liquid-phase compositions.

are based on the use of analytical equations of an empirical or a semi-theoretical description of the activity coefficient's variation with liquid phase composition. These methods enable us to calculate the composition and boiling point temperature of a ternary azeotrope on the basis of experimental vapor–liquid equilibrium data for the contributing binary systems.

The equations for the liquid–vapor equilibrium of a binary azeotropic system can be generalized to multicomponent systems in a relatively straightforward manner. To illustrate this, consider a ternary system obeying

$$\Delta \bar{G}^{ex}_{123} = X_1 X_2 A_{12} + X_1 X_3 A_{13} + X_2 X_3 A_{23}$$

and see how binary data might be used to predict the possible existence of azeotropes in higher-order multicomponent systems.

Using Eq. (9.29) and the constraint $y_i = X_i$, the vapor pressure of the ternary system at the azeotrope is expressed

$$P = P_1^{\bullet} \exp\left[\frac{A_{12}(X_2^{az})^2 + A_{13}(X_3^{az})^2 + (A_{12} + A_{13} - A_{23})X_2^{az}X_3^{az}}{RT^{az}}\right] \quad (9.56a)$$

$$P = P_2^{\bullet} \exp\left[\frac{A_{12}(X_1^{az})^2 + A_{23}(X_3^{az})^2 + (A_{12} + A_{23} - A_{13})X_1^{az}X_3^{az}}{RT^{az}}\right] \quad (9.56b)$$

$$P = P_3^{\bullet} \exp\left[\frac{A_{13}(X_1^{az})^2 + A_{23}(X_2^{az})^2 + (A_{13} + A_{23} - A_{12})X_1^{az}X_2^{az}}{RT^{az}}\right] \quad (9.56c)$$

$$X_1^{az} + X_2^{az} + X_3^{az} = 1$$

in terms of the three binary interactions, the azeotropic temperature (T^{az}), and the azeotropic compositions (X_2^{az}, X_3^{az}). A prior knowledge of the three binary interaction parameters from binary liquid–vapor equilibrium data enables one to estimate T^{az} and X_i^{az} by solving the four equations simultaneously. Inherent in this method is the basic assumption that the constants A_{ij} are temperature independent.

Malesinski (1965) related the compositions of a ternary azeotrope to the compositions of the constituent binary azeotropes in an elegant manner. By combining Eqs. (9.56a) and (9.56b) it is possible to write

$$1 = (P_1^{\bullet}/P_2^{\bullet})\exp\{[A_{12}(1 - 2X_1^{az})/RT^{az}] \\ + [(A_{13} - A_{23} - A_{12})X_3^{az}/RT^{az}]\} \quad (9.57)$$

Using the Clausius–Clapeyron relationship one can rewrite Eq. (9.57) as

$$0 = \Delta \bar{S}_1^{vap}(T^{az} - T_1) - \Delta \bar{S}_2^{vap}(T^{az} - T_2) \\ + A_{12}(1 - 2X_1^{az}) + (A_{13} - A_{23} - A_{12})X_3^{az} \quad (9.58)$$

F. Prediction of Ternary Azeotropes

Adding and subtracting the quantity $(\Delta \bar{S}_1^{\text{vap}} + \Delta \bar{S}_2^{\text{vap}})(T^{\text{az}})_{12}$ from the right-hand side of Eq. (9.58) gives

$$\{\Delta \bar{S}_1^{\text{vap}}[(T^{\text{az}})_{12} - T_1]/2A_{12}\} - \{\Delta \bar{S}_2^{\text{vap}}[(T^{\text{az}})_{12} - T_2]/2A_{12}\}$$
$$+ 0.5 - X_1^{\text{az}} + \{(\Delta \bar{S}_1^{\text{vap}} - \Delta \bar{S}_2^{\text{vap}})[T^{\text{az}} - (T^{\text{az}})_{12}]/2A_{12}\}$$
$$+ \{(A_{13} - A_{23} - A_{12})X_3^{\text{az}}/2A_{12}\} = 0 \qquad (9.59)$$

where $(T^{\text{az}})_{12}$ is the azeotropic temperature of the binary mixture containing components 1 and 2. Careful inspection of Eq. (9.59) with $X_3^{\text{az}} = 0$ reveals that the 1–2 binary mixture would obey

$$\frac{\Delta \bar{S}_1^{\text{vap}}[(T^{\text{az}})_{12} - T_1]}{2A_{12}} - \frac{\Delta \bar{S}_2^{\text{vap}}[(T^{\text{az}})_{12} - T_2]}{2A_{12}} + 0.5 - (X_1^{\text{ax}})_{12} = 0 \quad (9.60)$$

at its azeotropic composition. Combination of these two expressions yields

$$(X_1^{\text{az}})_{12} - X_1^{\text{az}} + \{(\Delta \bar{S}_1^{\text{vap}} - \Delta \bar{S}_2^{\text{vap}})[T^{\text{az}} - (T^{\text{az}})_{12}]/2A_{12}\}$$
$$+ (A_{13} - A_{23} - A_{12})X_3^{\text{az}}/2A_{12} = 0 \qquad (9.61)$$

Three sets of equations can be produced in this manner, with their use being dictated by the availability of binary information. For example, if the azeotropic compositions between components (1, 2) and (2, 3) are available, then the following two equations may be solved for the ternary azeotropic compositions X_1^{az} and X_3^{az}:

$$(X_1^{\text{az}})_{12} - X_1^{\text{az}} + [(A_{13} - A_{23} - A_{12})/2A_{12}]X_3^{\text{az}} = 0 \qquad (9.62)$$

$$(X_3^{\text{az}})_{23} - X_3^{\text{az}} + [(A_{13} - A_{23} - A_{12})/2A_{12}]X_1^{\text{az}} = 0 \qquad (9.63)$$

The mole fraction composition of the third component is obtained from a mass balance relationship. The term involving the azeotropic temperature has been omitted because its contribution to the ternary azeotropic compositions is generally less than 1%.

Eduljee and Tiwari (1979) used Eqs. (9.62) and (9.63) to predict the azeotropic compositions of 16 ternary systems. The results of their calculations are shown in Table II for 13 different mixtures. Numerical values of the binary interaction parameters A_{ij} were based on an empirically derived classification scheme that tries to take into account the hydrogen bonding characteristics of the individual molecules. (The authors' classification scheme is outlined in Table III.) Systems 11, 12, and 13 listed in Table II are non-azeotropic and are correctly predicted as such. Generally, there is a fairly good agreement between the experimental and predicted compositions, particularly in light of the simplifying assumptions involving the binary interaction parameters.

TABLE II
Comparison between the Experimental Ternary Azeotropic Compositions and the Values Calculated Via Eqs. (9.62) and (9.63).[a]

System	Components	$(X^{az})^{calc}$	$(X^{az})^{exp}$	$(T^{az})^{exp}$ (°C)
1	Chloroform	0.62	0.76	
	Ethyl Formate	0.15	0.08	62
	2-Bromopropane	0.23	0.16	
2	Chloroform	0.30	0.24	
	Methanol	0.41	0.44	57.5
	Acetone	0.29	0.32	
3	Chloroform	0.56	0.52	
	Ethanol	0.08	0.13	55
	Acetone	0.36	0.35	
4	Methanol	0.26	0.30	
	Methyl acetate	0.42	0.31	46
	n-Hexane	0.32	0.39	
5	n-Butanol	0.03	0.04	
	Benzene	0.51	0.50	77.4
	Cyclohexane	0.46	0.46	
6	Isopropanol	0.28	0.29	
	2-Butanone	0.23	0.17	68.9
	Cyclohexane	0.49	0.54	
7	Propanol	0.16	0.20	
	Benzene	0.40	0.30	73.8
	Cyclohexane	0.44	0.50	
8	Ethanol	0.39	0.44	
	Benzene	0.19	0.09	64.7
	Cyclohexane	0.42	0.47	
9	Methanol	0.18	0.17	
	Acetone	0.62	0.55	51.1
	Cyclohexane	0.20	0.28	
10	Water	0.33	0.20	
	2-Butanone	0.22	0.35	63.6
	Cyclohexane	0.45	0.45	
11	Chloroform	0.18		
	Methanol	−0.28	Nonazeotropic	
	2-Butanone	1.10		
12	2-Butanone	−0.02		
	Ethyl acetate	0.40	Nonazeotropic	
	n-Hexane	0.62		
13	Trichloroethylene	−0.02		
	Benzene	0.52	Nonazeotropic	
	Cyclohexane	0.50		

[a] Reprinted with permission from Eduljee and Tiwari, (1979). Copyright 1979 Pergamon, Oxford.

TABLE III.A
Eduljee and Tiwari's Classification of Organic Liquids According to Hydrogen-Bonding Potentiality[a]

Class	Characteristics	Examples
I	Three-dimensional network of strong hydrogen bonds	Water, glycol, glycerol, amino-alcohols, amides, polyphenols, etc.
I	Both donor atoms and active hydrogen bonds in the same molecule	Alcohols, acids, phenols, primary and secondary amines, nitro-compounds with active hydrogen atoms, etc.
III	Only donor atoms	Ethers, ketones, aldehydes, esters, tertiary amines, nitro-compounds without active hydrogen atoms, etc.
IV	Only active hydrogen atoms	Halogenated hydrocarbons like $CHCl_3$, CH_2Cl_2, CH_3CHCl_2, etc.
V	No hydrogen bond forming capabilities	Hydrocarbons, sulphides, mercaptans, halogenated hydrocarbons not listed in class IV.

TABLE III.B
Numerical Values for the Binary Interaction Parameter for Various Class Combinations

Class	A_{ij} (cal/mol)	Class	A_{ij} (cal/mol)
I–V	1915	III–IV	−460
II–V[b]	885	IV–IV	144
II–V[c]	1237	I–III	1565
III–V	585	II–III	670
IV–V	545	III–III	190
V–V	288	I–II	—
I–IV	1745	II–II	760
II–IV	925	I–I	—

[a] Reprinted with permission from Eduljee and Tiwari, (1979). Copyright 1979 Pergamon, Oxford.
[b] Excluding formic acid azeotropes.
[c] Formic acid azeotropes.

Problems

9.1 In older literature, binary vapor–liquid equilibrium data are sometimes represented by equations of the form

$$\alpha_{12} = (y_1/X_1/y_2/X_2) = (1 + a_{12}X_{12} + a_{122}X_2^2)/(1 + a_{21}X_1 + a_{211}X_1^2)$$

where α is the relative volatility of the mixture and the a's are empirical constants. Hala el al. (1968) gave the following values for the constants for the carbon disulfide (1) + acetone (2) system at 35.17°C:

$$a_{12} = 7.1875 \qquad a_{122} = -2.4064$$
$$a_{21} = 1.4685 \qquad a_{211} = 2.5891$$

The vapor pressures of the pure components at 35.17°C are approximately

$$P_1^\bullet = 526 \text{ torr} \quad \text{and} \quad P_2^\bullet = 355 \text{ torr}$$

Using the data just given, estimate the azeotropic composition at 35.17°C.

9.2 Under conditions such that the isothermal vapor–liquid equilibrium in a binary system is properly described by Eqs. (9.1) and (9.2), show that

$$(dP/dX_1)_{X_1=0} \geq -P_2^\bullet$$

and

$$(dP/dX_1)_{X_1=1} \leq P_1^\bullet$$

9.3 For a binary system in vapor–liquid equilibrium described by Eqs. (9.1) and (9.2), prove that the slope of the bubble-point curve at constant temperature is given by

$$dP/dX_1 = [1 + X_1(d \ln \gamma_1/dX_1)](\gamma_1 P_1^\bullet - \gamma_2 P_2^\bullet)$$

Under what conditions does $dP/dX_1 = 0$?

9.4 A rare type of binary vapor–liquid equilibrium behavior is that of multiple azeotropy, in which the dew- and bubble-point curves are

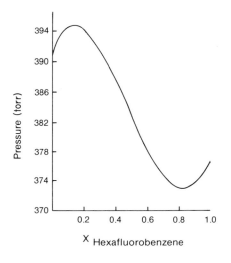

Fig. 9.1 Liquid–vapor equilibria for the Hexafluorobenzene + Benzene system at 60°C. The experimental data were taken from Gaw and Swinton, (1968).

S-shaped, thus yielding at different compositions both a minimum pressure and maximum pressure azeotrope. A well-documented example is the hexafluorobenzene + benzene system, depicted in Fig. 9.1. Assuming the applicability of Eqs. (9.1) and (9.2), determine under what circumstances multiple azeotropy is likely to occur.

9.5 Traditional methods for determining infinite dilution activity coefficients often require graphical extrapolation of finite concentration data and fitting of binary vapor–liquid equilibrium data to an expression for the excess Gibbs free energy. Whereas such methods do provide estimates of γ^∞, the numerical values depend on the choice of analytical expression used to mathematically represent $\Delta \bar{G}_{12}^{ex}$. Differential ebulliometry, however, enables the experimental determination of infinite dilution activity coefficients from knowledge of pure component properties and the limiting slope of the isobaric boiling point temperature with respect to the liquid-phase composition $(\partial T/\partial X_1)_P^\infty$. Assuming vapor-phase ideality, show that

$$\gamma_1^\infty = \frac{P_2^\bullet - (1 - P_2^\bullet \bar{V}_2/RT)(\partial P_2^\bullet/\partial T)(\partial T/\partial X_1)_P^\infty}{P_1^\bullet \exp[(P_2^\bullet - P_1^\bullet)\bar{V}_1/RT]}$$

9.6 As mentioned at the beginning of this chapter, it is possible for one to estimate activity coefficients from total pressure measurements by assuming a mathematical form for $\Delta \bar{G}_{12}^{ex}$. Using the experimental data

TABLE IV
Vapor–Liquid Equilibrium Data for Binary
Mixtures of Benzene (1) +
1,1,2-Trichlorotrifluoroethane (2)
at 25°C[a]

X_1	P (torr)
0.000	333.1
0.260	291.9
0.348	278.5
0.359	276.8
0.510	250.2
0.650	221.8
0.674	215.8
0.776	188.7
1.000	95.2

$\bar{V}_1 = 89.41$ cm^3/mol, $\bar{V}_2 = 119.9$ cm^3/mol

[a] Experimental data reproduced with permission from Linford and Hildebrand, (1969). Copyright 1969 American Chemical Society.

in Table IV and the Regular Solution model, estimate the activity coefficients of benzene and 1,1,2 trichlorotrifluoroethane, and $X_{\text{benzene}} = 0.510$. (Hint: Find the numerical value of A_{12} that best describes the total pressure P)

$$P = P_1 + P_2$$
$$\ln P_1 = \ln P_1^{\bullet} + \ln X_1 + \bar{V}_1 \phi_2^2 A_{12}/RT$$
$$\ln P_2 = \ln P_2^{\bullet} + \ln X_2 + \bar{V}_2 \phi_1^2 A_{12}/RT$$

9.7 The methanol (1) + benzene (2) system has a minimum boiling point azeotrope $T^{\text{az}} = 58.3°C$ at 1 atm. Use the Prigogine equation to predict the azeotropic composition. How does the predicted value compare to the experimental value $X_1^{\text{az}} = 0.610$ as determined by Nagata (1969)? Properties needed in the calculation include the boiling points of the two pure components, $T_1 = 64.7°C$ and $T_2 = 80.1°C$, and the enthalpies of vaporization, $\Delta \bar{H}_1^{\text{vap}} = 35 \text{ kJ/mol}$ and $\Delta \bar{H}_2^{\text{vap}} = 34 \text{ kJ/mol}$.

10
Solubility Behavior of Nonelectrolytes

The excess properties of a solute near infinite dilution have been useful in studying solution nonideality effects. Measurements of the enthalpies of solution of hydrogen-bonding acids in mixtures containing a hydrogen-bonding acceptor and an inert co-solvent are used to determine equilibrium constants and enthalpies of formation of hydrogen-bonded complexes (Arnett et al., 1967, 1970; Bertrand and Burchfield, 1974; Duer and Bertrand, 1970). This high dilution technique is applied also to complexation reactions involving Lewis acid–base adducts (Bolles and Drago, 1965) and proton transfer reactions (Arnett et al., 1974). Gas–liquid chromatographic studies on a binary liquid phase provide another method for determining thermodynamic properties of a solute near infinite dilution, and for investigating association complexes between the solute and one of the solvent components.[†] Measurements of excess partial molar enthalpies (Burchfield, 1977; Burchfield and Bertrand, 1975; Delmas et al., 1964) and excess chemical potentials (Acree, 1981; Acree and Bertrand, 1977, 1981, 1983; Acree and Rytting, 1982a,b, 1983) of a third component at infinite dilution in binary solvent mixtures of nonreacting components demonstrated that this type of measurement can provide valuable insight into the underlying causes of solution nonideality.

Solubility is a strong function of the intermolecular forces between the solute and solvent molecules. The well-known adage *similis similibus solvantur* (like dissolves like) serves merely as an empirical statement of fact that, in the absence of specific interactions, the intermolecular forces between chemically similar species lead to a smaller endothermic enthalpy of solution than those between dissimilar species. Because dissolution must be accompanied by a decrease in the Gibbs free energy, a low endothermic enthalpy is preferable to a large one. Factors other than the intermolecular forces

---

[†] See Acree and Bertrand (1979), Ashworth and Hooker (1977), Bruno (1981), Conder and Young (1979), Chien et al. (1981), Harbison et al. (1979), Laub and Pecsok (1978), Laub and Purnell (1976), Laub et al. (1978), Mathiasson and Jonsson (1974), Meyer and Meyer (1981), Perry and Tiley (1978), and Queignec and Cabanetos–Queignec (1981).

between the solute and solvent, however, play an important role in determining the solubility of a solid.

The solubility of a solid substance may be considered to arise from three, and in some cases four, contributions:

(a) The breaking of solute–solute interactions in the crystalline lattice;

(b) the breaking of solvent–solvent interactions; often referred to as *cavity formation*;

(c) the formation of solvent–solvent interactions; and

(d) the perturbation of solvent–solvent interactions in the immediate vicinity of the solute, as in *solvent structuring*.

Each of these four contributions may be further divided into specific and nonspecific interactions. To illustrate, consider the solubilities of two isomers, phenanthrene and anthracene, as given in Table I. The solubility of phenanthrene in benzene is approximately 25 times greater than that of anthracene, even though both solids are chemically similar to each other. The reason for this large difference in solubility follows from something that is all too often overlooked, that is, the solubility depends not only on the activity coefficient of the solute (which reflects the intermolecular forces between the solute and solvent), but also depends on the fugacity of the standard state to which the activity coefficient refers and on the fugacity of the pure solid. Let the solute be designated by the subscript 3. Then the

TABLE I
Solubility of Anthracene and Phenanthrene in Various Solvents at 25°C[a]

Solvent	X_3^{sat}
Anthracene	
Ethanol	0.0008
Benzene	0.0081
Carbon disulfide	0.0109
Carbon tetrachloride	0.0063
Diethyl ether	0.0059
n-Hexane	0.0018
Phenanthrene	
Ethanol	0.0125
Benzene	0.2068
Carbon disulfide	0.2554
Carbon tetrachloride	0.1850
Diethyl ether	0.1514
n-Hexane	0.0423

[a] Experimental solubilities were taken from Hildebrand *et al.*, (1917). Printed 1917 American Chemical Society.

A. Solid–Liquid Equilibrium in an Ideal Solution

equation of equilibrium is

$$f_{3(ps)} = f_{3(ss)} \tag{10.1}$$

or

$$f_{3(ps)} = \gamma_3^{sat} X_3^{sat} f_3^o \tag{10.2}$$

where ps is the pure solid, ss stands for solute in saturated solution, X_3^{sat} is the mole fraction solubility of the solute in the solvent, γ_3^{sat} is the liquid-phase activity coefficient, and f_3^o is the standard state fugacity to which γ_3^{sat} refers.

The selection of the standard state f_3^o is arbitrary, the only thermodynamic requirement being that it must be at the same temperature as the saturated solution. For convenience, it is advantageous to define the standard state fugacity as the fugacity of the pure supercooled liquid at the solution's temperature and at some specified pressure. Although this is a hypothetical standard state, it is one whose properties can be calculated with a fair degree of accuracy provided the solution's temperature is not far removed from the triple point of the solute.

A. Solid–Liquid Equilibrium in an Ideal Solution

To demonstrate the utility of Eq. (10.1) consider first a simple case. Assume that the vapor pressure above both the pure solid and supercooled liquid is small, so that vapor pressures can be substituted for fugacities without introducing large errors. This simplifying assumption is an excellent one in the majority of cases commonly encountered. It is further stipulated that the solvent and solute (as a supercooled liquid) form an ideal solution ($\gamma_3 = 1$). Within the limitations of these two assumptions Eq. (10.1) becomes

$$X_3^{id} = X_3^{sat} = f_{3(ps)}/f_{3(pscl)} = P_{3(ps)}/P_{3(pscl)} \tag{10.3}$$

where X_3^{id} refers to the ideal solubility and pscl the pure supercooled liquid. We also define the activity of the solid solute a_3^{solid} as

$$a_3^{solid} = f_{3(ps)}/f_{3(pscl)} \tag{10.4}$$

the ratio of the fugacity of the solid to the fugacity of the pure supercooled liquid.

The significance of these expressions can be best seen by referring to the pressure–temperature diagram depicted in Fig. 10.1. If the solute exists as a solid, then the solution temperature must be below the triple point temperature T_{TP}. The vapor pressure of the solid at the solution's temperature can be obtained graphically from the solid–vapor pressure curve. The supercooled liquid's vapor pressure, however, must be found by extrapolating the experimental liquid–vapor pressure curve from the triple-point temperature

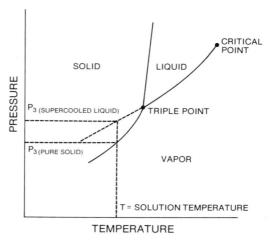

Fig. 10.1 Graphical extrapolation of the vapor pressure for the supercooled liquid solute from its pressure–temperature diagram.

to the solution temperature. Since the slope of the solid–vapor pressure is always larger than that of the extrapolated liquid–vapor pressure curve, it follows from Eq. (10.3) that the saturation solubility of the solid in an ideal solution must always be less than unity, except at the triple-point temperature where $P_{3(ps)} = P_{3(pscl)}$.

Equation (10.3) explains why phenanthrene and anthracene have different solubilities in simple hydrocarbon solvents such as benzene of hexane. Because of structural differences, the triple-point temperatures of the two solids are considerably different. Consequently, the pure-component fugacity ratios at the same temperature T also differ for the two isomeric solids.

The extrapolation indicated in Fig. 10.1 is fairly easy to make when the solution temperature is fairly close to the triple-point temperature. As the temperature interval increases, the uncertainty in this extrapolation also increases and in certain instances the uncertainty may become prohibitively large. It is therefore imperative to establish a systematic method for performing the desired extrapolation. Fortunately, a systematic extrapolation for the fugacity of the supercooled liquid (rather than its vapor pressure) can be readily derived from basic thermodynamic principles.

To develop a systematic extrapolational method, construct the following three-step thermodynamic cycle

$$\text{Component 3(solid, } T) \xrightarrow{\Delta G_\text{I}} \text{Component 3(solid, } T_{TP})$$

$$\text{Component 3(solid, } T_{TP}) \xrightarrow{\Delta G_\text{II}} \text{Component 3(liquid, } T_{TP})$$

$$\text{Component 3(liquid, } T_{TP}) \xrightarrow{\Delta G_\text{III}} \text{Component 3(supercooled liquid, } T)$$

A. Solid–Liquid Equilibrium in an Ideal Solution

with the overall process being

$$\text{Component 3(solid, } T) \xrightarrow{\Delta G_{IV}} \text{component 3(supercooled liquid, } T)$$

and

$$\Delta G_{IV} = RT \ln(f_{3(\text{pscl})}/f_{3(\text{ps})}) = \Delta G_I + \Delta G_{II} + \Delta G_{III} \tag{10.5}$$

The change in the Gibbs free energy for the overall process ΔG_{IV} is also related to the corresponding enthalpy and entropy changes

$$\Delta G_{IV} = \Delta H_{IV} - T \Delta S_{IV} \tag{10.6}$$

Because both enthalpy and entropy are state functions independent of path, it is permissible to express the total property as the sum of the individual steps

$$\Delta H_{IV} = \Delta H_I + \Delta H_{II} + \Delta H_{III} \tag{10.7}$$

$$\Delta S_{IV} = \Delta S_I + \Delta S_{II} + \Delta S_{III} \tag{10.8}$$

In the first step, the solid is heated at constant pressure[†] from temperature T to the triple-point temperature T_{TP}. The enthalpy and entropy accompanying this process are calculated by

$$\Delta \bar{H}_I = \int_T^{T_{TP}} \left(\frac{\partial \Delta \bar{H}}{\partial T}\right)_P dT = \int_T^{T_{TP}} \bar{C}_{p(\text{solid})} \, dT \tag{10.9}$$

$$\Delta \bar{S}_I = \int_T^{T_{TP}} \left(\frac{\partial \Delta \bar{S}}{\partial T}\right)_P dT = \int_T^{T_{TP}} \frac{\bar{C}_{p(\text{solid})}}{T} \, dT \tag{10.10}$$

Step II involves a phase change at the triple-point temperature, and

$$\Delta \bar{H}_{II} = \Delta \bar{H}^{\text{fus}} \tag{10.11}$$

$$\Delta \bar{S}_{II} = \Delta \bar{H}^{\text{fus}}/T_{TP} \tag{10.12}$$

In step III, the solute (as a supercooled liquid) is returned to its original temperature

$$\Delta \bar{H}_{III} = \int_{T_{TP}}^T \bar{C}_{p(\text{pscl})} \, dT \tag{10.13}$$

$$\Delta \bar{S}_{III} = \int_{T_{TP}}^T \frac{\bar{C}_{p(\text{pscl})}}{T} \, dT \tag{10.14}$$

Substituting Eqs. (10.6–10.14) into Eq. (10.5), and assuming that the difference in heat capacities between the solid and supercooled liquid $\Delta \bar{C}_p$

[†] Equations (10.9–10.14) neglect the effect of pressure on the properties of the solid and supercooled liquid. Unless the pressure is large, this effect is negligible.

remains constant over the temperature range $T \to T_{TP}$ gives[†]

$$\ln \frac{f_{3(\text{pscl})}}{f_{3(\text{ps})}} = \frac{\Delta \bar{H}^{\text{fus}}}{RT}\left(1 - \frac{T}{T_{TP}}\right) - \frac{\Delta \bar{C}_p(T_{TP} - T)}{T} + \frac{\Delta \bar{C}_p}{R}\ln\frac{T_{TP}}{T} \quad (10.15)$$

where

$$\Delta \bar{C}_p = \bar{C}_{p(\text{psl})} - \bar{C}_{p(\text{ps})}$$

Two simplifications in Eq (10.15) are frequently made, but these usually introduce only a slight error. First, for most substances there is little difference between the triple-point temperature and the normal melting temperature; also the differences in heats of fusion at these two temperatures is often negligible. Therefore, in practice it is common to substitute the normal melting temperature for T_{TP} and to use the heat of fusion at the melting temperature for $\Delta \bar{H}^{\text{fus}}$. Second, it is often necessary to remove the $\Delta \bar{C}_p$ terms from Eq. (10.15), as experimental values of the heat capacity of the supercooled liquid are rarely known. Two alternative approximations for $\Delta \bar{C}_p$ are commonly used: $\Delta \bar{C}_p = 0$, and $\Delta \bar{C}_p = \Delta \bar{S}^{\text{fus}} = \Delta \bar{H}^{\text{fus}}/T_{TP}$. These approximations lead to the following simplifications:

If $\Delta \bar{C}_p = 0$

$$\ln a_3^{\text{solid}} = \ln(f_{3(\text{ps})}/f_{3(\text{pscl})}) = -\Delta \bar{S}^{\text{fus}}(T_{TP} - T)/RT \quad (10.16)$$

If $\Delta \bar{C}_p = \Delta \bar{S}^{\text{fus}}$

$$\ln a_3^{\text{solid}} = \ln(f_{3(\text{ps})}/f_{3(\text{pscl})}) = (-\Delta \bar{S}^{\text{fus}}/R)\ln(T_{TP}/T) \quad (10.17)$$

Equation (10.16) was used by Yalkowsky and co-workers [see Yalkowsky and Valvani (1979, 1980) and Yalkowsky et al. (1972, 1980)] to explain the solubility nonelectrolyte substances in aqueous solutions; whereas Eq. (10.17) was used by Martin et al. (1980) (see also Adjei et al. 1980) and was recommended by Hildebrand et al. (1970). Calculations using Eqs. (10.16) and (10.17) reveal that the numerical values of $(T_{TP} - T)/RT$ and $\ln(T_{TP}/T)/T$ do not differ for low melting point solutes ($T_{TP} < 500$ K) studied at room temperature. They can differ significantly from one another, however, at higher temperatures. Values of the two terms as a function of melting point temperature are shown in Table II for $T = 300$ K and $T = 400$ K. For the most part, Eq. (10.16) will be used in the numerical examples because it is the more widely used expression for describing solubilities in nonelectrolyte solvent systems.

[†] When a solid undergoes a phase transition, the expression for a_3^{solid} must include additional term(s). Weimer and Prausnitz (1965) and Choi and McLaughlin (1983) discussed the effect of a phase transition on the solubility of a solid.

A. Solid–Liquid Equilibrium in an Ideal Solution

TABLE II
Dependence of $(T_{TP} - T)/2.303\,RT$ and $\ln(T_{TP}/T)/2.303\,R$ on T_{TP} and T

	$T = 300$ K		$T = 400$ K	
	$(T_{TP} - T)$	$\ln(T_{TP}/T)$	$(T_{TP} - T)$	$\ln(T_{TP}/T)$
T_{TP}	$2.303\,RT$	$2.303\,R$	$2.303\,RT$	$2.303\,R$
300	0.000	0.000	—	—
400	0.015	0.015	0.000	0.000
500	0.030	0.027	0.013	0.012
600	0.045	0.036	0.026	0.021
700	0.061	0.044	0.039	0.029

Equation (10.18) provides a reliable method for estimating the ideal solubility of a crystalline solute in a liquid solvent

$$\ln a_3^{\text{solid}} = \ln(\gamma_3^{\text{sat}} X_3^{\text{sat}}) = \frac{-\Delta \bar{H}^{\text{fus}}(T_{TP} - T)}{RT\,T_{TP}} + \frac{\Delta \bar{C}_p(T_{TP} - T)}{RT} - \frac{\Delta \bar{C}_p}{R}\ln(T_{TP}/T) \quad (10.18)$$

as the activity coefficient of the solute γ_3^{sat} is unity for an ideal solution. Returning to the experimental solubilities of phenanthrene and anthracene in benzene at 25°C, one finds that the experimental values differ from the solubilities calculated via Eq. (10.18) by no more than 20%. This is remarkable considering the properties of anthracene had to be extrapolated from 217° to 25°C.

Equation (10.18) immediately provides two useful conclusions regarding the solubility of solids in liquids. Although these conclusions rigorously apply to ideal solutions, they serve as useful guides for other solutions that do not deviate excessively from ideal behavior:

(a) For a given solid–solvent system, the solubility increases with increasing temperature. The rate of increase is approximately proportional to the enthalpy of fusion and, to a first approximation, does not depend on the melting point (triple point) temperature.

(b) For a given solvent and at a fixed temperature, if two solids have a similar entropy of fusion then the solid with the lower melting point temperature has the higher solubility. Similarly, if two solids have about the same melting point temperature then the one with the lower enthalpy of fusion has the higher solubility.

A typical application of Eq. (10.18) is depicted in Fig. 10.2 for the solubilities of eight aromatic hydrocarbons in benzene within the temperature

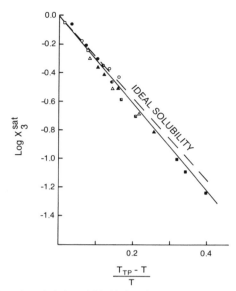

Fig. 10.2 Solubility of naphthalene (○), biphenyl (●), phenanthrene (◐), 1,3,5-triphenylbenzene (■), fluorene (▲), o-terphenyl (□), fluoranthene (△), and acenaphthene (◨) in benzene as a function of $(T_{TP} - T)/T$. The dashed line (— —) represents ideal solubility. The experimental values were taken from a paper by McLaughlin and Zainal (1959). Reproduced with permission of the copyright owners.

interval 30–70°C. With the aid of Eq. (10.18) these data can be correlated in a relatively simple manner. Because the melting points of the eight solutes are not more than 150°C above the temperatures of the saturated solutions, the terms involving $\Delta \bar{C}_p$ may be neglected. Also, it is permissible to substitute melting point temperatures for T_{TP}. For the eight solutes considered here, the entropies of fusion do not vary much and an average value is about 54.6 J/mol K. Therefore, a plot of log X_3^{sat} vs. $(T_{TP} - T)/T$ should be fairly linear with a slope approximately equal to $-54.6/2.303R$ and with intercept log $X_3^{sat} = 1$ when $(T_{TP} - T)/T = 0$. Such a plot is shown by the dashed line in Fig. 10.2 and it is evident that this line gives a good mathematical representation of the experimental data. As a result, the assumption of an ideal solution is appropriate for these systems. For more precise work, however, the assumption of ideality is only an approximation as the solutions show slight positive deviations from ideality. The observed solubilities are slightly below those calculated using Eq. (10.18) with $\Delta \bar{C}_p = 0$. The continuous line in Fig. 10.2 was determined empirically to represent the best fit of the data and has a slope equal to $-58.0/2.303R$.

Equation (10.18) gives the ideal solubility of solid 3 in solvent 1. By interchanging the subscripts, one may use the same equation to calculate the

B. The Scatchard–Hildebrand Model

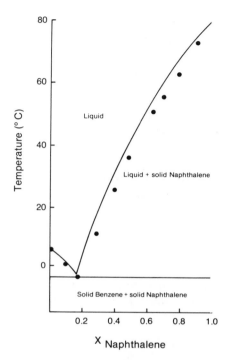

Fig. 10.3 Comparison between experimental and calculated (Eq. (10.18) with $\Delta \bar{C}_p = 0$) solid–liquid phase diagram for binary mixtures containing benzene and naphthalene. The experimental data were taken from a compilation by Timmermans, (1959).

ideal solubility of solid 1 in solvent 3. Repetition of such calculations at different temperatures enables one to estimate the freezing diagram of the entire binary system as a function of composition, assuming that the liquid phase is an ideal solution and there is complete immiscibility in the solid phase. Such a diagram is shown in Fig. 10.3 for the benzene + naphthalene system. The left-hand side of the diagram describes the equilibrium between the liquid mixture and solid benzene whereas the right-hand side describes the equilibrium between the liquid mixture and solid naphthalene. At the point of intersection, called the eutectic point, all three phases are in equilibrium.

B. The Scatchard–Hildebrand Solubility Parameter Model

Although Eq. (10.18) was used to calculate the ideal solubility ($\gamma_3^{\text{sat}} = 1$) of crystalline solutes in liquid solvents, it is also applicable to nonideal solutions. Whenever there is a sufficient difference in the nature and size of the solute and solvent molecules, it may be expected that γ_3^{sat} will not be equal to unity. In nonpolar solutions where only nonspecific interactions are important, γ_3^{sat} is generally larger than unity (and thus, the actual solubility is less than the ideal solubility), but in solutions where specific interactions are important, the activity coefficients may well be less than unity with correspondingly higher solubilities.

In Fig. 10.2 it was noted that the solubilities of eight aromatic hydrocarbons in benzene could be adequately described assuming ideal solution behavior. Figure 10.4 shows the solubilities of the same or similar aromatic hydrocarbons in carbon tetrachloride, the experimental data taken from a paper by McLaughlin and Zainal (1960). The dashed line denotes the ideal solubility curve using $\Delta \bar{S}^{fus} = 54.4$ J/mol K; the continuous line represents the best linear least-squares fit of the data and has a slope of $-66.8/2.303R$.

Comparison of Figs. 10.2 and 10.4 reveals that at the same temperature, solubilities in carbon tetrachloride are lower than those in benzene. In other words, the activity coefficients of the solutes in carbon tetrachloride are larger than those in benzene.

As in the case of liquid mixtures, there is no general method for predicting activity coefficients of solid solutes in liquid solvents. For nonpolar solutes and solvents, however, a reasonable estimate can frequently be made using the Scatchard–Hildebrand solubility parameter model discussed previously.

$$\ln \gamma_3^{sat} = \bar{V}_3^\circ (\delta_{solv} - \delta_3)^2 (1 - \phi_3^{sat})^2 \qquad (10.19)$$

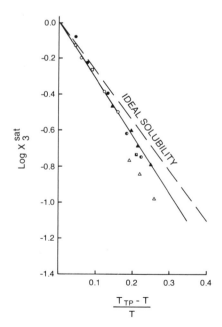

Fig. 10.4 Solubility of naphthalene (○), biphenyl (●), phenanthrene (◐), fluorene (▲), o-terphenyl (□), fluoranthene (△), and acenaphthene (■) in carbon tetrachloride as a function of $(T_{TP} - T)/T$. The dashed line (———) represents ideal solubility. The experimental values were taken from a paper by McLaughlin and Zainal (1960). Reprinted with permission of the copyright owners.

B. The Scatchard–Hildebrand Model

Combining Eqs. (10.18) and (10.19) gives the solubility as

$$\ln X_3^{\text{sat}} = \ln a_3^{\text{solid}} - \bar{V}_3^{\circ}(\delta_{\text{solv}} - \delta_3)^2(1 - \phi_3^{\text{sat}})^2/RT \tag{10.20}$$

where δ_{solv} and δ_3 refer to the solubility parameters of the solvent and supercooled liquid solute, $\bar{V}_3^{\circ}$ is the molar volume of the supercooled liquid solute, X_3^{sat} is the saturation mole fraction solubility, and ϕ_3^{sat} is the solute's volume fraction solubility calculated using the ideal molar volume approximation. For solvent components, the liquid molar volumes and solubility parameters are often tabulated in the literature (for example, see Hoy, 1970), and when not available, they can be calculated directly from density and vapor pressure measurements on the pure liquid. Molar volumes of supercooled liquid solutes, $\bar{V}_3^{\circ}$, on the other hand, are estimated either by group contribution methods or by the experimentally determined apparent partial molar volumes in the solvent of interest. Solubility parameters of the supercooled liquid can be obtained indirectly from solubility measurements and several calculational methods (Acree et al., 1981; Cave et al. 1980; Martin and Carstensen, 1981) are suggested for determining the best value of δ_3.

To illustrate the applicability of Eq. (10.20), consider the solubility of white phosphorous in n-heptane at 25°C. The melting point of white phosphorous is 44.2°C; the enthalpy of fusion ($\Delta \bar{H}^{\text{fus}} = 251.5$ J/mol) and the heat capacities of the solid ($\bar{C}_p = 89.79 + 12.02 \times 10^{-2}t$; J/mol°C) and liquid ($\bar{C}_p = 102.38 - 39.84 \times 10^{-3}t - 16.43 \times 10^{-5}t^2$; J/mol°C) were measured by Young and Hildebrand (1942). From Eq. (10.18) the ideal solubility at 25°C is $a_3^{\text{solid}} = 0.942$. A much better approximation of the actual solubility can be obtained from the solubility parameter theory as given by Eq. (10.20). Using the solubility parameter ($\delta_3 = 13.1$)[†] and the molar volume ($\bar{V}_3^{\circ} = 70.4$ cm³/mol) of the supercooled liquid phosphorous, extrapolated to 25°C using the thermal and volumetric properties of liquid phosphorous at temperatures slightly above its normal melting point and combined with the solubility parameter of n-heptane ($\delta_1 = 7.4$), one calculates a saturation solubility of $X_3^{\text{sat}} = 0.022$, which is strikingly different from the value obtained by assuming ideal liquid-phase behavior. The experimentally observed solubility is $X_3^{\text{sat}} = 0.0124$, in reasonably good agreement with the predictions based on Eq. (10.20).

The solubility parameter of a binary solvent mixture is (Scott, 1949; Smith et al., 1950)

$$\delta_{\text{solv}} = (\phi_1 \delta_1 + \phi_2 \delta_2)/(\phi_1 + \phi_2) = \phi_1^{\circ} \delta_1 + \phi_2^{\circ} \delta_2 \tag{10.21}$$

$$\phi_1^{\circ} = 1 - \phi_2^{\circ} = n_1 \bar{V}_1/(n_1 \bar{V}_1 + n_2 \bar{V}_2)$$

[†] Solubility parameters have the units of cal$^{1/2}$/cm$^{3/2}$. To convert from calories to joules, multiply by 4.184.

a volume fraction average of the solubility parameters of the two pure solvents δ_1 and δ_2. The superscript (o) indicates that the solvent composition is calculated as if the solute were not present.

Figure 10.5 compares the solubility parameter predictions to the observed solubilities for phenanthrene in binary mixtures containing cyclohexane and methylene iodide (Gordon and Scott, 1952). Numerical values used in the predictive expressions include the molar volumes and solubility parameters of cyclohexane ($\bar{V}_1 = 108$ cm^3/mol, $\delta_1 = 8.2$), methylene iodide ($\bar{V}_2 = 80$ cm^3/mol, $\delta_2 = 11.8$), phenanthrene ($\bar{V}_3^o = 150$ cm^3/mol, $\delta_3 = 9.8$), and the activity of the solute $a_3^{solid} = 0.256$ at 300°C. Examination of Fig. 10.5 reveals that phenanthrene exhibits a maximum mole fraction solubility at $\phi_{cyclo} = 0.38$, which is expected based on the calculations of Eqs. (10.20) and (10.21). The solubility parameter theory predicts that a maximum mole fraction solubility will occur whenever the solubility parameter of the solute lies between the solubility parameters of the pure solvents, i.e., $\delta_1 < \delta_3 < \delta_2$. Gordon and Scott (1952) explained the discrepancy between the ideal solubility and the observed maximum solubility by stating that the optimum solvent composition "is near enough to the solvent–solvent critical composition and temperature that extensive clustering must occur; consequently, the solute is actually dissolved not in a homogeneous *ideal* solvent, but rather in a solvent which consists of microscopic regions richer either in

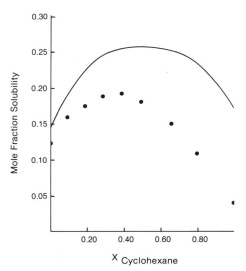

Fig. 10.5 Comparison between the experimental solubilities (●) and the solubility parameter predictions for phenanthrene in the cyclohexane + methylene iodide solvent system at 300 K. Experimental values were taken from a curve given by Gordon and Scott (1952).

B. The Scatchard–Hildebrand Model

cyclohexane or methylene iodide than the overall composition would indicate." Hildebrand, et al. (1970) later attributed the lack of agreement between X_3^{sat} and a_3^{solid} to the (quantitative) failure of the solubility parameter theory to describe the binary interactions between phenanthrene and cyclohexane, $A_{13} > (\delta_1 - \delta_3)^2$.

The popularity of the solubility parameter approach arises primarily because it relates the solubility of a solute in a given solvent, either pure or mixed, to the bulk properties of the pure components. Whereas this application of the solubility parameter approach has certain practical advantages in that it requires only a minimal number of experimental observations, a more flexible expression for binary solvent systems can be developed by replacing the individual δ_is with the measured solubilities in the pure solvents and the measured thermodynamic excess properties of the solvent mixture. This overcomes the difficulties associated with determining the best value of δ_3 and removes the mathematical restrictions imposed by $A_{ij} = (\delta_i - \delta_j)^2 \geq 0$. Ironically, Scott (1949) in his development of the solubility parameter of a binary solvent mixture started with the most general ternary mixing model and imposed the requirement that $A_{ij} = (\delta_i - \delta_j)^2$ as his last step. Essentially, Scott's set of mathematical manipulations is being reversed to estimate the binary interaction parameters A_{ij} with direct experimental observations rather than solubility parameters.

To incorporate direct experimental observations into the basic solubility parameter model, first substitute Eq. (10.21) into Eq. (10.20)

$$(1 - \phi_3^{sat})^{-2} RT \ln(a_3^{solid}/X_3^{sat}) = \bar{V}_3^\circ(\phi_1^\circ \delta_1 + \phi_2^\circ \delta_2 - \delta_3)^2 \qquad (10.22)$$

and actually multiply out the squared term

$$\begin{aligned}(1 - \phi_3^{sat})^{-2} & RT \ln(a_3^{solid}/X_3^{sat}) \\ &= \bar{V}_3^\circ[(\phi_1^\circ \delta_1)^2 + 2\phi_1^\circ \delta_1 \phi_2^\circ \delta_2 + (\phi_2^\circ \delta_2)^2 - 2\delta_3(\phi_1^\circ \delta_1 + \phi_2^\circ \delta_2) + \delta_3^2] \\ &= \bar{V}_3^\circ[\phi_1^\circ(\delta_1^2 - 2\delta_1 \delta_3 + \delta_3^2) + \phi_2^\circ(\delta_2^2 - 2\delta_2 \delta_3 + \delta_3^2) - \phi_1^\circ \phi_2^\circ(\delta_1^2 - 2\delta_1 \delta_2 + \delta_2^2)] \\ &= \bar{V}_3^\circ[\phi_1^\circ(\delta_1 - \delta_3)^2 + \phi_2^\circ(\delta_2 - \delta_3)^2 - \phi_1^\circ \phi_2^\circ(\delta_1 - \delta_2)^2] \end{aligned} \qquad (10.23)$$

Inspection of Eq. (10.23) reveals that for model systems obeying this expression, the saturation solubility of the solute in a pure solvent would be described by

$$\begin{aligned}(\Delta \bar{G}_3^{ex})_{X_i = 1}^\infty &= (1 - \phi_3^{sat})^{-2} RT \ln(a_3^{solid}/X_3^{sat}) \\ &= \bar{V}_3^\circ(\delta_i - \delta_3)^2 \qquad i = 1, 2 \end{aligned} \qquad (10.24)$$

The small subscripts $X_i = 1$ are used to distinguish the properties of the solute in a pure solvent from those in the binary solvent mixture. One should recall previous discussions of the solubility parameter approach; that the excess molar Gibbs free energy of the binary solvent mixture can also be described

$$\Delta \bar{G}_{12}^{ex} = (X_1^o \bar{V}_1 + X_2^o \bar{V}_2)\phi_1^o \phi_2^o (\delta_1 - \delta_2)^2 \qquad (10.25)$$

in terms of solubility parameters.

Combining Eqs. (10.23–10.25) one finds that the solubility of a solute in binary solvent mixtures containing only nonspecific interactions is

$$RT \ln(a_3^{solid}/X_3^{sat}) = (1 - \phi_3^{sat})^2 [\phi_1^o (\Delta \bar{G}_3^{ex})_{X_1=1}^\infty + \phi_2^o (\Delta \bar{G}_3^{ex})_{X_2=1}^\infty \\ - \bar{V}_3^o (X_1^o \bar{V}_1 + X_2^o \bar{V}_2)^{-1} (\Delta \bar{G}_{12}^{ex})] \qquad (10.26)$$

a volume fraction average of the solute's properties in the two pure solvents $[(\Delta \bar{G}_3^{ex})_{X_1=1}^\infty$ and $(\Delta \bar{G}_3^{ex})_{X_2=1}^\infty]$ and a contribution due to the unmixing of the solvent pair by the presence of the solute. Enhancement of the unmixing term by a large solute molecule can lead to predictions of maximum or minimum solubilities. There is no mathematical requirement, however, that the maximum mole fraction solubility must equal the activity of the supercooled liquid solute. The predictions of Eq. (10.26) will be compared later (in Table III) to the experimental solubilities of naphthalene, p-dibromobenzene, iodine, stannic iodide, benzil, and p-benzoquinone in simple binary solvent mixtures.

C. Stoichiometric Complexation Model of Higuchi

Stoichiometric complexation models are used frequently to quantitatively explain enhanced solubilities of a polar organic solute in binary solvent mixtures containing an inert hydrocarbon and a polar co-solvent. The basic model assumes complexation between the solute (component A) and an interacting cosolvent (component C) (Anderson, 1977; Anderson et al., 1980; Fung, 1970; Fung et al., 1971):

$$A + C \rightleftharpoons AC \xrightarrow{+C} AC_2 \xrightarrow{+C} \cdots AC_n$$

in which each reaction is described by an appropriate molarity-based equilibrium constant

$$K_{AC}^c = C_{AC}/(C_A^{sat})_{X_B=1} C_C^{free} \qquad (10.27)$$

$$K_{AC_n}^c = C_{AC_n}/C_{AC_{n-1}} C_C^{free} \qquad (10.28)$$

where $(C_A^{sat})_{X_B=1}$ is the saturation solubility of the solute (mol/l) in pure inert hydrocarbon (assumed to also represent the free solute concentration

C. Stoichiometric Complexation Model of Higuchi

in binary mixtures), and C_C^{free} is the free (uncomplexed) ligand. The particular model just given assumes only a single solute molecule is present in each complex, but the mathematical form of the resulting equations is not significantly altered by additional solute molecules per complex. The total solubility of solute in any system C_A^{sat} can be expressed as

$$C_A^{sat} = (C_A^{sat})_{X_B=1} + K_{AC}^c (C_A^{sat})_{X_B=1} C_C^{free} + K_{AC}^c K_{AC_2}^c (C_A^{sat})_{X_B=1} (C_C^{free})^2 + \cdots \quad (10.29)$$

and the concentration of complexing agent C_C is

$$C_C = C_C^{free} + K_{AC}^c (C_A^{sat})_{X_B=1} C_C^{free} + 2 K_{AC}^c K_{AC_2}^c (C_A^{sat})_{X_B=1} (C_C^{free})^2 + \cdots \quad (10.30)$$

In the absence of solute, the total concentration of interactive co-solvent C_C is equal to C_C^{free} only if the extent of self-association is negligible. Furthermore, the mathematical form of Eq. (10.29) predicting plots of solubility versus co-solvent should be concave upward in solvents incapable of self-association.

If only 1:1 complexes are present, Eqs. (10.29) and (10.30) can be combined to give

$$\text{Fractional change in solubility} = [C_A^{sat} - (C_A^{sat})_{X_B=1} / (C_A^{sat})_{X_B=1}]$$
$$= K_{AC}^c C_C / [1 + K_{AC}^c (C_A^{sat})_{X_B=1}] \quad (10.31)$$

and a plot of the fractional change in solubility versus ligand added gives a straight line. The solute–solvent equilibrium constant can be determined from the slope via Eq. 10.31.

Direct graphical evaluation of equilibrium constants is also possible for systems having both 1:1 and 1:2 solute–solvent complexation. Suitable mathematical manipulation of Eqs. (10.29) and (10.30) results in

$$\frac{C_A^{sat} - (C_A^{sat})_{X_B=1}}{C_C - 2[C_A^{sat} - (C_A^{sat})_{X_B=1}]} = \alpha + \beta \{C_C - 2[C_A^{sat} - (C_A^{sat})_{X_B=1}]\} \quad (10.32)$$

where

$$\alpha = \frac{K_{AC}^c (C_A^{sat})_{X_B=1}}{1 - K_{AC}^c (C_A^{sat})_{X_B=1}} \quad \text{and} \quad \beta = \frac{K_{AC_2}^c (C_A^{sat})_{X_B=1}}{[1 - K_{AC}^c (C_A^{sat})_{X_B=1}]^2}$$

Plots of the left-hand side of Eq. (10.32) versus $C_C - 2[C_A^{sat} - (C_A^{sat})_{X_B=1}]$ give a straight line. The K_{AC}^c and $K_{AC_2}^c$ can be easily calculated from the slope and intercept.

Experimental solubilities of carbazole in isooctane + n-butyl ether mixtures, determined by Anderson et al. (1980), were analyzed according to

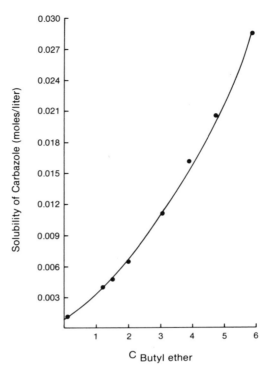

Fig. 10.6 Plot of the solubility of carbazole in isooctane + n-butyl ether mixtures as a function of butyl ether concentration. Experimental data points (●) were taken from a paper by Anderson et al., (1980). The solid line is used to connect the data points and does not represent predictions based on the theoretical model.

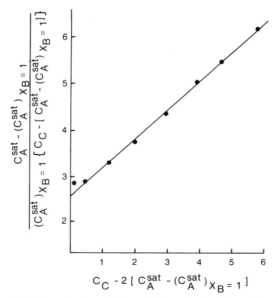

Fig. 10.7 Graphical determination of K_{AC} and K_{AC_2} for the interaction of carbazole with n-butyl ether.

C. Stoichiometric Complexation Model of Higuchi

Eq. (10.32) to demonstrate its utility. As shown in Fig. 10.6, the solubility curve for carbazole is slightly concaved upward, thus suggesting that both complexes are present. Numerical values of the equilibrium constants $K^c_{AC} \approx$ 2.5 liter/mol and $K^c_{AC_2} \approx 0.25$ liter/mol, are determined from the slope and intercept of Fig. 10.7.

Application of Eqs. (10.31) and (10.32) to systems not involving complexation theoretically should provide equilibrium constants equal to zero. Consequently, one might expect the model to be capable of distinguishing between complexing and noncomplexing systems through the relative magnitudes of the calculated equilibrium constants. As shown in Figs. 10.8 and

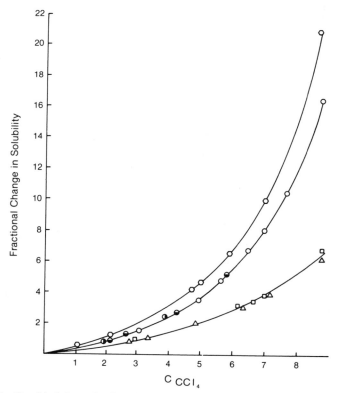

Fig. 10.8 Graphical determination of K_{AC} from plots of the fractional change in solubility versus carbon tetrachloride molarity in several binary solvents consisting of carbon tetrachloride and n-octane (○), n-heptane (◐), n-hexane (◑), isooctane (○), cyclohexane (△), and cyclooctane (□). The basic model requires additional solute–solvent complexes to explain the nonlinear behavior. This figure has been reproduced from Acree and Rytting, (1982b) with the permission of the copyright owners.

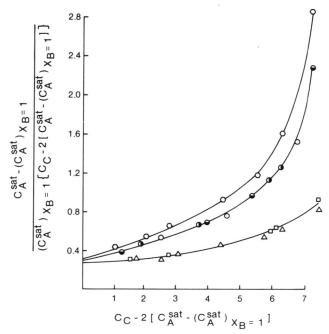

Fig. 10.9 Graphical determination of K_{AC} and K_{AC_2} for benzil in several binary solvents at 25°C (See Eq. 10.32). The binary mixtures contain carbon tetrachloride and n-octane (○), n-heptane (◓), n-hexane (◐), isooctane (○), cyclohexane (△), and cyclooctane (□).

10.9, however, important Ks can be obtained in systems where complexation is not expected to occur.

In Figs. 10.8 and 10.9 the graphical determination of potential equilibrium constants are shown, calculated from Eqs. (10.31) and (10.32) for benzil solubilities in binary mixtures containing carbon tetrachloride. Inspection of the two graphs show that the various co-solvents appear to be grouped along three different curves; one for isooctane, a second for the three normal hydrocarbons (n-hexane, n-heptane, and n-octane), and a third curve for the two cyclic hydrocarbons (cyclohexane and cyclooctane). Presumed equilibrium constants calculated in mixtures containing cyclohexane are not identical to values calculated in isooctane mixtures. More importantly, the model requires several constants to mathematically represent the experimental data, and yet there is little (if any) experimental evidence to suggest complexation between carbon tetrachloride and benzil. In fact, dipole moment measurements of benzil in benzene and carbon tetrachloride were interpreted as evi-

C. Stoichiometric Complexation Model of Higuchi

dence for specific interactions of benzil with benzene but not with carbon tetrachloride (Hopkins et al., 1971). Gas–liquid chromatographic retention behavior of carbon tetrachloride and chloroform on a benzil stationary phase was explained by Vernon (1971) in terms of London or simple dispersion forces in the case of carbon tetrachloride and in terms of hydrogen bonding in the case of chloroform.

A second flaw in the basic model of Higuchi et al. becomes apparent upon writing the solubility expressions in terms of $(C_A^{sat})_{X_B=1}$ and $(C_A^{sat})_{X_B=1}$. The complete description of experimental solubility in the pure complexing solvent through Eq. (10.31) requires

$$\frac{[(C_A^{sat})_{X_C=1} - (C_A^{sat})_{X_B=1}]}{(C_A^{sat})_{X_B=1}} = \frac{K_{AC}^c C_C^*}{1 + K_{AC}^c (C_A^{sat})_{X_B=1}} \tag{10.33}$$

where C_C^* refers to the concentration of pure complexing solvent in the saturated solution. Within limitations of the approximate relationships

$$C_i = 10^3 X_i^o / (X_B \bar{V}_B + X_C \bar{V}_C) \tag{10.34}$$

$$(C_A^{sat})_{X_i=1} = 10^3 (X_A^{sat})_{X_i=1} / \bar{V}_i \quad i = B, C \tag{10.35}$$

combination of Eqs. (10.31–10.35) enables the solubility in binary mixtures to be expressed as a mole fraction average of values in the two pure solvents

$$X_A^{sat} = X_B^o (X_A^{sat})_{X_B=1} + X_C^o (X_A^{sat})_{X_C=1} \tag{10.36}$$

where

$$X_i^o = X_i / (X_B + X_C)$$

Predictions of Eq. (10.36) are off by as much as 50% for *p*-benzoquinone in *n*-heptane + carbon tetrachloride mixtures (Acree and Rytting, 1982a), and off by a factor of two for benzil in the isooctane + carbon tetrachloride system (Acree and Rytting, 1982b). It is difficult to attribute the failure of Eq. (10.29) to specific solute–solvent interactions between the solutes and carbon tetrachloride or to the departure from infinite dilution, as the Nearly Ideal Binary Solvent (NIBS) approach (to be discussed in the next section) describes the experimental solubilities to within a maximum deviation of 6% without introducing a single equilibrium constant.

Inspection of Eq. (10.36) also reveals that it is incapable of describing systems containing either a maximum or minimum mole fraction solubility. Classic examples are found in studies of phenanthrene (Gordon and Scott, 1952) and 2-nitro-5-methylphenol (Buchowski et al., 1979) in cyclohexane + methylene iodide, where the observed solubilities show maximum values that are almost twice that predicted by Eq. (10.36).

D. The Nearly Ideal Binary Solvent Theory (NIBS), Solubility in Binary Solvents of Nonspecific Interactions

The NIBS approach was developed by Bertrand and co-workers[†] for describing the thermodynamic properties of a solute in binary solvent mixtures, and has been successful in predicting enthalpies of solution and solubilities in mixed solvents containing only nonspecific interactions. The basic principles of this method as they pertain to the chemical potential or the partial molar Gibbs free energy of a solute will be reviewed to identify the assumptions made in the derivation of the predictive equations.

In the NIBS approach, expressions for the partial molar excess properties of the solute near infinite dilution in binary solvent mixtures were developed for a model ternary system obeying a general mixing equation

$$\Delta Z^{ex}_{123} = (n_1\Gamma_1 + n_2\Gamma_2 + n_3\Gamma_3)^{-1}$$
$$\times (n_1\Gamma_1 n_2\Gamma_2 A_{12} + n_1\Gamma_1 n_3\Gamma_3 A_{13} + n_2\Gamma_2 n_3\Gamma_3 A_{23}) \quad (10.37)$$

in which Z represents any extensive thermodynamic property, Γ_i is the weighting factor for component i, and A_{ij} is a binary interaction parameter that is independent of composition. Through differentiation of Eq. (10.37) the corresponding partial molar excess properties of the solute (component 3) can be expressed in terms of a weighted mole fraction average of the properties in the pure solvents $[(\Delta\bar{Z}^{ex}_3)^\infty_{X_1=1}$ and $(\Delta\bar{Z}^{ex}_3)^\infty_{X_2=1}]$ and a contribution because of the unmixing of the binary solvent pair by the presence of the solute

$$(\Delta\bar{Z}^{ex}_3)^\infty = f^o_1(\Delta\bar{Z}^{ex}_3)^\infty_{X_1=1} + f^o_2(\Delta\bar{Z}^{ex}_3)^\infty_{X_2=1} - \Gamma_3(X^o_1\Gamma_1 + X^o_2\Gamma_2)^{-1}(\Delta\bar{Z}^{ex}_{12})$$
(10.38)

where

$$f^o_1 = 1 - f^o_2 = n_1\Gamma_1/(n_1\Gamma_1 + n_2\Gamma_2)$$

and $X^o_1 = 1 - X^o_2 = n_1/(n_1 + n_2)$.

In Eq. (10.38) and subsequent equations, the superscript (∞) indicates an extrapolated value for the infinitely dilute solution ($f_3 = 0$). Most of the specific elements of the model Eq. (10.37) were removed; only the weighting factors remain. If reasonable estimates for the weighting factors can be developed, the thermodynamic excess properties of the solute in binary solvent mixtures can be predicted and compared to experimental data.

[†] See Burchfield (1977), Burchfield and Bertrand (1975), Acree (1981), and Acree and Bertrand (1977, 1981, 1983).

D. NIBS, Solubility in Binary Solvents

Weighting factors represent a measure of the skew of the binary thermodynamic excess property from mole fraction symmetry, and can be evaluated, in a relative sense, as the ratio of two weighting factors (Γ_i/Γ_j). Several methods were proposed previously for the evaluation of these weighting factors from the thermodynamic properties of binary mixtures (Burchfield, 1977; Burchfield and Bertrand, 1975). To avoid calculating weighting factors, several simple approximations can be made: Approximating the weighting factors for each component by its molar volume, equating the weighting factors for each component, or approximating the weighting factors for each component by the surface area of the molecule.

Thermodynamic excess properties are relatively simple for directly observed excess properties such as volume and enthalpy. But in the case of Gibbs free energy, thermodynamic excess properties are more complex because the total free energy of mixing is experimentally determined and the excess value must be calculated as the difference between the observed value and the value of an ideal solution

$$\Delta G^{\text{mix}} = RT \sum_{i=1}^{N} n_i \ln X_i + \Delta G^{\text{ex}} \tag{10.39}$$

For mixtures containing molecules with considerable differences in molar volumes, general mixing equations possessing the mathematical form of Eq. (10.37) more accurately describe differences between the total free energy of mixing and that predicted using the Flory–Huggins expression

$$\Delta G^{\text{mix}} = RT \sum_{i=1}^{N} n_i \ln \phi_i + \Delta G^{\text{fh}} \tag{10.40}$$

For a binary mixture, the excess molar Gibbs free energy over the predictions of the Flory–Huggins equation is related to the defined excess free energy by

$$\Delta \bar{G}^{\text{fh}}_{12} = \Delta \bar{G}^{\text{ex}}_{12} + RT[\ln(X_1 \bar{V}_1 + X_2 \bar{V}_2) - X_1 \ln \bar{V}_1 - X_2 \ln \bar{V}_2] \tag{10.41}$$

The mathematical treatment of these general mixing equations leads to two general expressions for estimating the partial molar excess Gibbs free energy of a solute near infinite dilution in a binary solvent

$$\Delta \bar{G}^{\text{ex}}_3 = (1 - f_3)^2 [f_1^\circ (\Delta \bar{G}^{\text{ex}}_3)^\infty_{X_1=1} + f_2^\circ (\Delta \bar{G}^{\text{ex}}_3)^\infty_{X_2=1} \\ - \Gamma_3 (X_1^\circ \Gamma_1 + X_2^\circ \Gamma_2)^{-1} (\Delta \bar{G}^{\text{ex}}_{12})] \tag{10.42}$$

and

$$\Delta \bar{G}^{\text{fh}}_3 = (1 - f_3)^2 [f_1^\circ (\Delta \bar{G}^{\text{fh}}_3)^\infty_{X_1=1} + f_2^\circ (\Delta \bar{G}^{\text{fh}}_3)^\infty_{X_2=1} \\ - \Gamma_3 (X_1^\circ \Gamma_1 + X_2^\circ \Gamma_2)^{-1} (\Delta \bar{G}^{\text{fh}}_{12})] \tag{10.43}$$

The term $\Delta \bar{G}^{\text{fh}}_3$ represents an excess partial molar free energy of the solute relative to an ideal mixing equation based on volume fractions rather than mole fractions.

Through basic thermodynamic relationships the chemical potential of the solute in binary solvent mixtures can be related to the solubility

$$\Delta \bar{G}_3^{ex} = RT \ln(a_3^{solid}/X_3^{sat}) \tag{10.44}$$

$$\Delta \bar{G}_3^{fh} = RT\{\ln(a_3^{solid}/\phi_3^{sat}) - [1 - (\bar{V}_3^{\circ}/\bar{V}_{sol})]\} \tag{10.45}$$

in which $\bar{V}_3^{\circ}$ is the molar volume of the solute in the liquid state at the desired temperature, and a_3^{solid} is the activity of the solute referred to the hypothetical liquid supercooled below the normal melting point. The numerical value of a_3^{solid} is calculated from the fusion heat of the solute and the heat capacities of this solid and of its supercooled liquid via Eq. (10.18).

The application of Eqs. (10.42) and (10.44) or Eqs. (10.43) and (10.45) to solubilities enables the partial molar thermodynamic excess properties of the solute in the pure solvents to be calculated if the appropriate weighting factors are known. The evaluation of weighting factors from the experimental data, as was done in the prediction of enthalpies of solution in binary solvent mixtures (Burchfield, 1977; Burchfield and Bertrand, 1975), is not mathematically feasible because of the nature of solubility measurements. Hence, simple approximations must be employed. Furthermore, weighting factors calculated from the excess free energy of mixing of binary solutions are often found to be strongly dependent upon temperature. This condition probably arises because the interaction parameters for the free energy functions are usually small differences between much larger interaction parameters for enthalpy and entropy. Small changes in the interaction parameters and weighting factors for entropy and enthalpy can show up as large relative changes in the corresponding parameters for the free energy.

Acree and Bertrand (1977) derived the following three expressions:

$$RT \ln(a_3^{solid}/X_3^{sat}) = (1 - X_3^{sat})^2 [X_1^{\circ}(\Delta\bar{G}_3^{ex})_{X_1=1}^{\infty} + X_2^{\circ}(\Delta\bar{G}_3^{ex})_{X_2=1}^{\infty} - \Delta\bar{G}_{12}^{ex}] \quad (XX)$$

$$RT \ln(a_3^{solid}/X_3^{sat}) = (1 - \phi_3^{sat})^2 [\phi_1^{\circ}(\Delta\bar{G}_3^{ex})_{X_1=1}^{\infty} + \phi_2^{\circ}(\Delta\bar{G}_3^{ex})_{X_2=1}^{\infty}$$
$$- \bar{V}_3^{\circ}(X_1^{\circ}\bar{V}_1 + X_2^{\circ}\bar{V}_2)^{-1}\Delta\bar{G}_{12}^{ex}] \quad (XV)$$

and

$$RT\{\ln(a_3^{solid}/\phi_3^{sat}) - (1 - \phi_3^{sat})[1 - \bar{V}_3^{\circ}/(X_1^{\circ}\bar{V}_1 + X_2^{\circ}\bar{V}_2)]\}$$
$$= (1 - \phi_3^{sat})^2 [\phi_1^{\circ}(\Delta\bar{G}_3^{fh})_{X_1=1}^{\infty} + \phi_2^{\circ}(\Delta\bar{G}_3^{fh})_{X_2=1}^{\infty}$$
$$- \bar{V}_3^{\circ}(X_1^{\circ}\bar{V}_1 + X_2^{\circ}\bar{V}_2)^{-1}\Delta\bar{G}_{12}^{fh}] \quad (VV)$$

based on different approximations for weighting factors. Equation (XX) is based on Eq. (10.42) with all weighting factors equal. Equation (XV) is based on Eq. (10.42) using molar volumes as weighting factors. Equation (VV) is

D. NIBS, Solubility in Binary Solvents

based on Eq. (10.43) using molar volumes as weighting factors. When the molar volumes of all components are approximately equal, the volume fraction of each component approaches its mole fraction and the three equations are almost identical.

As was shown in the first section of this chapter, Eq. (XV) can be derived from the Scatchard–Hildebrand solubility parameter theory by eliminating the three δs with the actual experimental solubilities in the pure solvents and the thermodynamic excess properties of the binary solvent. The NIBS treatment is more general, however, and does not place any restrictions on the numerical values of the binary interaction parameters. Equation (8.16), on the other hand, relates the A_{ij} terms to solubility parameters via

$$A_{ij} = (\delta_i - \delta_j)^2$$

and thus, cannot be applied to systems having negative deviations from Raoult's law.

To illustrate the predictive application of Eq. (XX), assume that one wants to estimate the solubility of naphthalene in binary carbon tetrachloride (1) + n-hexane (2) mixtures, at $X_2^o = 0.5971$ and 25°C, from the measured solubilities in the pure solvents $(X_3^{sat})_{X_1=1} = 0.2591$, and $(X_3^{sat})_{X_2=1} = 0.1168$. Begin by calculating $(\Delta \bar{G}_3^{ex})_{X_1=1}^{\infty}$ and $(\Delta \bar{G}_3^{ex})_{X_2=1}^{\infty}$ using $a_3^{solid} = 0.312$

$$(\Delta \bar{G}_3^{ex})_{X_1=1}^{\infty} = (8.314)(298.15)(1 - 0.2591)^{-2} \ln(0.312/0.2591)$$
$$= 838.93 \text{ J/mol}$$

$$(\Delta \bar{G}_3^{ex})_{X_2=1}^{\infty} = (8.314)(298.15)(1 - 0.1168)^{-2} \ln(0.312/0.1168)$$
$$= 3122.17 \text{ J/mol}$$

These values are then combined with the experimental excess Gibbs free energy of the binary solvent $\Delta \bar{G}_{12}^{ex} = 141.8$ J/mol, to yield

$$(8.314)(298.15) \ln(0.312/X_3^{sat})$$
$$= (1 - X_3^{sat})^2 [(0.5971)(838.93) + (0.4029)(3122.17) - 141.8]$$

The equation can be solved in a reiterative manner. Letting $(1 - X_3^{sat})^2 = 1$, generate a first approximation of $X_3^{sat} = 0.1625$, which is then used to calculate $(1 - X_3^{sat})^2$ for the second approximation. Convergence to a constant value of X_3^{sat} generally takes three or four iterations, depending upon the saturation solubility.

Benzil solubilities in binary solvent mixtures containing carbon tetrachloride should provide a definitive test of the various weighting factor approximations as the mole fraction solubilities encompass up to a 14-fold range. Relatively small differences in Γ_i can result in large errors in the predicted values provided that the thermodynamic excess properties of the solute are greatly different in each pure solvent. Graphical comparison

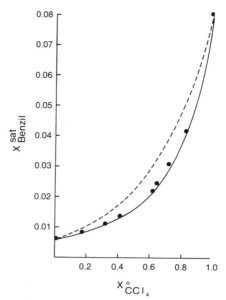

Fig. 10.10 Comparison between experimental solubilities (●) and the NIBS predictions for benzil in binary mixtures of carbon tetrachloride + isooctane using Eq. (XX) (— — —) and Eqs. (XV) and (VV) (———). The excess free energy of mixing data for the binary solvent ($\Delta \bar{G}_{12}^{ex}$) is taken from measurements by Battino, (1968). This figure has been reproduced from Acree and Rytting (1982b) with permission of the copyright owners.

between experimental and calculated values are shown in Figs. 10.10 and 10.11 for benzil in carbon tetrachloride + isooctane and carbon tetrachloride + n-octane mixtures. As indicated by the solid lines, Eqs. (XV) and (VV) are comparable and predict the experimental solubilities to within a maximum deviation of 5%. In comparison, Eq. (XX) provides rather poor predictions, with several of the deviations as large as 20%. These observations are in complete agreement with those of Burchfield and Bertrand (1975) who concluded that approximation of weighting factors with molar volumes gave better predictions for enthalpies of solution in binary solvent mixtures than the somewhat simpler approximation of equating all weighting factors.

Because these solubilities can be adequately described by the NIBS model suggests that specific solute–solvent interactions (i.e., complexation) does not contribute significantly to the large solubility enhancement. Rather, the 14-fold range of benzil solubilities in carbon tetrachloride + isooctane mixtures is a result of differences in nonspecific interactions. Stoichiometric complexation models that attribute all solubility enhancement to the formation of molecular complexes ignore differences in nonspecific interactions.

D. NIBS, Solubility in Binary Solvents

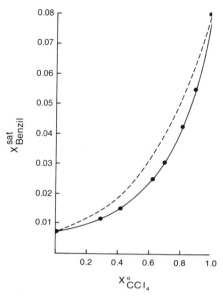

Fig. 10.11 Comparison between experimental solubilities (●) and the NIBS predictions for benzil in binary mixtures of carbon tetrachloride + n-octane using Eq. (XX) (— — —) and Eqs. (XV) and (VV) (———). The excess Gibbs free energy for the binary solvent ($\Delta \bar{G}^{ex}_{12}$) was taken from measurements by Jain and Yadav, (1971). This figure has been reproduced from Acree and Rytting (1982b) with permission of the copyright owners.

Why should one expect differences in nonspecific interactions, shown to be present in several mixtures containing two noncomplexing solvent components, to suddenly vanish whenever one of the *inert* solvents is replaced by a complexing solvent?

The predictive abilities of Eqs. (XX), (XV), and (VV) are summarized in Table III for 58 systems for which solubility data and thermodynamic mixing data of the binary solvent were available at or near the same temperature. Molar volumes and surface areas used in the predictions are listed in Table IV. For the most part, Eqs. (XV) and (VV) provide reasonable estimates ($\pm 5\%$) for the experimental solubilities; the exceptions being the anthracene solubilities in solvent mixtures containing benzene. (These systems will be discussed in more detail later). Equation (XX) provides only fair predictions (± 15–25%) for several of the benzil, p-benzoquinone, and iodine systems.

The superiority of expressions based on molar volumes suggests that the relative sizes of the molecules are an important consideration. The use of surface areas as weighting factors may be revealing because surface area represents a different measure of molecular size. Introduction of molecular

TABLE III
Solubility of Nonelectrolyte Solutes in Binary Solvent Mixtures

Binary solvent system	Data reference[c]	RMS deviations (%)[b] of calculated values			$\Delta \bar{G}_{12}^{ex}$ reference[c]
		(XX)	(XV)	(VV)	
Napthalene					
Benzene + cyclohexane	1	−1.4	−1.2	−1.1	2
Benzene + carbon tetrachloride	1	+1.5	+1.7	+1.9	2
Benzene + n-hexane	1	+2.3	1.4	1.4	3
Cyclohexane + hexadecane	4	−4.0	+2.9	+1.4	5
n-Hexane + n-hexadecane	4	−6.8	+1.8	0.8	6
Carbon tetrachloride + cyclohexane	4	0.3	−0.5	−0.6	2
Carbon tetrachloride + n-hexadecane	4	+8.0	+4.5	+2.4	7
Benzene + n-hexadecane	1	+9.2	+2.4	+0.7	7
Carbon tetrachloride + n-hexane	4	+3.2	1.3	0.5	8
Cyclohexane + n-hexane	4	0.8	0.6	0.6	9
Benzene + toluene	1	+0.5	0.5	+0.6	10
Carbon tetrachloride + toluene	11	−1.5	−1.5	−1.5	12
Cyclohexane + toluene	11	−1.8	−0.9	−0.8	13
Ethylbenzene + carbon tetrachloride	11	0.5	0.7	0.6	14
Ethylbenzene + cyclohexane	11	−1.7	0.3	0.3	14
p-Dibromobenzene					
Carbon tetrachloride + n-hexadecane	15	+5.4	+1.8	−0.8	7
n-Hexane + n-hexadecane	15	−8.5	+1.6	0.6	6
Carbon tetrachloride + cyclohexane	15	−0.6	−1.5	−1.5	2
Stannic Iodide					
Benzene + cyclohexane	16	2.2	2.8	3.1	2
Cyclohexane + carbon tetrachloride	17	−0.8	−1.3	−1.2	2

Iodine					
Cyclohexane + n-hexane	18	+1.7	−1.5	−2.5	9
Cyclohexane + n-hexane	19	1.2	−2.0	−3.0	9
Cyclohexane + n-hexane	20	+1.6	−1.9	−2.0	9
Carbon tetrachloride + n-hexane	18	+5.0	−1.3	−2.0	8
Carbon tetrachloride + n-hexane	20	+5.8	−0.8	−1.3	8
Cyclohexane + n-heptane	21	+2.1	−0.5	−0.8	13
n-Heptane + n-hexadecane	17	−8.2	2.4	−1.0	22
Isooctane + n-hexadecane	17	−8.1	1.8	−0.9	10
Cyclohexane + carbon tetrachloride	20	+0.9	−0.7	−0.7	2
Cyclohexane + OMCTS[a]	20	+5.3	+4.2	−3.0	23
Carbon tetrachloride + OMCTS	20	+8.6	+5.3	−4.5	24
n-Heptane + carbon tetrachloride	20	+5.4	−0.6	−1.3	25
Benzil					
n-Hexane + cyclohexane	20	−0.4	−1.3	−1.2	9
n-Hexane + carbon tetrachloride	20	+13.0	−3.4	−3.1	8
n-Heptane + carbon tetrachloride	20	+17.2	−3.1	−2.5	25
n-Hexane + n-heptane	20	−0.6	0.3	0.3	10
Cyclohexane + carbon tetrachloride	20	+1.5	−2.2	−2.4	2
Benzene + toluene	20	0.2	−0.3	−0.3	10
Isooctane + cyclohexane	26	+2.4	−1.8	−1.2	27
Cyclohexane + cyclooctane	26	−2.2	0.3	0.4	28
n-Octane + carbon tetrachloride	26	+25.9	−1.8	0.9	29
Cyclohexane + n-Octane	26	+1.8	0.6	0.3	29
Cyclohexane + n-heptane	26	+2.0	0.5	+0.7	30
Isooctane + carbon tetrachloride	26	+25.7	−4.2	−3.4	31
p-Benzoquinone					
n-Octane + carbon tetrachloride	32	+15.7	−3.1	−4.2	29
n-Heptane + carbon tetrachloride	32	+12.4	−3.8	−4.4	25
n-Heptane + n-dodecane	32	−1.4	+1.3	0.2	10

TABLE III (continued)

Binary solvent system	Data reference[c]	RMS deviations (%)[b] of calculated values			$\Delta \bar{G}_{12}^{ex}$ reference[c]
		(XX)	(XV)	(VV)	
Cyclohexane + isooctane	32	0.3	0.6	−1.1	27
Cyclohexane + cyclooctane	32	−0.5	0.5	0.4	28
Cyclohexane + n-heptane	32	+0.4	−0.5	−0.8	30
Anthracene					
Cyclohexane + n-heptane	33	1.0	0.6	0.6	30
Cyclohexane + cyclooctane	33	−1.4	0.9	0.9	28
Cyclohexane + n-octane	33	−1.3	+0.6	+0.7	29
Cyclohexane + isooctane	33	+1.9	−1.2	−1.0	27
Cyclohexane + n-hexane	33	−1.2	−1.2	−1.1	10
Benzene + n-heptane	33	+10.6	2.1	1.6	34
Benzene + n-heptane	33	+7.9	−4.6	−3.8	35
Benzene + cyclohexane	33	−6.9	−7.7	−7.5	2
Benzene + cyclohexane	36	−6.2	−6.8	−6.6	2
Benzene + carbon tetrachloride	36	−3.1	−2.1	−2.0	2
Benzene + n-hexane	33	+2.7	−6.0	−5.4	3
Benzene + cyclooctane	33	1.8	−8.3	−7.7	37
Benzene + isooctane	33	+10.5	−11.6	−10.7	38

[a] Octamethylcyclotetrasiloxane

[b] RMS Deviations (%) = $(100/N^{\frac{1}{2}})\{\sum^N [\ln(X_{\text{calc}}^{\text{sat}}/X_{\text{exp}}^{\text{sat}})]^2\}^{\frac{1}{2}}$. The algebraic sign indicates the deviations were all of the same sign.

[c] References: (1) Heric and Posey (1964a); (2) Goates et al. (1959); (3) Smith and Robinson (1970); (4) Heric and Posey (1965); (5) Gomez-Ibanez and Shieh (1965); (6) McGlashan and Williamson (1961); (7) Jain and Lark (1973); (8) Jain et al. (1970); (9) Li et al. (1973); (10) Rowlinson (1959); (11) Heric and Posey (1964b); (12) Rastogi et al. (1967); (13) Katayama et al. (1965); (14) Jain and Yadav (1974); (15) Berryman and Heric (1967); (16) Purkayastha and Walkley (1972); (17) Acree and Bertrand (1977); (18) Sytilin (1974); (19) Nakanishi and Asakura (1977); (20) Acree (1981); Kortum and Vogel (1955); (22) Bronsted and Koefoed (1946); (23) Tomlins and Marsh (1976); (24) Marsh (1968); (25) Bissell and Williamson (1975); (26) Acree and Rytting (1982b); (27) Battino (1966); (28) Ewing and Marsh (1974); (29) Jain and Yadav (1971); (30) Young et al. (1977); (31) Battino (1968); (32) Acree and Rytting (1982a); (33) Acree and Rytting (1983); (34) Jain et al. (1973); (35) Harris and Dunlop (1970); (36) Smutek et al. (1967); (37) Mitra et al. (1973); (38) Weissman and Wood (1960).

D. NIBS, Solubility in Binary Solvents

TABLE IV
Properties Used in the NIBS Predictions

Solvents Component	$\bar{V}_i$ (cm³/mol)	$\bar{A}_i{}^a$ (Å²/mol)
Benzene	89.41	109.5
Carbon tetrachloride	97.08	—
Cyclohexane	108.76	120.8
n-Hexadecane	294.12	323.2
n-Heptane	147.48	160.3
n-Hexane	131.51	142.1
Cyclooctane	134.88	148.8
n-Octane	163.46	178.4
Isooctane	166.09	163.1
Toluene	106.84	126.5
Ethylbenzene	123.06	144.9
OMCTS	314.00	—

Solutes Component	$\bar{V}_i$ (cm³/mol)	$\bar{A}_i$ (Å²/mol)	a_3^{solid}
Naphthalene	123.0	155.8	0.312
p-Dibromobenzene	118.0	156.6	0.248
Anthracene	150.0	202.2	0.01049
p-Benzoquinone	82.1	—	0.182
Benzil	183.0	—	0.224
Iodine	59.591	—	0.258
Stannic iodide	151.0	—	0.1127

[a] Valvani et al. (1976), Yalkowsky and Valvani (1976, 1979), Yalkowsky et al. (1979).

surface areas ($\bar{A}_i$) into the model leads to the development of two more predictive expressions

$$RT \ln(a_3^{\text{solid}}) = (1-\theta_3^{\text{sat}})^2[\theta_1^o(\Delta\bar{G}_3^{\text{ex}})_{X_1=1}^\infty + \theta_2^o(\Delta\bar{G}_3^{\text{ex}})_{X_2=1}^\infty \\ - \bar{A}_3(X_1^o\bar{A}_1 + X_2^o\bar{A}_2)^{-1}\Delta\bar{G}_{12}^{\text{ex}}] \quad \text{(XA)}$$

and

$$RT\{\ln(a_3^{\text{solid}}/\phi_3^{\text{sat}}) - (1-\phi_3^{\text{sat}})[1-\bar{V}_3^o/(X_1^o\bar{V}_1 + X_2^o\bar{V}_2)]\} \\ = (1-\theta_3^{\text{sat}})^2[\theta_1^o(\Delta\bar{G}_3^{\text{fh}})_{X_1=1}^\infty + \theta_2^o(\Delta\bar{G}_3^{\text{fh}})_{X_2=1}^\infty - \bar{A}_3(X_1^o\bar{A}_1 + X_2^o\bar{A}_2)^{-1}\Delta\bar{G}_{12}^{\text{fh}}] \quad \text{(VA)}$$

depending on whether a Regular Solution model (Eq. XA) or a Flory–Huggins model (Eq. VA) is used to describe solution ideality.

The predictive abilities of these new equations are compared in Table V for 25 systems. The unavailability of molecular surface areas for benzil,

TABLE V
Solubility of Nonelectrolyte Solutes in Binary Solvent Mixtures[a]

Binary solvent system	RMS deviations (%) of calculated values[b]				
	(XX)	(XV)	(VV)	(XA)	(VA)
Naphthalene					
Benzene + cyclohexane	−1.4	−1.2	−1.1	+0.4	+0.6
Benzene + n-hexane	+2.3	1.4	1.4	+1.9	+2.2
Cyclohexane + n-hexadecane	−4.0	+2.9	+1.4	+2.8	+1.5
n-Hexane + n-hexadecane	−6.8	+1.8	+0.8	+1.7	+0.6
Benzene + n-hexadecane	+9.2	+2.4	+0.7	+3.3	+2.5
Cyclohexane + n-hexane	0.8	0.6	0.6	0.8	0.8
Benzene + toluene	+0.5	+0.5	+0.6	+0.5	+0.7
Cyclohexane + toluene	−1.8	−0.9	−0.8	+1.0	+1.0
Cyclohexane + ethylbenzene	−1.7	0.3	0.3	+1.4	+1.5
p-Dibromobenzene					
n-Hexane + n-hexadecane	−8.5	+1.6	0.6	+1.7	0.5
Anthracene					
Cyclohexane + n-heptane	1.0	0.6	0.6	0.6	+1.0
Cyclohexane + cyclooctane	−1.4	0.9	0.9	+1.0	+1.2
Cyclohexane + n-octane	−1.3	+0.6	+0.7	+0.9	+1.5
Cyclohexane + isooctane	+1.9	−1.2	−1.0	+0.5	+1.1
Cyclohexane + n-hexane	−1.2	−1.2	−1.1	0.3	0.3
Benzene + n-heptane	+10.6	2.1	1.6	+5.9	+6.9
Benzene + n-heptane	+7.9	−4.6	−3.8	+2.1	+3.1
Benzene + cyclohexane	−6.9	−7.7	−7.5	−1.8	−1.7
Benzene + cyclohexane	−6.2	−6.8	−6.6	1.1	1.1
Benzene + carbon tetrachloride	−3.1	−2.1	−2.0	—	—
Benzene + n-hexane	+2.7	−6.0	−5.4	+2.6	+3.1
Benzene + cyclooctane	1.8	−8.3	−7.7	−3.6	3.0
Benzene + isooctane	+10.5	−11.6	−10.7	+1.8	+3.0

[a] Table V reproduced from Acree and Rytting, (1983) with permission of the copyright owners.
[b] RMS Deviations (%) = $(100/N^{1/2})\{\sum_{i=1}^{N}[\ln(X_{\text{case}}^{\text{sat}}/X_{\text{expl}}^{\text{sat}})]^2\}^{1/2}$; an algebraic sign indicates all deviations were of the same sign.

iodine, p-benzoquinone, and carbon tetrachloride prevented the application of Eqs. (XA) and (VA) to the remaining 33 systems considered in Table III. Equation (XA), based on surface areas as weighting factors for the excess free energy relative to Raoult's law, is seen to be the most generally applicable equation with an overall average (rms) deviation of 1.7% and a maximum error for a single data point of 7.5%. This maximum deviation occurs in a system (benzene + n-heptane) in which conflicting values of $\Delta \bar{G}_{12}^{\text{ex}}$ were reported. As shown in Table V, deviations between the predicted and observed solubilities depend to a large extent on which literature source is used for the solvent properties. This leads to two sets of predicted anthracene solubilities that differ from each other by as much as 6%. Discrepancies in the reported values of $\Delta \bar{G}_{12}^{\text{ex}}$ were not noted for the remaining binary systems listed in Tables III and V. The primary advantage of Eq. (XA) over expressions based on molar volumes Eqs. (XV) and (VV) is its applicability to anthracene solubilities in solvent mixtures containing benzene. If these five systems are excluded from the calculations, Eqs. (XV) and (VV) are slightly better than equations based on surface areas.

Unfortunately, these calculations do not clearly indicate whether weighting factors are better approximated with molar volumes or surface areas. From the standpoint of calculational simplicity and the ready availability of molar volumes, Eq. (XV) is preferred, and some support for this form can be found in its adaptability to the Scatchard–Hildebrand solubility parameter theory. Similar support for Eqs. (XA) and (VA) can be found in correlations of partition coefficients with surface areas and in several semi empirical expressions developed for predicting liquid–vapor equilibrium. Equation (VV), however, is also applicable to polymer solutions, and this form of the basic NIBS model is personally preferred because it is more ideally suited to molarity-based equilibrium constants and to gas–liquid partition coefficients.

E. The Nearly Ideal Binary Solvent Model: (NIBS) Monomeric and Dimeric Treatment of Carboxylic Acids.

Spectroscopic evidence indicates that monofunctional carboxylic acids exist in solutions of nonpolar solvents as both monomers and dimers, with the extent of dimerization depending upon the characteristics of the solvent (Allen *et al.*, 1967; Glasoe *et al.*, 1971; Krishnan *et al.*, 1979). The NIBS theory should be applicable for predicting carboxylic acid solubilities in binary solvent mixtures if the extent of dimerization does not vary appreciably between the two pure solvents and if the properties of the dimeric solute are known or can be estimated.

Application of the monomeric and dimeric forms of the NIBS expressions is relatively straightforward in that the molecular weight, molar volume, and enthalpy of fusion[†] of the dimeric solute are taken to be twice the values for

---

[†] The enthalpy of fusion is actually for the transition of the solid to a mixture of monomers and dimers. For consistency in the NIBS calculations, however, the melt is considered to be completely monomeric for the monomer model and completely dimeric for the dimer model.

the monomeric solute. This leads to

$$a_{3,\text{dim}}^{\text{solid}} = (a_{3,\text{mon}}^{\text{solid}})^2 \qquad (10.46)$$

$$X_{3,\text{dim}}^{\text{sat}} = X_{3,\text{mon}}^{\text{sat}}/(2 - X_{3,\text{mon}}^{\text{sat}}) \qquad (10.47)$$

$$\bar{V}_{3,\text{dim}} = 2\bar{V}_{3,\text{mon}} \qquad (10.48\text{a})$$

$$\phi_{3,\text{dim}}^{\text{sat}} = \phi_{3,\text{mon}}^{\text{sat}} \qquad (10.48\text{b})$$

where dim is dimer and mon is monomer. Acree and Bertrand (1981) demonstrated that the solubility of benzoic acid in several binary solvent mixtures could be reasonably estimated via Eqs. (XV) and (VV) provided that the acid was considered to be completely dimerized. Calculations treating benzoic acid as completely monomeric, on the other hand, gave a rather poor description of the carboxylic acid solutions with the predicted solubilities differing from the experimental values by as much as 11% for a single data point. Results of these calculations are summarized in Table VI along with the activities and molar volumes of benzoic acid.

Examination of Table VI reveals that the accuracy of the dimeric treatment is comparable to that of the monomer model for solutes such as iodine, naphthalene, p-dibromobenzene, benzil, and p-benzoquinone, which are considered incapable of self-association. Benzoic acid, however, undergoes self-

TABLE VI

Comparison between the Monomer and Dimer Treatment of Benzoic Acid in the Predictions of Solubilities in Binary Solvent Systems at 25°C

Solvent system	Model[a]	RMS deviations (%) of calculated values*		
		(XX)	(XV)	(VV)
Cyclohexane + n-heptane	Monomer	−2.5	−3.0	−3.0
	Dimer	−2.6	−1.0	0.5
Cyclohexane + n-hexane	Monomer	−2.5	−3.5	−3.5
	Dimer	−2.6	−0.8	0.7
Cyclohexane + carbon tetrachloride	Monomer	−3.2	−6.8	−6.8
	Dimer	−2.3	−2.9	−2.9
n-Hexane + carbon tetrachloride	Monomer	+2.6	−8.6	−8.6
	Dimer	+3.4	−2.4	−1.7
n-Heptane + carbon tetrachloride	Monomer	+7.1	−8.0	−8.2
	Dimer	+8.0	−2.3	−1.0

[a] Monomeric Model: $a_{3,\text{mon}}^{\text{solid}} = 0.2275$ and $\bar{V}_{3,\text{mon}} = 104.38$ cm^3/mol Dimeric Model: $a_{3,\text{dim}}^{\text{solid}} = 0.0518$ and $\bar{V}_{3,\text{dim}} = 208.76$ cm^3/mol.

* RMS Deviations (%) = $(100/N^{1/2})\{\sum_{i=1}^{N} [\ln(X_{\text{calc}}^{\text{sat}}/X_{\text{exp}}^{\text{sat}})]^2\}^{1/2}$; an algebraic sign indicates all deviations were of the same sign.

F. NIBS with Solute–Solvent Complexation

association, and the superiority of the dimeric treatment is particularly gratifying because calculations using reported dimerizations constants in cyclohexane (Allen et al., 1967) and carbon tetrachloride (Allen et al., Glasoe et al., 1971) indicate that less than 4% of the benzoic acid molecules are monomers at saturation at 25°C. A realistic interpretation of carboxylic acid solutions requires the presence of both monomeric and dimeric species, but even this may be an oversimplification as Goldman and co-workers (Krishnan et al., 1979) presented evidence to support the exis-tence of trimers of benzoic acid in benzene.

F. Extension of the Nearly Ideal Binary Solvent Model to Systems Having Solute–Solvent Complexation

The success of Eq. (XV) in predicting solubilities in binary solvent mixtures with molar volumes covering a three-fold range and for solutes encompassing up to a 14-fold range of mole fraction solubilities suggests that this expression should provide adequate estimates of the physical contributions to nonideality in systems containing chemical interactions such as those between the solute and a complexing solvent. Application of the Flory–Huggins form of Eq. (10.37) to the quaternary system (A_1, B, C_1, AC)

$$A_1 + C_1 \rightleftharpoons AC$$

$$K_{AC}^{\phi} = (\hat{\phi}_{AC}/\hat{\phi}_{A_1}\hat{\phi}_{C_1}) \tag{10.49}$$

takes the form

$$\Delta G^{\text{mix}} = RT[\hat{n}_{A_1} \ln \hat{\phi}_{A_1} + \hat{n}_B \ln \hat{\phi}_B + \hat{n}_{C_1} \ln \hat{\phi}_{C_1} + \hat{n}_{AC} \ln \hat{\phi}_{AC}]$$
$$+ (\hat{n}_{A_1}\bar{V}_A + \hat{n}_B\bar{V}_B + \hat{n}_{C_1}\bar{V}_C + \hat{n}_{AC}\bar{V}_{AC})[\hat{\phi}_{A_1}\hat{\phi}_B A_{A_1 B} + \hat{\phi}_{A_1}\hat{\phi}_{C_1} A_{A_1 C_1}$$
$$+ \hat{\phi}_{A_1}\hat{\phi}_{AC} A_{A_1 AC} + \hat{\phi}_B\hat{\phi}_{C_1} A_{BC_1} + \hat{\phi}_B\hat{\phi}_{AC} A_{BAC} + \hat{\phi}_{C_1}\hat{\phi}_{AC} A_{C_1 AC}] \tag{10.50}$$

the only assumption being that the molar volume of the AC-complex equals the sum of the molar volumes of components A and C, that is, $\bar{V}_{AC} = \bar{V}_A + \bar{V}_C$. As in earlier chapters, the superscript (^) denotes the *true* quaternary solution. The chemical potentials of the individual components relative to the pure liquids ($\mu_i^{\bullet}$) are obtained through the appropriate differentiation

$$\hat{\mu}_{A_1} - \mu_{A_1}^{\bullet} = RT[\ln \hat{\phi}_{A_1} + 1 - (\bar{V}_A/\hat{V}_{\text{sol}})]$$
$$+ \bar{V}_A[\hat{\phi}_B(1 - \hat{\phi}_{A_1})A_{A_1 B} + \hat{\phi}_{C_1}(1 - \hat{\phi}_{A_1})A_{A_1 C_1}$$
$$+ \hat{\phi}_{AC}(1 - \hat{\phi}_{A_1})A_{A_1 AC} - \hat{\phi}_B\hat{\phi}_{C_1} A_{BC_1}$$
$$- \hat{\phi}_B\hat{\phi}_{AC} A_{BAC} - \hat{\phi}_{C_1}\hat{\phi}_{AC} A_{C_1 AC}] \tag{10.51}$$

$$\hat{\mu}_B - \mu_B^{\bullet} = RT[\ln \hat{\phi}_B + 1 - (\bar{V}_B/\hat{V}_{sol})]$$
$$+ \bar{V}_B[\hat{\phi}_{A_1}(1 - \hat{\phi}_B)A_{A_1B} + \hat{\phi}_{C_1}(1 - \hat{\phi}_B)A_{BC_1}$$
$$+ \hat{\phi}_{AC}(1 - \hat{\phi}_B)A_{BAC} - \hat{\phi}_{A_1}\hat{\phi}_{C_1}A_{A_1C_1}$$
$$- \hat{\phi}_{C_1}\hat{\phi}_{AC}A_{C_1AC} - \hat{\phi}_{A_1}\hat{\phi}_{AC}A_{A_1AC}] \quad (10.52)$$

and

$$\hat{\mu}_{C_1} - \mu_{C_1}^{\bullet} = RT[\ln \hat{\phi}_{C_1} + 1 - (\bar{V}_C/\hat{V}_{sol})]$$
$$+ \bar{V}_C[\hat{\phi}_{A_1}(1 - \hat{\phi}_{C_1})A_{A_1C_1} + \hat{\phi}_B(1 - \hat{\phi}_{C_1})A_{BC_1}$$
$$+ \hat{\phi}_{AC}(1 - \hat{\phi}_{C_1})A_{C_1AC} - \hat{\phi}_{A_1}\hat{\phi}_B A_{A_1B}$$
$$- \hat{\phi}_{A_1}\hat{\phi}_{AC}A_{A_1AC} - \hat{\phi}_B\hat{\phi}_{AC}A_{BAC}] \quad (10.53)$$

where $\hat{V}_{sol}$ is the molar volume of the *true* solution

$$1/\hat{V}_{sol} = \hat{\phi}_{A_1}/\bar{V}_A + \hat{\phi}_B/\bar{V}_B + \hat{\phi}_{C_1}/\bar{V}_C + \hat{\phi}_{AC}/(\bar{V}_A + \bar{V}_C) \quad (10.54)$$

As shown in Chapter 7, the chemical potential of stoichiometric component C (and also A) is equal to the chemical potential of the uncomplexed species in the solution

$$\mu_C = \hat{\mu}_{C_1}$$

Combining Eqs. (10.49–10.54), one can write the Gibbs free energy of mixing

$$\Delta G^{mix} = RT[n_A \ln \hat{\phi}_{A_1} + n_B \ln \hat{\phi}_B + n_C \ln \hat{\phi}_{C_1} + n_A + n_B + n_C$$
$$- (n_A\bar{V}_A + n_B\bar{V}_B + n_C\bar{V}_C)/\hat{V}_{sol}] + (n_A\bar{V}_A + n_B\bar{V}_B + n_C\bar{V}_C)$$
$$\times [\phi_A\phi_B A_{A_1B} + \phi_A\hat{\phi}_{C_1}A_{A_1C} + \phi_A\hat{\phi}_{AC}A_{A_1AC}$$
$$+ \phi_C\hat{\phi}_{A_1}A_{A_1C_1} + \phi_C\phi_B A_{BC_1} + \phi_C\hat{\phi}_{AC}A_{C_1AC} - \hat{\phi}_{A_1}\hat{\phi}_{C_1}A_{A_1C_1}$$
$$- \hat{\phi}_{A_1}\hat{\phi}_{AC}A_{A_1AC} - \hat{\phi}_{C_1}\hat{\phi}_{AC}A_{C_1AC}] \quad (10.55)$$

where

$$n_A = \hat{n}_{A_1} + \hat{n}_{AC} \quad \text{and} \quad n_C = n_{C_1} + n_{AC}$$

Treatment of all interaction parameters involving the AC-complex in terms of solubility parameters leads to

$$A_{A_1AC} = \bar{V}_C^2(\bar{V}_A + \bar{V}_C)^{-2}A_{A_1C_1}$$
$$A_{C_1AC} = \bar{V}_A^2(\bar{V}_A + \bar{V}_C)^{-2}A_{A_1C_1}$$

Substitution of these approximations into Eq. (10.55), after suitable mathematical manipulation, yields the following expression for the Gibbs free energy:

$$\Delta G^{mix} = RT[n_A \ln \hat{\phi}_{A_1} + n_B \ln \hat{\phi}_B + n_C \ln \hat{\phi}_{C_1} + n_A + n_B + n_C$$
$$- (n_A\bar{V}_A + n_B\bar{V}_B + n_C\bar{V}_C)/\hat{V}_{sol}] + (n_A\bar{V}_A + n_B\bar{V}_B + n_C\bar{V}_C)$$
$$\times [\phi_A\phi_B A_{A_1B} + \phi_B\phi_C A_{BC_1} + \phi_A\phi_C A_{A_1C_1}] \quad (10.56)$$

F. NIBS with Solute–Solvent Complexation

Using the equilibrium condition defined by Eq. (10.48), it can be easily shown that the chemical potential of the solid solute (component A) at saturation is

$$\mu_A - \mu_A^\bullet = RT \ln a_A^{\text{solid}} = RT[\ln \hat{\phi}_{A_1}^{\text{sat}} + 1 - \bar{V}_A^\circ/\hat{V}_{\text{sol}}] + \bar{V}_A^\circ(1 - \phi_A^{\text{sat}})^2$$
$$[\phi_B^\circ A_{A_1B} + \phi_C^\circ A_{A_1C_1} - \phi_B^\circ \phi_C^\circ A_{BC_1}] \qquad (10.57)$$

Inspection of Eq. (10.57) reveals that for model systems obeying this expression, the A_{A_1B} and $A_{A_1C_1}$ interaction parameters can be eliminated from the basic model via the saturation solubilities in the pure solvents, and the A_{BC_1} parameter can be eliminated via the excess Gibbs free energy of the binary solvent mixture relative to the Flory–Huggins model. Performing these substitutions gives

$$RT[\ln(a_3^{\text{solid}}/\hat{\phi}_{A_1}^{\text{sat}}) - 1 + \bar{V}_A^\circ/\hat{V}_{\text{sol}}]$$
$$= (1 - \phi_A^{\text{sat}})^2[\phi_B^\circ(\Delta\bar{G}_A^{\text{fh}})_{X_B=1}^\infty + \phi_C^\circ(\Delta\bar{G}_A^{\text{fh}})_{X_C=1}^\infty$$
$$- \bar{V}_A^\circ(X_B^\circ\bar{V}_B + X_C^\circ\bar{V}_C)^{-1}\Delta\bar{G}_{BC}^{\text{fh}}] \qquad (10.58)$$

where

$$(\Delta\bar{G}_A^{\text{fh}})_{X_B=1}^\infty = (1 - \phi_A^{\text{sat}})^{-2}RT[\ln(a_A^{\text{solid}}/\phi_A^{\text{sat}}) - (1 - \phi_A^{\text{sat}})(1 - \bar{V}_A^\circ/\bar{V}_B)]$$

and

$$(\Delta\bar{G}_A^{\text{fh}})_{X_C=1}^\infty = (1 - \phi_A^{\text{sat}})^{-2}RT[\ln(a_A^{\text{solid}}/\hat{\phi}_{A_1}^{\text{sat}}) - 1 + \bar{V}_A^\circ(\hat{\phi}_{A_1}^{\text{sat}}/\bar{V}_A^\circ + \phi_C/\bar{V}_C)]$$

Calculation of the solute–solvent equilibrium constants using Eq. (10.58) is relatively straightforward. The quantities $(\Delta\bar{G}_A^{\text{fh}})_{X_B=1}^\infty$ and $(\Delta\bar{G}_A^{\text{fh}})_{X_C=1}^\infty$ are calculated from the volume fraction of the solute

$$\phi_A^{\text{sat}} = \hat{\phi}_{A_1}^{\text{sat}}[1 + \bar{V}_A^\circ K_{AC}^\phi \hat{\phi}_{C_1}/(\bar{V}_A^\circ + \bar{V}_C)]$$

in the pure solvents using an assumed value for the equilibrium constant. These quantities, along with the excess Gibbs free energy of the binary solvent mixture, are then used in Eq. (10.58) to calculate $\hat{\phi}_A^{\text{sat}}$ via a reiterative approach. One continues to vary K_{AC}^ϕ until one obtains a numerical value that best describes the experimental solubilities in a particular binary solvent mixture.

When the solubility is sufficiently small, $\phi_A^{\text{sat}} \approx 0$ and $1 - \phi_A^{\text{sat}} \approx 1$, reasonable estimates of K_{AC}^ϕ are often obtainable from a simplified form of Eq. (10.58) relating the overall solute solubility in the binary solvent mixture to the solubility in the two pure solvents $(\phi_A^{\text{sat}})_{X_B=1}$ and $(\phi_A^{\text{sat}})_{X_C=1}$

$$\ln(\phi_A^{\text{sat}}) = \phi_B^\circ \ln(\phi_A^{\text{sat}})_{X_B=1} + \phi_C^\circ \ln(\phi_A^{\text{sat}})_{X_C=1} + \ln[1 + \bar{V}_A^\circ K_{AC}^\phi \phi_C^\circ/(\bar{V}_A^\circ + \bar{V}_C)]$$
$$- \phi_C^\circ \ln[1 + \bar{V}_A^\circ K_{AC}^\phi/(\bar{V}_A^\circ + \bar{V}_C)] + [\bar{V}_A^\circ \Delta\bar{G}_{BC}^{\text{fh}}/RT(X_B^\circ\bar{V}_B + X_C^\circ\bar{V}_C)] \qquad (10.59)$$

To calculate the equilibrium constant, substitute the solute solubility at a particular solvent composition (i.e., $\phi_C^\circ \approx 0.5$) into Eq. (10.59) and solve the resulting mathematical expression for K_{AC}^ϕ. For example, to evaluate the anthracene–chloroform association constant from the anthracene solubility in the n-hexane + chloroform system at $X_C^\circ = 0.7371$ ($\phi_C^\circ = 0.6322$), one would need to solve

$$\ln 0.01165 = (0.3678) \ln 0.001471 + (0.6322) \ln 0.01998$$
$$+ \ln[1 + K_{AC}^\phi(150.00)(0.6322)/(230.63)]$$
$$- (0.6322) \ln[1 + K_{AC}^\phi(150.00)/(230.63)]$$
$$+ [(150.00)(308.0)/(8.314)(298.15)(94.00)]$$

obtaining the numerical value of $K_{AC}^\phi = 5.53$. Volume fraction compositions of anthracene used in the preceding example were

$$(\phi_A^{sat})_{X_C=1} = \frac{(0.01084)(150.00)}{(0.01084)(150.00) + (0.98916)(80.63)} = 0.01998$$

$$(\phi_A^{sat})_{X_B=1} = \frac{(0.00129)(150.00)}{(0.00129)(150.00) + (0.99871)(131.51)} = 0.001471$$

$$\phi_A^{sat} = \frac{(0.00733)(150.00)}{(0.00733)(150.00) + (0.99267)(94.00)} = 0.01165$$

calculated from the experimental solubilities listed in Table VII using the ideal molar volume approximation. The experimental vapor–liquid measurements of Bissell and Williamson (1975) served as the literature reference for $\Delta \bar{G}_{BC}^{fh} = 308.0 \, J/mol$.

TABLE VII
Mole Fraction Solubility of Anthracene at 25°C in n-Hexane (1) + Chloroform (2) Mixtures[a]

X_1°	X_3^{sat}
0.0000	0.00129
0.2227	0.00232
0.4019	0.00354
0.5199	0.00460
0.6120	0.00563
0.7371	0.00733
0.8815	0.00914
1.0000	0.01084

[a] Data source: Acree, unpublished results.

G. Solubility Predictions Using UNIFAC

Readers are reminded that the calculation of association constants for a *presumed* anthracene–chloroform complex does not imply that the author actually believes such a complex exists in solution. Rather, the experimental solubilities are used to illustrate the calculation of solute–solvent equilibrium constants from solubility data. As in all cases, the presence of molecular complexes should be supported by independent measurements involving spectroscopy, calorimetry, and so forth.

Before ending this discussion of solubility in complexing systems, it should be mentioned that equilibrium constants are only one approach for describing systems having specific interactions. Purkayastha and Walkley (1972) employed an expression similar to Eq. (10.38), with modifications of weighting factors based on the solubility parameters of the solute and solvents, to treat the solubility of iodine in binary solvent mixtures containing benzene. Nakanishi and Asakura (1977) used a similar approach with weighting factors based on the enthalpy of formation of the specific complex. Nitta and Katayama (1973, 1974, 1975) and Nitta et al. (1973, 1980) approximated weighting factors by molar volumes and included the equilibrium constant for complex formation in their development of a predictive equation for solubilities in complexing systems.

G. Solubility Predictions Using the UNIFAC Group Contribution Method

Although it is theoretically possible to use any of the solution models discussed in Chapters 6–8 to correlate activity coefficients as a function of liquid-phase composition, the nature of solubility determinations will restrict the number of adjustable parameters (or binary interaction parameters) to no more than one per binary mixture, as there is only one data point associated with the saturation solubility of a solute in a pure solvent. Keeping this restriction in mind, one realizes that the more sophisticated solution models such as the Wilson or UNIQUAC models are of little use when it comes to predicting solubility in binary solvent mixtures. Group contribution methods derived from these solution models, however, are ideally suited to solubility predictions because the various group parameters need not be determined from solubility measurements.

For example, the predictive expression of the UNIFAC model

$$\ln(a_3^{\text{solid}}/\phi_3^{\text{sat}}) = (z/2)q_3 \ln(\theta_3^{\text{sat}}/\phi_3^{\text{sat}}) + \ell_3 - (\theta_3^{\text{sat}}/X_3^{\text{sat}}) \sum_j X_j \ell_j$$

$$+ \sum_{\text{groups}} v_k^{(3)}[\ln \Gamma_k - \ln \Gamma_k^{(3)}] \qquad (10.60)$$

where the summation extends over all groups in the solution and

$$\ln \Gamma_k = Q_k \left[1 - \ln \sum_m \theta_m^g \psi_{mk} - \sum_m \theta_m^g \psi_{km} \Big/ \sum_n \theta_n^g \psi_{nm} \right] \quad (10.61)$$

requires only a prior knowledge of a_3^{solid}, the pure component structural parameters r_i and q_i, and the group interaction parameters that are listed in Table IV. The liquid-phase compositions and group compositions (θ_m^g and θ_n^g) used in Eqs. (10.60) and (10.61) correspond to those of the saturated solution. (A more complete description of the UNIFAC method can be found in Section VI.C.)

Martin et al. (1981) compared the predictions of the UNIFAC method to the experimental solubilities of naphthalene in a wide range organic solvents. As is seen in Table VIII, the UNIFAC model provides reasonable predictions, with many of the deviations around 10–20%.

TABLE VIII
Comparison between UNIFAC Predictions and Experimental Naphthalene Solubilities in Pure Solvents at 40°C[a]

Solvent	$(X_3^{\text{sat}})^{\text{exp}}$	(X_3^{sat}) Eq. (10.60)	Deviation (%)[b]
n-Hexane	0.222	0.2629	+16.9
Carbon tetrachloride	0.395	0.4071	+3.0
Toluene	0.422	0.4425	−4.7
Benzene	0.428	0.4499	+5.0
Chloroform	0.467	0.4695	+0.5
Chlorobenzene	0.444	0.3979	−11.0
Acetone	0.378	0.3628	−4.1
Carbon disulfide	0.494	0.4197	−16.3
1,1-Dibromoethane	0.456	0.3837	−17.3
Sec-Butanol	0.112	0.1132	+1.1
Tert-Butanol	0.1009	0.0819	−20.9
Cyclohexanol	0.232	0.2080	−10.9
Aniline	0.306	0.2689	−12.9
1-Butanol	0.116	0.1124	−3.2
Isopropanol	0.076	0.0948	+22.1
Ethylene dibromice	0.439	0.3832	−13.6
1-Propanol	0.094	0.0939	−0.1
Acetic acid	0.117	0.1267	+8.0
Ethanol	0.0726	0.0552	−27.4
Methanol	0.0437	0.0489	+11.2

[a] Table VII has been reproduced from Martin et al. (1981) with the permission of the copyright owners.
[b] Deviations (%) = $100 \ln(X_3^{\text{calc}}/X_3^{\text{exp}})$.

G. Solubility Predictions Using UNIFAC

Extension of the UNIFAC model to binary solvent mixtures is relatively straightforward, and Table IX compares the UNIFAC predictions to the experimental solubilities of Heric and Posey (1964a,b, 1965) for naphthalene in 16 different solvent mixtures. Included in this comparison is the corresponding predictions of Eq. (VV) derived from the NIBS model. Inspection of the last two columns of Table IX reveals that the NIBS predictions are superior to those of the UNIFAC method. It should be remembered, however, that the NIBS model was developed specifically for systems containing nonspecific interactions, and all of the systems listed in Table IX are believed to be free of specific solute–solvent and solvent–solvent interactions. The NIBS model cannot predict solubilities in complexing systems nor can it be used to predict solubilities in the pure solvents. The UNIFAC model, on the other hand, provides reasonable predictions of activity coefficients in several complexing systems, and one naturally expects the model to be fairly successful in providing first approximations of solubilities in complex systems. As seen in Table VII, the UNIFAC model predicted naphthalene

Table IX

Summarized Comparison between the UNIFAC Predictions, NIBS Predictions, and Experimental Solubilities for Naphthalene in Several Binary Solvent Mixtures

Solvent system	RMS deviations (%) of calculated values[a]	
	UNIFAC	NIBS
Benzene + cyclohexane	+15.8	−1.1
Benzene + carbon tetrachloride	+4.8	+1.9
Benzene + n-hexane	+15.3	1.4
Cyclohexane + n-hexadecane	12.7	+1.4
n-Hexane + n-hexadecane	12.7	+0.8
Carbon tetrachloride + cyclohexane	+16.7	−0.6
Benzene + n-hexadecane	7.6	+0.7
Carbon tetrachloride + n-hexane	+15.6	+0.5
Cyclohexane + n-hexane	+24.6	0.6
Benzene + toluene	+ 5.1	+0.6
Carbon tetrachloride + toluene	+2.7	−1.5
Cyclohexane + toluene	+14.9	−0.8
Carbon tetrachloride + n-hexadecane	7.1	−2.4
Toluene + n-hexane	+14.3	—
Toluene + n-hexadecane	6.9	—
Ethylbenzene + benzene	3.4	—

[a] RMS Deviation (%) = $(100/N^{1/2})\{\Sigma[\ln(X_3^{cal}/X_3^{exp})]^2\}^{1/2}$; an algebraic sign indicates all deviations were of the same sign.

solubilities in alcohol solvents to within 10–28%, and these deviations are comparable in magnitude to that in *n*-hexane (17%).

In using the UNIFAC model, one must remember that errors in calculating a_3^{solid} are reflected in the solubility predictions. Inspection of Eq. (10.18) reveals that these errors are greatest for solutes far removed from their normal melting point temperatures. For this reason, anthracene solubilities were not included in Table IX because deviations between experimental and predicted values represent not only the limitations of the UNIFAC model, but also errors in calculating the activity of the supercooled liquid solute, a_3^{solid}. It should be stated, however, that deviations for anthracene are much larger than for naphthalene.

Problems

10.1 (a) The heat of fusion of naphthalene is 146.9 J/g and its melting point is 80°C. Estimate the ideal mole fraction solubility of naphthalene in benzene at 25°C and compare it with the experimental value $(X_3^{\text{sat}})_{\text{ben}} = 0.296$.

(b) Estimate the ideal mole fraction solubility of napthalene in ethylbenzene at 25°C and compare it with the experimentally determined value of $(X_3^{\text{sat}})_{\text{et ben}} = 0.2926$.

(c) For anthracene, $\Delta H^{\text{fus}} = 161.9$ J/g and the melting point is 217°C. Estimate the ideal solubility of anthracene in benzene at 60°C.

10.2 Assuming ideal solution behavior and constant heats of fusion, estimate the freezing-point composition diagram for the biphenyl + *p*-dichlorobenzene system. Compare the calculations to the experimentally determined values given in Table X. Values needed in the calculations include $\Delta \bar{H}^{\text{fus}} = 17.72$ kJ/mol and $T_{TP} = 69.0°C$ for biphenyl; and $\Delta \bar{H}^{\text{fus}} = 17.24$ kJ/mol and $T_{TP} = 52.9°C$ for *p*-dichlorobenzene.

10.3 The experimental mole fraction solubility of iodine in carbon disulfide $(X_3^{\text{sat}} = 0.0558)$ is six times greater than in cyclohexane $(X_3^{\text{sat}} = 0.00918)$. Predict the solubilities of iodine in binary mixtures having an initial cyclohexane composition of $X_{\text{cyclo}}^{\circ} = 0.2, 0.4, 0.6, 0.8$, and 1.0. The excess Gibbs free energy of the cyclohexane (1) + carbon disulfide (2) is given by

$$\Delta \bar{G}_{12}^{\text{ex}}(J/\text{mol}) = X_2(1-X_2)[844.3 + 205.9(2X_2-1) + 71.5(2X_2-1)^2]$$

10.4 Vitoria and Walkley (1969) measured the solubility of tetraphenyltin in cyclohexane $(X_3^{\text{sat}} = 0.000208)$ and carbon tetrachloride $(X_3^{\text{sat}} = $

TABLE X
Crystallization Temperatures of Binary Biphenyl (1) + p-Dichlorobenzene (2) Mixtures[a]

X_1	T (°C)
0.000	52.9
0.038	51.5
0.119	47.1
0.186	43.4
0.200	42.8
0.291	37.0
0.402	29.3
0.425	27.7
0.496	34.1
0.575	41.2
0.692	49.8
0.802	57.4
0.859	61.4
0.923	65.0
1.000	69.0

[a] Reproduced with permission from Warner et al., (1934).

TABLE XI
Saturation Solubility of Benzil in Various Solvents at 298.15 K

Solvent	X_3^{sat}
Cyclohexane	0.01072
n-Hexane	0.00570
n-Heptane	0.00654
Carbon tetrachloride	0.0804
n-Octane	0.00726
Isooctane	0.00587
Benzene	0.1804
Toluene	0.1504

0.00106) at 25°C. Using the NIBS model, estimate the solubility of tetraphenyltin in binary carbon tetrachloride (1) + cyclohexane (2) mixtures having a composition of $X_1^\circ = 0.25$, 0.50, and 0.75. Values needed in the calculations include $\bar{V}_3^\circ = 314.4$ cm^3/mol, $\Delta \bar{H}^{fus} = 37.2$ kJ/mol and $T_{TP} = 229$°C. The excess Gibbs free energy of mixing at 25°C is given by

$$\Delta \bar{G}_{12}^{ex}(J/mol) = X_1 X_2 [284.1 + 18.1(X_1 - X_2) - 4.2(X_1 - X_2)^2]$$

10.5 Calculate the best value for the solubility parameter of benzil from the data given in Table XI. Numerical values of $\bar{V}_i$ and a_3^{solid} can be found in Table IV.

10.6 Treating biphenyl as an aromatic hydrocarbon having 10 ACH and 2 AC groups, use the UNIFAC model to predict the solubility of biphenyl in the pure solvents n-heptane and carbon tetrachloride at 40°C. Warner et al. (1934) reported biphenyl solubilities of $X_3^{sat} = 0.2735$, and $X_3^{sat} = 0.518$ in n-heptane and carbon tetrachloride, respectively. The numerical value of a_3^{solid} is 0.5708 at 40°C. What additional information is required for the prediction of biphenyl solubilities via the Solubility Parameter model?

10.7 Predict the solubility of anthracene in binary n-hexane (1) + chloroform (2) mixtures at 25°C using the NIBS model, the Solubility Parameter approach, and the UNIFAC method. For the UNIFAC predictions, treat anthracene as an aromatic hydrocarbon having 10 ACH and 4 AC groups. Which of these three solution models best predicts the experimental solubilities listed in Table VII? Properties needed in the calculations include $\delta_3 = 9.9$ cal$^{1/2}$/cm$^{3/2}$ and the excess Gibbs free energy of n-hexane + chloroform mixtures (Bissell and Williamson, 1975):

$$\Delta \bar{G}^{ex}_{12}/RT = X_2(1 - X_2)[0.55099 + 0.07957(2X_2 - 1)]$$

10.8 A charge transfer complex between iodine and benzene is fairly well documented in the chemical literature. Using Eq. (10.59) and the experimental iodine solubilities listed in Table XII, estimate the association constant K^ϕ_{AC} for the iodine–benzene complex. Spectroscopic studies by Showmik (1971) provide an independent determination of the iodine–benzene equilibrium constant in cyclohexane ($K^c_{AC} = 0.260$ M^{-1}), methylcyclohexane ($K^c_{AC} = 0.252$ M^{-1}), and n-heptane ($K^c_{AC} = 0.242$ M^{-1}). How does the solubility-based association constant com-

TABLE XII
Mole Fraction Solubilities of Iodine in n-Heptane (1) + Benzene (2) Mixtures at 25°C

X^o_1	X^{sat}_3
0.0000	0.04852
0.3115	0.02868
0.3898	0.02478
0.4796	0.02108
0.5451	0.01849
0.6700	0.01456
0.7673	0.01188
0.8695	0.00938
0.9470	0.00801
1.0000	0.00691

TABLE XIII
Mole Fraction Solubilities of Copper Acetylacetonate in Binary Benzene (1) + n-Heptane (2) Mixtures at 25°C^a

X^o_1	X^{sat}_3
0.00	6.59 × 10^{-6}
0.05	9.48 × 10^{-6}
0.15	1.37 × 10^{-5}
0.29	2.44 × 10^{-5}
0.41	3.82 × 10^{-5}
0.52	6.90 × 10^{-5}
0.62	8.64 × 10^{-5}
0.71	1.14 × 10^{-4}
0.79	1.46 × 10^{-4}
0.87	1.92 × 10^{-4}
0.94	2.34 × 10^{-4}
1.00	2.75 × 10^{-4}

^{a} Experimental data are taken from Koshmura (1977). Reproduced with permission of copyright owners.

pare to the spectroscopically determined values? The excess molar Gibbs free energy of the n-heptane (1) + benzene (2) system is given by Harris and Dunlop (1970)

$$\Delta \bar{G}_{12}^{\text{ex}}/RT = X_1 X_2 (0.5667)/[1 + 0.268(X_1 - X_2)]$$

10.9 Using Eq. (VV) of the NIBS model, predict the solubility of copper acetylacetonate in binary benzene + n-heptane mixtures at the nine compositions listed in Table XIII. If solute–solvent complexation is suggested, use Eq. (10.59) to calculate the equilibrium constant. Properties needed in the calculations include

$$\bar{V}_3^{\circ} = 193 \text{ cm}^3/\text{mol} \quad \text{and} \quad a_3^{\text{solid}} = 0.00182$$

11

Liquid–Liquid Equilibrium: Distribution of a Solute between Two Immiscible Liquid Phases

Often, liquid mixtures exist not as a single phase but as two or more equilibrium phases. This phenomenon of partial immiscibility is important in many separation processes, especially extraction and extractive distillation. It is also a stringent test of solution theories, because liquid–liquid immiscibility is a manifestation of extreme nonidealities in the liquid mixture.

Commonly, a partially miscible system might behave like the carbon disulfide + methanol system, shown in Fig. 11.1. At ambient temperatures the two liquids are only partly soluble in each other and this mutual solubility increases with increasing temperature until complete miscibility occurs at

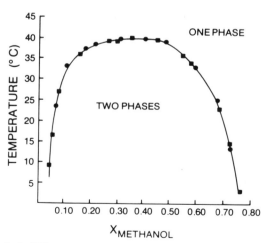

Fig. 11.1 Partial miscibility and UCST exhibited by the methanol + carbon disulfide system. Experimental data were taken from Rothmund, (1898), and Drucker, (1923).

11 Liquid–Liquid Equilibrium

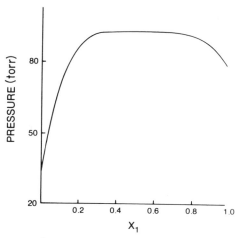

Fig. 11.2 Total pressure of vapor in equilibrium with the partially miscible isooctane (1) + nitroethane (2) system at 35°C.

about 40°C. The point at which the two liquid phases merge is called the consolute point, or upper critical solution temperature (UCST). Virtually all binary mixtures will display a UCST, though frequently it will not appear because it exists far below the freezing point of one of the pure components.

At any temperature below the UCST there exists two liquid phases in complete equilibrium with the common gas phase above the solution. For example, the vapor pressure of the isooctane + nitroethane system below its UCST is shown in Fig. 11.2. Note first the large positive deviations from Raoult's law, always typical of immiscible mixtures. The flat portion of the pressure curve, between about $X_1 = 0.31$ and $X_1 = 0.74$, represents the miscibility gap, and the vapor composition is invariant over this concentration range.

Though not as common, many examples also exist (primarily in hydrogen-bonded systems) of lower critical solution temperatures (LCST); an example being the water + triethylamine system depicted in Fig. 11.3. A few examples exist of systems having both critical points. For instance, the water + 2-methylpiperidine system (Fig. 11.4) has a closed loop extending over a 100°C temperature interval within which partial miscibility occurs. Conversely, but rarely, the sulfur + benzene solution (Fig. 11.5) has its LCST at a significantly higher temperature than the UCST.

Thermodynamic analysis of partially miscible mixtures stems from the basic principle of minimization of free energy to achieve a stable equilibrium. Consider the process of mixing two liquids. If the liquids mix completely to

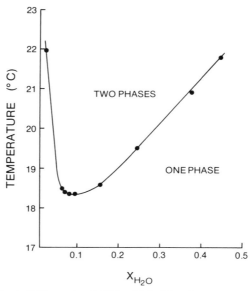

Fig. 11.3 Partial miscibility and a LCST exhibited by the water + triethylamine system. The experimental data were taken from the work of Kohler and Rice (1957).

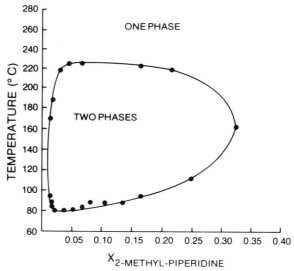

Fig. 11.4 Phase boundary curve for the water + 2-methyl-piperdine system that forms a closed loop for the partially miscible system with the UCST approximately 140°C higher than the LCST. The experimental data were taken from Flaschner and Macewen (1908).

11 Liquid–Liquid Equilibrium

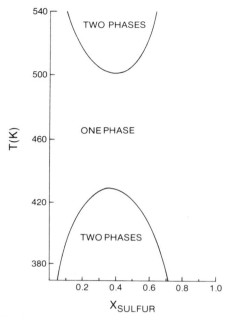

Fig. 11.5 Partial miscibility exhibited by the sulfur + benzene system.

form a single solution, as do benzene and cyclohexane, the Gibbs free energy of the solution must be a minimum for the system. Alternatively, if the two liquids are only slightly miscible in each other, then the Gibbs free energy of the system as it exists, with two different phases at different composition, must be lower than the Gibbs free energy for any single phase of intermediate composition.

Consider a graphical representation of the thermodynamic properties of a binary mixture (see Fig. 11.6). In the upper plot the Gibbs free energy of mixing is shown to have a slope that increases monotonically in mole fraction. On the other hand, for a partially miscible system the Gibbs free energy curve has points of inflection, so that a portion of the curve (dotted line) represents a metastable or unstable equilibrium state. The true stable-equilibrium Gibbs free energy is given by the tangent line (solid line), that clearly represents a condition of lower Gibbs free energy. Along this tangent line the system exists as two equilibrium phases of composition X_1^α and X_1^β, determined by the two points of tangency. From this diagram it can be seen that phase splitting occurs whenever the Gibbs free energy curve has a point of inflection, so that two points on the curve can be connected by a straight line that is lower in energy than a portion of the curve itself.

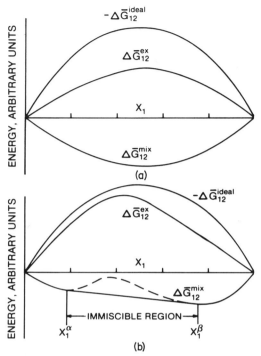

Fig. 11.6 Gibbs free energy functions for totally miscible (Fig. 11.6a) and partially miscible (Fig. 11.6b) binary liquid mixtures.

To illustrate this point, consider a binary mixture obeying

$$\Delta \bar{G}_{12}^{mix} = RT[X_1 \ln X_1 + X_2 \ln X_2] + X_1 X_2 A_{12} \tag{11.1}$$

Plugging in various values for the interaction parameter, it is found $A_{12}/RT \geq 2$ for phase separation to occur, as is indicated in Fig. 11.7.

Mathematically, the condition for partial immiscibility is

$$(\partial^2 \Delta \bar{G}^{mix}/\partial X^2) = 0 \tag{11.2}$$

An equivalent relationship is given in terms of activity a_i of one component

$$(\partial a_i/\partial X) = 0$$

Most partially immiscible systems, especially those interacting primarily by dispersion forces, become more miscible as the temperature increases. If the vapor–liquid critical point of either component is not exceeded, the two liquids eventually become completely soluble in each other. The point

11 Liquid–Liquid Equilibrium

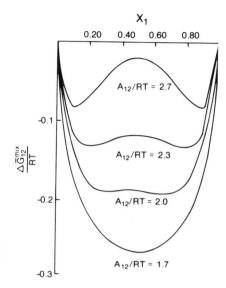

Fig. 11.7 Effect to the A_{12} term on the Gibbs free energy of mixing (and thus, liquid–liquid immiscibility) of binary mixtures.

at which the two phases merge is called the upper critical solution point, or consolute point. The most common type of behavior is that shown in Fig. 11.1, a T–X plot showing the single phase region and the two-phase region, separated by the co-existence curve, or locus of phase composition co-existing at equilibrium. The lower temperature limit of the co-existence is the UCST. The upper and lower plots in Fig. 11.6 represent systems above and below the UCST, respectively. The Gibbs free energy of mixing at the UCST is intermediate between the two previous conditions, exhibiting a region of zero curvature. The mathematical condition for the liquid–liquid critical point is

$$\partial^2 \Delta \bar{G}^{\text{mix}}/\partial X^2 = 0 \qquad (11.3a)$$

$$\partial^3 \Delta \bar{G}^{\text{mix}}/\partial X^3 = 0 \qquad (11.3b)$$

$$\partial^4 \Delta \bar{G}^{\text{mix}}/\partial X^4 > 0 \qquad (11.3c)$$

Thus, if a valid expression is available for the Gibbs free energy of mixing, it can be differentiated with respect to mole fraction and the result used in conjunction with Eqs. (11.3a–c) to determine the critical solution point. In addition, the entire coexistence curve may be determined from plots of the Gibbs free energy of mixing as a function of composition at various temperatures below the critical solution temperature.

A. Liquid–Liquid Equilibrium in Binary Systems

Returning to Eq. (11.1), one can see under what conditions partial miscibility occurs. Differentiation of Eq. (11.1) with respect to X_1 gives

$$\partial \Delta \bar{G}_{12}^{mix}/\partial X_1 = RT + RT \ln X_1 - RT$$
$$- RT \ln X_2 + X_2 A_{12} - X_1 A_{12} \qquad (11.4)$$

$$\partial^2 \Delta \bar{G}_{12}^{mix}/\partial X_1^2 = RT/X_1 + RT/X_2 - 2A_{12} \qquad (11.5)$$

Setting the second derivative equal to zero, one finds

$$2A_{12}/RT = 1/X_1 X_2 \qquad (11.6)$$

or that $A_{12}/RT \geq 2$ for phase separation to occur, because 4 is the smallest value that $1/(X_1 X_2)$ can take on. This agrees with earlier observations based on Fig. 7.

The UCST may be found from the Gibbs free energy of mixing from criteria defined by Eqs. (11.3a–c). The second and third derivatives of the Gibbs free energy of mixing with respect to composition are set equal to zero, and the simultaneous solution of these two equations yields the UCST for the binary mixture in terms of two unknowns: The temperature (T^c) and the composition (X_1^c) of the consolute point

$$X_1^c = 0.5$$
$$T^c = A_{12}/2R$$

The remainder of the phase boundary is determined from

$$RT \ln(X_1^\alpha/X_1^\beta) = [(X_2^\beta)^2 - (X_2^\alpha)^2] A_{12} \qquad (11.7)$$

$$RT \ln(X_2^\alpha/X_2^\beta) = [(X_1^\beta)^2 - (X_1^\alpha)^2] A_{12} \qquad (11.8)$$

with the restrictions that in each phase $X_1 + X_2 = 1$.

In general, the solution model just presented is more successful in predicting the shape of a T–X phase-boundary curve than it is in predicting the exact equilibrium compositions and UCST. An improved description of the liquid–liquid equilibrium can be obtained by replacing Eq. (11.1) with the Regular Solution model. Readers will have the opportunity to derive similar equations for the UCST based on the Regular Solution model in problem 11.1.

It should be stated that not all solution models will describe partially miscible liquid mixtures. Taking the second derivative of the Wilson equation

B. A Solute between Two Immiscible Solvents

$$\Delta \bar{G}_{12}^{mix} = RT[X_1 \ln X_1 + X_2 \ln X_2 - X_1 \ln(X_1 + X_2\Lambda_{12}) - X_2 \ln(X_2 + X_1\Lambda_{21})]$$

$$\frac{\partial^2 \Delta \bar{G}_{12}^{mix}}{\partial X_1^2} = \frac{RT}{X_1} + \frac{RT}{X_2} - \frac{2RT(1-\Lambda_{12})}{(X_1 + X_2\Lambda_{12})} + \frac{2RT(\Lambda_{21}-1)}{(X_2 + X_1\Lambda_{21})}$$

$$+ \frac{RTX_1(1-\Lambda_{12})^2}{(X_1 + X_2\Lambda_{12})^2} + \frac{RTX_2(\Lambda_{21}-1)^2}{(X_2 + X_1\Lambda_{21})^2}$$

$$= \frac{\Lambda_{12}^2 RT}{X_1(X_1 + X_2\Lambda_{12})^2} + \frac{\Lambda_{21}^2 RT}{X_2(X_2 + X_1\Lambda_{21})^2} \quad (11.9)$$

one discovers that no numerical values exist for the Λ_{ij} parameters that will satisfy $\partial^2 \Delta \bar{G}_{12}^{mix}/\partial X_1^2 = 0$. Despite this limitation, the Wilson equation continues to be one of the more popular solution models for describing vapor–liquid equilibrium. Its multicomponent form provides reasonable predictions of activity coefficients and excess Gibbs free energies for a large number of completely miscible liquid mixtures.

B. Distribution of a Solute between Two Immiscible Solvents

The distribution of a solute between two phases in which it is soluble was an important research area for many years. Historically this technique has been used in some form to isolate natural products, such as the essence of flowers.

The first systematic study of distribution between two immiscible liquids that led to a theory having predictive capabilities was performed by Berthelot and Jungfleisch (1872). These investigators accurately measured the equilibrium concentrations of I_2 and Br_2 distributed between water and carbon disulfide. In addition, they determined the concentrations of various organic acids, H_2SO_4, HCl, and NH_3 distributed between water and ethyl ether. From these early investigations came the first appreciation that the ratio of the solute concentrations distributed between two immiscible solvents was a constant and did not depend on the relative volumes of the two solvents.

It was further concluded from these early observations that there are a small variation in partition coefficient with temperature, with the more volatile solvent being favored by a temperature increase. It was also evident that some systems, notably succinic acid partitioned between water and ether, did not obey these simple rules, even in dilute solutions, but the authors intuitively felt the rules would be justified nevertheless.

Nernst (1891) made the next significant contribution to the subject. He stressed the fact that the partition coefficient would be constant only if a single molecular species were being considered as partitioned between the two phases. Considered as such, partitioning could be treated by classical thermodynamics as an equilibrium process where the tendency of any single molecular species of solute to leave one solvent and enter another would be a measure of its activity in that solvent and would be related in the usual fashion to the other commonly measured activity functions such as partial pressure, osmotic pressure, and chemical potential. As the primary example of a more exact expression of the *Partition law*, it was noted that benzoic acid distributed itself between benzene and water so that

$$(C_A^{\text{org}})^{1/2}/C_A^{\text{water}} = \text{constant}$$

where C_A^{organic} and C_A^{water} refer to the molar concentration of benzoic acid in benzene (chiefly in dimeric form) and water, respectively. The constant combines the partition coefficient for benzoic acid monomer and the dimerization constant for the acid in benzene. Since benzoic acid exists largely as a dimer in benzene at the concentrations employed, the monomer concentration in benzene is proportional to the square root of its total concentration in that solvent. Nernst was also aware that, at low concentrations, he would have to correct the aqueous phase concentration for the ionization of benzoic acid.

The association and dissociation of solutes in different phases remains the most perplexing problem in studying partitioning coefficients. For a true partition coefficient, one must consider the same species in each phase. A precise definition of this in the strictest sense is impossible as the water and solvent molecules form bonds of varying degrees of firmness with different solutes. Any system more complex than inert gases in hydrocarbons and water becomes impossible to define sharply at the molecular level. Limited attention was given to the fact that solutes other than carboxylic acids may carry one or more bound water molecules into the nonaqueous phase. This is possible in solvents such as sec-butyl alcohol, which on a molar basis contains more water molecules in the butanol phase than butanol molecules.

During the 1900s, numerous carefully planned partition experiments were reported, most of which were carried out with the objective of determining the ionization constant in an aqueous medium of moderately ionized acids and bases. As a point of historical fact, the method did not fully live up to its early promise, partly because of unexpected association in the organic solvents and partly because of solvent changes.

After reliable ionization constants became available through other means, partitioning measurements were used to calculate the association constants of organic acids in the nonaqueous phase as a function of temperature. This

B. A Solute between Two Immiscible Solvents

yielded values of ΔH, ΔS, and ΔG for the association reaction (Banewicz et al., 1957; Davies and Griffiths, 1954, 1955; Davies et al., 1951; Schrier et al., 1964). Any calculation of self-association constants from partition data, however, can be misleading when hydrate formation occurs (Fujii et al., 1978, 1981; Lassetre, 1937; Van Duyne et al. 1967).

Thermodynamically, the partitioning of a solute between two immiscible liquid phases requires equality of the solute's chemical potential (or fugacity) in both phases

$$f_i^\alpha = f_i^\beta \tag{11.10}$$

$$X_i^\alpha \gamma_i^\alpha (f_i^o)^\alpha = X_i^\beta \gamma_i^\beta (f_i^o)^\beta \tag{11.11}$$

where the f_i^os are the standard state fugacities. Adoption of the Raoult's law convention for the activity coefficients $(f_i^o)^\alpha = (f_i^o)^\beta = f_i^\bullet$ leads to the result that at equilibrium

$$X_i^\alpha \gamma_i^\alpha = X_i^\beta \gamma_i^\beta \tag{11.12}$$

Rearrangement of Eq. (11.12) gives the partition coefficient K_p^x

$$K_p^x = X_i^\alpha / X_i^\beta = \gamma_i^\beta / \gamma_i^\alpha \tag{11.13}$$

expressed in terms of mole fractions.

To derive the partition coefficients in molar concentration ratios, one makes the approximation that near infinite dilution each mole fraction is equal to the product of the molar concentration C_i and the molar volume $\bar{V}$ of each phase. Substitution into Eq. (11.11) gives

$$C_i^\alpha \gamma_i^\alpha \bar{V}^\alpha = C_i^\beta \gamma_i^\beta \bar{V}^\beta \tag{11.14}$$

$$K_p^c = K_p^\phi = C_i^\alpha / C_i^\beta = (\gamma_i^\beta \bar{V}^\beta / \gamma_i^\alpha \bar{V}^\alpha) \tag{11.15}$$

For dilute solutions this latter partition coefficient (Eq. 11.15) differs from the coefficient based on mole fractions by a factor equal to the ratio of the solvent molar volumes.

Alternatively, one could express the solute fugacities via Henry's law

$$f_i = X_i (k_H)_x$$

$$\gamma_i = 1$$

in which case the partition coefficient becomes

$$X_i^\alpha (k_H^\alpha)_x = X_i^\beta (k_H^\beta)_x \tag{11.16}$$

$$K_p^x = X_i^\alpha / X_i^\beta = (k_H^\beta)_x / (k_H^\alpha)_x \tag{11.17}$$

Either description is valid, as is using Raoult's law for one phase and Henry's law for the other.

Green and Frank (1979) presented an example of how the partitioning of benzene between water and cyclohexane (or carbon tetrachloride) can be used to measure the Henry's law constant of benzene in water. Briefly, the method involves bringing water layers into equilibrium with solutions of benzene in cyclohexane (or carbon tetrachloride) and measuring the mole fraction of benzene in both phases. For the nonaqueous phase the authors calculated the fugacity of benzene from the vapor pressure measurements of Scatchard et al. (1939, 1940). This fugacity was then equated to that in the aqueous phase

$$X_{Ben}^{aq}(k_H^{aq})_x = X_{Ben}^{\alpha}\gamma_{Ben}^{\alpha} f_{Ben}^{\bullet}$$

$$\alpha = \text{cyclohexane, carbon tetrachloride}$$

where Ben is Benzene. Henry's law constants determined in this manner are 187 atm (α = cyclohexane) and 184 atm (α = carbon tetrachloride) at 15°C. Values determined by direct vapor pressure measurements range between 188.5 to 193.3 atm (Alexander, 1959; Bohon and Claussan, 1951; Franks et al., 1963).

Liquid–liquid equilibrium data are sometimes used to estimate the excess Gibbs free energies of binary systems. To illustrate this calculation, consider the equilibrium compositions of the two liquid phases in the ternary cyclohexane (1) + benzene (2) + water (3) system as determined by Arich et al. (1975). The experimental results are tabulated in Table I. At equilibrium, the activity of benzene (cyclohexane) in the organic phase must equal its activity in the aqueous phase

$$X_{Ben}^{org}\gamma_{Ben}^{org} f_{Ben}^{\bullet} = X_{Ben}^{aq}\gamma_{Ben}^{aq} f_{Ben}^{\bullet}$$

$$X_{Cyclo}^{org}\gamma_{Cyclo}^{org} f_{Cyclo}^{\bullet} = X_{Cyclo}^{aq}\gamma_{Cyclo}^{aq} f_{Cyclo}^{\bullet}$$

Raoult's law is used for both phases. Whereas γ_{Ben}^{org} and γ_{Cyclo}^{org} refer to the water-saturated organic phase, one would expect that their numerical values should approximate the activity coefficients of the binary cyclohexane + benzene system because the solubility of water is on the order of $X_{aq}^{org} = 0.001$ to 0.003. Furthermore, the low solubility of the organic solutes in the aqueous phase suggests, to a first approximation, that the aqueous-phase activity coefficients are independent of organic-phase composition. Within these limitations, one estimates γ_{Ben}^{aq} and γ_{Cyclo}^{aq} from the solubility of *pure* benzene in water $X_{Ben}^{aq} = 4.03 \times 10^{-4}$ (Amidon and Anik, 1981), water in *pure* benzene $X_W^{ben} = 3.10 \times 10^{-3}$ (Goldman and Krishnan, 1976), *pure* cyclohexane in water $X_{Cyclo}^{aq} = 1.18 \times 10^{-5}$ (McAuliffe, 1966) and water in *pure* cyclohexane $X_W^{cyclo} = 3.34 \times 10^{-4}$ (Goldman and Krishnan, 1976):

$$\gamma_{Ben}^{aq} = (1.00000 - 0.00310/4.10 \times 10^{-4}) = 2.47 \times 10^3$$

$$\gamma_{Cyclo}^{aq} = (1.00000 - 0.000334/1.18 \times 10^{-5}) = 8.44 \times 10^4$$

B. A Solute between Two Immiscible Solvents

TABLE I
Concentrations of the Two Liquid Phases at Equilibrium for the Ternary Cyclohexane (1) + Benzene (2) + Water (3) System at 25°C[a]

Organic Phase			Aqueous Phase		
X_1	X_2	X_3	X_1	X_2	X_3
0.1114140	0.8860360	0.0025500	0.0000020	0.0003672	0.9996308
0.1845540	0.8130360	0.0024100	0.0000031	0.0003417	0.9996566
0.2624050	0.7353350	0.0022600	0.0000041	0.0003153	0.9996806
0.3792240	0.6187360	0.0020400	0.0000054	0.0002766	0.9997180
0.4591440	0.5389960	0.0018600	0.0000062	0.0002497	0.9997441
0.5540840	0.4442660	0.0016500	0.0000072	0.0002161	0.9997767
0.6440710	0.3544890	0.0014400	0.0000080	0.0001817	0.9998103
0.7391120	0.2596880	0.0012000	0.0000090	0.0001414	0.9998496
0.8191800	0.1798200	0.0010000	0.0000098	0.0001035	0.9998867

[a] Reproduced with permission from Arich et al., (1975). Copyright Pergamon, Oxford.

These values, when combined with the ternary liquid–liquid equilibrium data given in Table II permit $\gamma_{\text{Ben}}^{\text{org}}$ and $\gamma_{\text{Cyclo}}^{\text{org}}$ to be calculated. As one can see from Fig. 11.8, the liquid–liquid equilibrium-based excess Gibbs free energies compare favorably with values determined from vapor pressure measurements.

TABLE II
Comparison between the Experimental and Predicted Partition Coefficients of Copper Acetylacetonate in the Benzene (1) + n-Heptane (2) Mixed Solvent[a]

X_1	$K_{P(BC)}^{\phi}$ Experimental	$K_{P(BC)}^{\phi}$ Eq. (11.24)[b]	$K_{P(BC)}^{\phi}$ Eq. (11.24)[c]
0.00	0.13	—	—
0.15	0.25	0.21	0.23
0.30	0.50	0.36	0.39
0.41	0.83	0.55	0.59
0.52	1.32	0.85	0.89
0.62	1.95	1.28	1.33
0.71	2.82	1.90	1.94
0.79	3.72	2.70	2.75
0.87	4.79	3.89	3.91
0.94	6.03	5.36	5.38
1.00	7.08	—	—

[a] Partition coefficients were determined by Koshimura, (1977).
[b] The $\Delta \bar{G}_{BC}^{fh}$ was taken from Harris and Dunlap (1970).
[c] The $\Delta \bar{G}_{BC}^{fh}$ was taken from Jain et al., (1973).

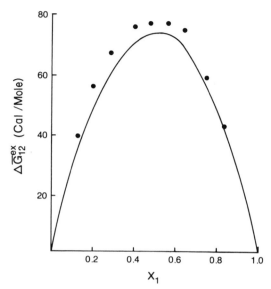

Fig. 11.8 Comparison between the excess Gibbs free energies calculated from liquid–liquid equilibria data (●) and vapor pressure measurements (— —) for the cyclohexane (1) + benzene (2) system at 25°C.

Methods based on liquid–liquid equilibrium have two distinct advantages over classical vapor–liquid equilibrium techniques:

(a) all measurements are performed at constant temperature and pressure, and

(b) it is easier to attain equilibrium between two liquids than between vapor and liquid.

One encounters experimental difficulties as extremely low solubilities are required. Hence, liquid–liquid equilibrium methods cannot replace conventional vapor–liquid equilibrium methods. It can provide, however, an alternative method for determining activity coefficients, and allows measurements to be made when vapor–liquid equilibrium is difficult to realize.

C. Partitioning of a Solute between a Binary Organic Phase and Water

Of the many attempts to introduce basic concepts into the treatment of partition coefficients, the most general approach is based on the NIBS model. According to Eqs. (10.42) and (10.43), the chemical potential of a solute is

C. A Solute in a Binary Organic Phase and Water

given by

$$\mu_A - \mu_A^\bullet = RT \ln X_A + \bar{V}_A \phi_{Solv}^2 A_{ASolv} \qquad (11.18)$$

or in the case of molecules of different sizes

$$\mu_A - \mu_A^\bullet = RT[\ln \phi_A + \phi_{Solv}(1 - \bar{V}_A/\bar{V}_{Solv})] + \bar{V}_A \phi_{Solv}^2 A_{ASolv} \qquad (11.19)$$

When an inert solute ($\phi_A \approx 0$) is partitioned between a binary organic phase and an aqueous phase, Eq. (11.19) can be used to express the solute's chemical potential in the organic phase (solvent components B and C)

$$\mu_A^{org} - \mu_A^\bullet = RT[\ln \phi_A^{org} + 1 - \bar{V}_A/(X_B^o \bar{V}_B + X_C^o \bar{V}_C)]$$
$$+ \bar{V}_A \phi_B^o A_{BC} + \bar{V}_A \phi_C^o A_{AC} - \bar{V}_A \phi_B^o \phi_C^o A_{BC} \qquad (11.20)$$

and in the aqueous phase

$$\mu_A^{aq} - \mu_A^\bullet = RT[\ln \phi_A^{aq} + 1 - \bar{V}_A/\bar{V}_{aq}] + \bar{V}_A A_{AWa} \qquad (11.21)$$

provided the two phases are completely immiscible with each other.
The partition coefficient $K_p^\phi = C_A^{org}/C_A^{aq} = \phi_A^{org}/\phi_A^{aq}$, is

$$-RT \ln K_p^\phi = RT \ln(\phi_A^{aq}/\phi_A^{org})$$
$$= RT[\bar{V}_A/\bar{V}_{aq} - \bar{V}_A/(X_B^o \bar{V}_B + X_C^o \bar{V}_C)]$$
$$+ \phi_B^o \bar{V}_A A_{AB} + \phi_C^o \bar{V}_A A_{AC} - \phi_B^o \phi_C^o \bar{V}_A A_{BC} - \bar{V}_A A_{AWa} \qquad (11.22)$$

obtained by combining Eqs. (11.20), (11.21) and the equilibrium requirement $\mu_A^{org} = \mu_A^{aq}$.

The predictive application of Eq. (11.22) can take several different forms. For example, a solute's partition coefficient can be predicted by approximating the various A_{ij} parameters with solubility parameters $A_{ij} = (\delta_i - \delta_j)^2$ or in the case of solid solutes, by evaluating interaction parameters from experimental solubilities

$$RT[\ln(a_A^{solid}/\phi_A^{sat}) - (1 - \phi_A^{sat})(1 - \bar{V}_A/\bar{V}_{Solv})] = (1 - \phi_A^{sat})^2 \bar{V}_A A_{ASolv} \qquad (11.23)$$

Incorporation of the pure solvent partition coefficients $K_{p(B)}^\phi$ and $K_{p(C)}^\phi$ into Eq. (11.22) leads to

$$\ln K_{p(BC)} = \phi_B^o \ln K_{p(B)} + \phi_C^o \ln K_{p(C)} + \bar{V}_A \Delta \bar{G}_{BC}^{fh}/(X_B^o \bar{V}_B + X_C^o \bar{V}_C)RT \qquad (11.24)$$

a third predictive method if one's interest is in binary organic mixtures.

In Table II is a comparison of the predictions of Eq. (11.24) to the experimental partition coefficients of copper acetylacetonate between water and binary benzene + n-heptane mixtures. The relatively poor agreement between the predicted and experimental values is not surprising as the solubility data (see problem 10.9) suggested the existence of specific interactions between copper acetylacetonate and benzene. From earlier discussions of the NIBS model, recall that the solute's chemical potential in systems having solute–solvent complexation

$$A + C \rightleftharpoons AC$$

$$K^{\phi}_{AC} = \hat{\phi}_{AC}/\hat{\phi}_{A_1}\hat{\phi}_{C_1}$$

is given by

$$\mu_A^{org} - \mu_A^{\bullet} = RT[\ln \hat{\phi}_{A_1} + 1 - \bar{V}_A/\bar{V}_{Sol}]$$
$$+ \bar{V}_A(1 - \phi_A^{sat})^2[\phi_B^o A_{A_1B} + \phi_C^o A_{A_1C_1} + \phi_B^o \phi_C^o A_{BC_1}] \quad (11.25)$$

Combination of Eqs. (11.21) and (11.25) enables the development of a calculational method for the solute–solvent equilibrium constant

$$\ln K^{\phi}_{p(BC)} = \phi_B^o \ln K^{\phi}_{p(B)} + \phi_C^o \ln K^{\phi}_{p(C)} + \ln[1 + \bar{V}_A K^{\phi}_{AC}\phi_C^o/(\bar{V}_A + \bar{V}_C)]$$
$$- \phi_C^o \ln[1 + \bar{V}_A K^{\phi}_{AC}/(\bar{V}_A + \bar{V}_C)]$$
$$+ [\bar{V}_A \Delta \bar{G}^{fh}_{BC}/(X_B^o \bar{V}_B + X_C^o \bar{V}_C)RT] \quad (11.26)$$

based on experimentally determined partition coefficients. Using the experimental value of $K^{\phi}_{p(BC)} = 1.95$ at $X_C = 0.62$, one calculates a value of $K^{\phi}_{AC} = 15.6$ or 15.7, depending upon which literature source is used for $\Delta \bar{G}^{fh}_{BC}$. It is unfortunate that the large solute molecule ($\bar{V}_A = 193$ cm^3/mol) magnifies the unmixing term, as conflicting values of $\Delta \bar{G}^{fh}_{BC}$ have been reported. Discrepancies in literature values, however, are a problem often encountered whenever the predictive method requires a prior knowledge of binary data.

There is, unfortunately, one problem inherent in the use of partition, and this problem arises because all solvents have some mutual solubility. Thus, the solvent in the aqueous phase is not pure water, but water saturated with the organic solvent, and similarly the solvent in the organic phase is the water-saturated organic solvent. Therefore, in principle at least, thermodynamic data obtained by partition methods may differ from data obtained by other methods. One should always be aware of this whenever interpreting partition coefficients at the molecular level.

Problems

11.1 Assuming that the Gibbs free energy of mixing obeys

$$\Delta \bar{G}^{mix}_{12} = RT[X_1 \ln X_1 + X_2 \ln X_2] + (X_1 \bar{V}_1 + X_2 \bar{V}_2)\phi_1 \phi_2 A_{12}$$

show that the UCST is defined by

$$X_1^c = [\bar{V}_1 - (\bar{V}_1^2 + \bar{V}_2^2 - \bar{V}_1 \bar{V}_2)^{1/2}]/(\bar{V}_1 - \bar{V}_2)$$
$$T^2 = [2X_1^c X_2^c \bar{V}_1^2 \bar{V}_2^2/R(X_1^c \bar{V}_1 + X_2^c \bar{V}_2)^3]A_{12}$$

11.2 The distribution coefficient for iodine between carbon tetrachloride and water, $K_p^c = C_{I_2}^{CCl_4}/C_{I_2}^{H_2O}$, is 85. Calculate:
(a) The number of moles of iodine remaining in 100 ml of an aqueous solution that was originally 0.01 M after extraction with two 50-ml portions of CCl_4.
(b) The number of moles of I_2 remaining if the aqueous solution in part (a) had been extracted with a single 100-ml portion of CCl_4 rather than the two 50-ml portions.
(c) The number of moles that would remain if the aqueous solution were shaken up successively with $100/s$-ml portions of CCl_4 and the procedure repeated s times so that in all, 100 ml of carbon tetrachloride was used. What is the limit of this expression as $s \to \infty$?

11.3 A solid, such as iodine, is distributed between two immiscible liquids such as water and carbon disulfide. Prove that if one phase is saturated with iodine the other phase must also be saturated. (Hint: No particular form of solution law need to be assumed for this proof.)

11.4 In the determination of the aqueous solubility of a single organic substance, an initially pure (neat) organic liquid is placed in direct contact with distilled water. Ethylbenzene and propylbenzene have *neat* molar solubilities of 0.00162 mol/liter and 0.000392 mol/liter, respectively. If an equimolar mixture of ethylbenzene and propylbenzene were allowed to equilibrate in direct contact with water, would the solubility of ethylbenzene still be 0.00162 mol/liter? Develop expressions for predicting how the solubility of ethylbenzene and propylbenzene varies with organic phase composition. Be sure to identify all approximations. Use the experimental solubilities of Tewari *et al.*, listed in Table III to test the predictive expressions.

TABLE III
Aqueous Solubilities of Mixtures of
n-Propylbenzene (1) + Ethylbenzene (2) at 25°C[a]

	Molar Solubilities (mol/l)	
ϕ_2	n-Propylbenzene	Ethylbenzene
0.091	0.000348	0.000176
0.250	0.000299	0.000459
0.500	0.000184	0.000879
0.750	0.0000869	0.00129
0.909	0.0000307	0.00149

[a] Reproduced with permission from Tewari, *et al.*, 1982.

11.5 In designing extraction processes, one needs to know the partition coefficient of the solute between two immiscible liquids. Many times it is impossible to find the desired values. Suppose one needs the partition coefficients of benzene distributed between water and alkanes (n-hexane, n-heptane, n-pentadecane, n-tetradecane, cyclohexane, and methylcyclohexane). Searching the chemical literature, one could find only the partition coefficient of benzene between water and n-heptane, $K_P^c = 195$. Estimate the remaining partition coefficients using the experimental excess Gibbs free energies of benzene + alkane mixtures. Information needed for the calculations include

Methylcyclohexane (1) + Benzene (2)
$$\Delta \bar{G}_{12}^{ex}/RT = X_1 X_2 [0.5053 - 0.1311(2X_1 - 1) + 0.0217(2X_1 - 1^2)]$$

Cyclohexane (1) + Benzene (2)
$$\Delta \bar{G}_{12}^{ex}/RT = X_1 X_2 [0.53045 - 0.05819(X_1 - X_2) + 0.06222(X_1 - X_2)^2]$$

n-Hexane (1) + Benzene (2)
$$\Delta \bar{G}_{12}^{ex}/RT = X_1 X_2 [0.61947 - 0.09652(X_1 - X_2) + 0.04754(X_1 - X_2)^2]$$

n-Heptane (1) + Benzene (2)
$$\Delta \bar{G}_{12}^{ex}/RT = X_1 X_2 [0.6945 - 0.0304(X_1 - X_2) + 0.1066(X_1 - X_2)^2]$$

Benzene (1) + n-Pentadecane (2)
$$\Delta \bar{G}_{12}^{ex}/RT = X_1 X_2 [0.1947 + 0.1744(2X_1 - 1) + 0.1751(2X_1 - 1)^2 + 0.1506(2X_1 - 1)^3]$$

Benzene (1) + n-Tetradecane (2)
$$\Delta \bar{G}_{12}^{ex}/RT = X_1 X_2 [0.3166 + 0.2735(2X_1 - 1) + 0.1765(2X_1 - 1)^2]$$

Hint: $RT \ln \gamma_i^{\infty} = \lim_{X_i \to 0} (\Delta \bar{G}_{ij}^{ex}/X_i X_j)$

12
Physio-Chemical Applications of Gas–Liquid Chromatography to Nonelectrolyte Solutions

Conventional methods for studying solution nonideality, which depend primarily on measurement of vapor pressure, suffer from the disadvantage that the relative experimental error of measurement at low concentrations is usually much greater than at higher concentrations. Furthermore, conventional static vapor pressure measurements on systems having a volatile solute + nonvolatile solvent are usually time consuming and require both solute and solvent in a state of high purity.

Gas–liquid chromatography, on the other hand, is a rapid method that requires only a pure solvent and not necessarily a pure solute. The accuracy obtainable at high dilution is generally superior to the accuracy of other experimental methods for mole fraction concentrations below 0.1 of solute.

The calculation of the activity coefficient from the retention behavior requires a knowledge of the solute's vapor pressure. Incorporation of gas-phase nonideality into our measurements also would require knowledge of the second virial coefficient of the pure solute and the mixed second virial coefficient of solute + carrier gas. These coefficients can usually be estimated with adequate accuracy or, if need arises, mixed second virial coefficient can be evaluated from gas chromatographic measurements over a wide range of pressures.

Gas–liquid chromatography should be regarded as an additional experimental technique for determining the activity coefficients of liquid mixtures and mixed virial coefficients of gaseous mixtures. It is by no means the only experimental technique available. From the aspect of solution theory modeling it is advantageous to have a detailed thermodynamic description of the system involving several different basic types of experimental methods, such as vapor pressure measurement, calorimetry, dilatometry and compressibility. Gas–liquid chromatography is essentially a technique of vapor pressure measurement.

Finally, before considering the various types of systems that have been investigated via chromatography, it is desirable to point out that thermodynamics is a collection of useful mathematical relationships between independently observable quantities. Thermodynamics, per se, cannot tell us anything about the microscopic origin of macroscopic properties. In order to interpret and predict the retention behavior of a solute one needs a thermodynamic description of the chromatographic system relating the experimentally determined properties (i.e., chromatographic partition coefficients, retention volumes and retention times) to the thermodynamic quantities used in solution studies, and a coherent theory that relates the properties of a microscopic model to the macroscopic thermodynamic quantities.

A. The Nearly Ideal Binary Solvent Theory: Gas–Liquid Partition Coefficients in Noncomplexing Systems

Gas–liquid chromatographic studies provide another method for determining partial molar thermodynamic properties of a solute near infinite dilution and investigating association complexes between the solute and of of the solvent components. The experimental data, expressed in terms of partition coefficients, can be employed to test modern solution theories, including NIBS theory. In their analysis of literature data for the infinite dilution partition coefficients of a solute (component A) distributed between a binary liquid phase (components B and C) and the vapor phase, Purnell and coworkers (Laub and Purnell, 1975, 1976a,b; Laub et al., 1977; Purnell and Vargas de Andrade, 1975) noted that the solute retention behavior could be described by a simple volume fraction average of the partition coefficients on the pure solvents ($K_{R(B)}$ and $K_{R(C)}$)

$$K_{R(BC)} = \phi_B^o K_{R(B)} + \phi_C^o K_{R(C)} \tag{12.1}$$

irrespective of the complexing nature of the system. They suggested that this linear relationship may imply "the prospect of a coherent theory of solutions of a generality not hitherto visualized (Laub et al., 1978). A slightly different form of Eq. (12.1)

$$K_{R(BC)} = K_{R(BC)}[1 + K_{AC} C_C^o] \tag{12.2}$$

$$K_{AC} = [K_{R(C)} - K_{R(B)}] \bar{V}_C / K_{R(B)} \tag{12.3}$$

is commonly used to estimate the solute–solvent equilibrium constant K_{AC} from chromatographic measurements. Despite the fact that Eq. (12.1) provides a reasonable mathematical representation of several hundred ternary

A. NIBS in Noncomplexing Systems

systems, it cannot be rigorously derived from conventional nonelectrolyte solution models such as the NIBS model discussed in Chapter 10. The expression can be readily derived, though, on the basis that the two liquid phases "exist in their macroscopic solutions as microscopically immiscible groups of like molecules" (Laub and Purnell, 1976a). Notice that this statement is completely different from the underlying assumptions of the two-liquid model used in the development of the Wilson, NRTL, and UNIQUAC equations.

Gas–liquid partition coefficients can be related to other thermodynamic properties through the definition (Conder and Young, 1979; Laub and Pecsok, 1978)

$$K_R = (RTQ/P_A^\bullet \gamma_A^\infty \bar{V}_{\text{solv}}) \tag{12.4}$$

in which $P_A^\bullet$ is the equilibrium vapor pressure of the solute at the specified temperature, γ_A^∞ is the activity coefficient (relative to Raoult's law) of the solute at infinite dilution (denoted by the superscript ∞), Q is a temperature-dependent term involving the molar volume of the pure liquid solute and the second virial coefficients of the solute and carrier gas, and solv stands for solvent. Using this relationship, Purnell and Vargas de Andrade (1975) showed that Eq. (12.1) is equivalent to

$$1/\gamma_A^\infty = X_B^o/(\gamma_A)_{X_B=1}^\infty + X_C^o/(\gamma_A)_{X_C=1}^\infty \tag{12.5}$$

with only the approximation that the excess molar volume of the solvent mixture is negligible. The nongenerality of Eq. (12.5) can be demonstrated by consideration of solubility in binary solvent mixtures.

The activity coefficient of the solute can be related to the solubility of a solid through the relationship

$$a_A^{\text{solid}} = X_A^{\text{sat}} \gamma_A^{\text{sat}} \tag{12.6}$$

in which the activity of the solid depends only on temperature and is determined relative to the pure supercooled liquid. If the solubility is sufficiently small, the activity coefficient at infinite dilution may be approximated directly as the activity coefficient at saturation, or through the approximation

$$\ln \gamma_A^{\text{sat}} = (1 - X_A^{\text{sat}})^2 \ln \gamma_A^\infty \tag{12.7}$$

Combination of Eqs. (12.5–12.7) gives

$$X_A^{\text{sat}} = X_B^o (X_A^{\text{sat}})_{X_B=1} + X_C^o (X_A^{\text{sat}})_{X_C=1} \tag{10.36}$$

which is identical to an equation developed from the stoichiometric complexation model of Higuchi *et al.* Experimental solubilities of benzil and *p*-benzoquinone exhibit considerable deviations from Eq. (10.36), with several deviations as large as 100%. Whereas it might be argued by some that the failure of Eq. (10.36) in these two instances is a result of complexation between the solutes and carbon tetrachloride or is a result of departure from infinite dilution, it should be noted that the NIBS model described the benzil and *p*-benzoquinone solubilities to within a maximum deviation of 6% without introducing a single equilibrium constant. Furthermore, Purnell and coworkers observed that Eq. (12.1), from which Eq. (10.36) was just derived, applied to complexing and noncomplexing systems.

Expression for estimating the partition coefficients of the solute in binary solvent mixtures can be developed through either of the five predictive equations of the NIBS theory. The nature of gas–liquid chromatographic studies, of small solute and large solvent molecules, however, suggests the Flory–Huggins expression might be more appropriate for calculating thermodynamic excess properties. Using the definitions

$$(\Delta \bar{G}_A^{fh})^\infty = RT \ln (\gamma_A^{fh}) \tag{12.8}$$

$$\ln \gamma_A^{fh} = \ln(a_A/\phi_A) - (1 - \phi_A)[1 - \bar{V}_A(X_B^o \bar{V}_B + X_C^o \bar{V}_C)^{-1}] \tag{12.9}$$

$$(\gamma_A^{fh})^\infty = \lim_{X_A \to 0} \gamma_A^{fh} \tag{12.10}$$

the equation for the properties of a solute at infinite dilution in noncomplexing binary solvent mixtures (Eq. (VV), page 218) takes the form

$$RT \ln(\gamma_A^{fh})^\infty = RT[\phi_B^o \ln(\gamma_A^{fh})^\infty_{X_B = 1} + \phi_C^o \ln(\gamma_A^{fh})^\infty_{X_C = 1}]$$
$$- \bar{V}_A(X_B^o \bar{V}_B + X_C^o \bar{V}_C)^{-1} \Delta \bar{G}_{BC}^{fh} \tag{12.11}$$

in which $\Delta \bar{G}_{BC}^{fh}$ represents the excess Gibbs free energy of the binary solvent pair calculated over the Flory–Huggins predictions, and the mole fraction and volume fraction compositions are now calculated as if the solute were not present. The ordinary activity coefficient of the solute γ_A^∞ is related to $(\gamma_A^{fh})^\infty$ by

$$\gamma_A^\infty = (\gamma_A^{fh})^\infty \bar{V}_A(X_B^o \bar{V}_B + X_C^o \bar{V}_C)^{-1} \exp[1 - \bar{V}_A(X_B^o \bar{V}_B + X_C^o \bar{V}_C)^{-1}] \tag{12.12}$$

and the combination of Eqs. (12.4), (12.11), and (12.12) yields

$$RT \ln K_{R(BC)} = RT[\phi_B^o \ln K_{R(B)} + \phi_C^o \ln K_{R(C)}]$$
$$+ \bar{V}_A(X_B^o \bar{V}_B + X_C^o \bar{V}_C)^{-1} \Delta \bar{G}_{BC}^{fh} \tag{12.13}$$

where $K_{R(i)}$ is the solute's partition coefficient in solvent *i*.

Application of Eq. (12.13) in predicting the partition coefficients in mixed solvents requires knowledge of the thermodynamic excess properties of the binary solvent, which for many systems may not be readily available. If the molar volume of the solute is small in comparison to the molar volumes of the solvent molecules (as is the usual case in gas–liquid chromatographic studies) and/or the excess free energy of the solvent pair is small, the last term

B. The Kretschmer–Wiebe Association Model

on the right-hand side of Eq. (12.13) becomes negligible, giving

$$\ln K_{R(BC)} = \phi_B^o \ln K_{R(B)} + \phi_C^o \ln K_{R(C)} \qquad (12.14)$$

In general, Eq. (12.14) predicts curvature in plots of $K_{R(BC)}$ vs. ϕ_B^o, the extent of curvature depends on the ratios of $K_{R(C)}/K_{R(C)}$ and the magnitude of the contribution due to the *unmixing* of the solvent pair by the presence of the solute. For example, with ratios of $K_{R(B)}/K_{R(C)}$ near unity and $\Delta \bar{G}_{BC}^{fh} = 0$, the values of $K_{R(BC)}$ calculated using Eq. (12.13) are almost identical to the values generated by Eq. (12.1), which has been shown to be applicable to a large number of ternary systems.

B. The Kretschmer–Wiebe Association Model: Gas–Liquid Partition Coefficients of Alcohol Solutes on Binary Solvent Mixtures of Inert Hydrocarbons

Additional support for a logarithmic relationship between partition coefficient and solvent composition can be found by looking at the chromatographic behavior of alcohol solutes on stationary phases containing two saturated (inert) hydrocarbons. Using the Kretschmer–Wiebe association model coupled with the Scatchard–Hildebrand model for nonspecific interactions, the excess Gibbs free energy of a ternary alcohol plus two inert hydrocarbon mixture can be written as

$$\Delta \bar{G}_{ABC}^{ex} = RT[X_A \ln(\hat{\phi}_{A_1}/\hat{\phi}_{A_1}^* X_A) + X_B \ln(\phi_B/X_B)$$
$$+ X_C \ln(\phi_C/X_C) + X_A K_A(\hat{\phi}_{A_1}/\hat{\phi}_{A_1}^*)]$$
$$+ (X_A \bar{V}_A + X_B \bar{V}_B + X_C \bar{V}_C)[\phi_A \phi_B A_{A_1B} + \phi_A \phi_C A_{A_1C} + \phi_B \phi_C A_{BC}]$$

with the activity coefficient of the alcohol solute being given by

$$RT \ln \gamma_A = RT[\ln(\hat{\phi}_{A_1}/\hat{\phi}_{A_1}^* X_A) + (1 - K_A \hat{\phi}_{A_1})(\phi_B + \phi_C)$$
$$+ K_A(\hat{\phi}_{A_1} - \hat{\phi}_{A_1}^*) - (\bar{V}_A \phi_B/\bar{V}_B) - (\bar{V}_A \phi_C/\bar{V}_C)]$$
$$+ \bar{V}_A[\phi_B(1 - \phi_A)A_{A_1B} + \phi_C(1 - \phi_A)A_{A_1C} - \phi_B \phi_C A_{BC}] \qquad (12.15)$$

Because chromatographic partition coefficients are related to γ_A^∞ rather than γ_A, the limit of Eq. (12.15) as $n_A \to 0$ needs to be examined.

$$RT \ln \gamma_A^\infty = RT \left[\ln \frac{\bar{V}_A}{(X_B^o \bar{V}_B + X_C^o \bar{V}_C)\hat{\phi}_{A_1}^*} + 1 - K_A \hat{\phi}_{A_1}^* - \frac{\bar{V}_A \phi_B^o}{\bar{V}_B} - \frac{\bar{V}_A \phi_C^o}{\bar{V}_C} \right]$$
$$+ \bar{V}_A[\phi_B^o A_{A_1B} + \phi_C^o A_{A_1C} - \phi_B^o \phi_C^o A_{BC}] \qquad (12.16)$$

The A_{A_1B} and A_{A_1C} parameters can be removed via the infinite dilution activity coefficients in the pure solvents $[(\gamma_A)_{X_B=1}^\infty$ and $(\gamma_A)_{X_C=1}^\infty]$

$$RT \ln \gamma_A^\infty = RT\left[\phi_B^\circ \ln(\gamma_A)_{X_B=1}^\infty + \phi_C^\circ \ln(\gamma_A)_{X_C=1}^\infty + \ln \frac{(\bar{V}_B)^{\phi_B^\circ}(\bar{V}_C)^{\phi_C^\circ}}{(X_B^\circ \bar{V}_B + X_C^\circ \bar{V}_C)}\right]$$
$$- \bar{V}_A \phi_B^\circ \phi_C^\circ A_{BC} \quad (12.17)$$

Expressed in terms of partition coefficients, Eq. (12.17) becomes

$$RT \ln K_{R(BC)} = RT[\phi_B^\circ \ln K_{R(B)} + \phi_C^\circ \ln K_{R(C)}] + \bar{V}_A \phi_B^\circ \phi_C^\circ A_{BC} \quad (12.18)$$

which is identical to Eq. (12.13) whenever the Gibbs free energy of the binary solvent pair

$$\Delta \bar{G}_{12}^{fh} = (X_B^\circ \bar{V}_B + X_C^\circ \bar{V}_C)\phi_B^\circ \phi_C^\circ A_{BC}$$

is used to replace the A_{BC} interaction parameter.

Although this discussion has been limited to the Kretschmer–Wiebe association model, it can be shown that Eq. (12.13) is derivable from several other self-association models provided that the physical interactions are described by the Scatchard–Hildebrand model.

C. The Nearly Ideal Binary Solvent Theory: Gas–Liquid Partition Coefficients in Systems Containing Solute–Solvent Complexation

The majority of gas–liquid chromatographic studies undertaken during recent years have involved binary solvent systems where complexation occurs between the solute and one of the solvent components

$$A + C \rightleftharpoons AC$$

The Flory–Huggins mixing model leads to an equilibrium constant of the form

$$K_a = [\hat{\phi}_{AC}(\gamma_{AC}^{fh})^\infty / \hat{\phi}_{A_1} \hat{\phi}_C (\gamma_A^{fh})^\infty (\gamma_C^{fh})^\infty] \exp[-1 - (\Delta \bar{V}/\bar{V}_{BC})] \quad (12.19)$$

in which $(\gamma_A^{fh})^\infty$ and $\hat{\phi}_{A_1}$ refer to the *uncomplexed* solute and $\Delta \bar{V}$ is the volume change for the formation of one mole of the complex, which should be negligible compared to the molar volumes of the solute and solvent molecules. If the ratio of activity coefficients in Eq. (12.19) is assumed to be constant in a given solvent pair, then an operational equilibrium constant K'_{AC} can be defined in terms of the moles of complex and uncomplexed solute $(\hat{n}_{AC}, \hat{n}_{A_1})$

$$K'_{AC} = \hat{n}_{AC}/(\hat{n}_{A_1}\phi_C) = [\bar{V}_A(\gamma_A^{fh})^\infty(\gamma_C^{fh})^\infty / \bar{V}_{AC}(\gamma^{fh})^\infty]K_a \quad (12.20)$$

C. NIBS in Solute–Solvent Complexation

and the fraction of the solute molecules that remain uncomplexed is given by

$$f_A = \hat{n}_{A_1}/(\hat{n}_{A_1} + \hat{n}_{AC}) = (1 + K'_{AC}\phi^o_C)^{-1} \quad (12.21)$$

The apparent activity coefficient for the solute $(\gamma^{fh}_A)'$ is then related to the activity coefficient of the uncomplexed solute in the binary solvent by

$$(\gamma^{fh}_A)^\infty = (\gamma^{fh}_A)'(1 + K'_{AC}\phi^o_C) \quad (12.22)$$

and in the pure complexing solvent by

$$(\gamma^{fh}_A)^\infty_{X_C=1} = (\gamma^{fh}_A)'_{X_C=1}(1 + K'_{AC}) \quad (12.23)$$

Substitution of Eqs. (12.22) and (12.23) into Eq. (12.11) gives an expression for the apparent activity coefficient of the solute in a binary solvent containing both an inert and complexing solvent molecule

$$RT \ln(\gamma^{fh}_A)' = RT[\phi^o_B \ln(\gamma^{fh}_A)_{X_B=1} + \phi^o_C \ln(\gamma^{fh}_A)'_{X_C=1} - \ln(1 + K'_{AC}\phi^o_C)]$$
$$- \bar{V}_A(X^o_B \bar{V}_B + X^o_C \bar{V}_C)^{-1} \Delta \bar{G}^{fh}_{BC} \quad (12.24)$$

Combination of Eqs. (12.4), (12.12), and (12.24) gives

$$RT \ln K_{R(BC)} = RT[\phi^o_B \ln K_{R(B)} + \phi^o_C \ln K'_{R(C)} + \ln(1 + K'_{AC}\phi^o_C)]$$
$$+ \bar{V}_A(X^o_B \bar{V}_B + X^o_C \bar{V}_C)^{-1} \Delta \bar{G}^{fh}_{BC} \quad (12.25)$$

in which $K'_{R(C)}$ is the partition coefficient of the *uncomplexed* solute in the pure complexing solvent C. Examination of Eq. (12.25) reveals that it reduces to a linear form

$$K_{R(BC)} = K_{R(B)}[1 + K'_{AC}\phi^o_C] \quad (12.26)$$

whenever $K_{R(B)}/K'_{R(C)}$ is unity and the $\Delta \bar{G}^{fh}_{BC}$ equals zero.

For complexing systems, in terms of the partition coefficients in the two pure solvents, Eq. (12.26) takes the one parameter (K'_{AC}) form

$$RT \ln K_{R(BC)} = RT[\phi^o_B \ln K_{R(B)} + \phi^o_C \ln K_{R(C)} + \ln(1 + K'_{AC}\phi^o_C)$$
$$- \phi^o_C \ln(1 + K'_{AC})]$$
$$+ \bar{V}_A(X^o_B \bar{V}_B + X^o_C \bar{V}_C)^{-1} \Delta \bar{G}^{fh}_{BC} \quad (12.27)$$

In many of the binary solvent systems used in chromatographic studies the low volatility of the solvent components prevents the direct experimental determination of the excess Gibbs free energy. The magnitude of this term, though, may be assessed through Eq. (12.17) for a solvent pair by measurement of partition coefficients of saturated hydrocarbons. Because these compounds are generally considered incapable of complex formation ($K'_{AC} = 0$), deviations from Eq. (12.18) can be attributed to the unmixing term in Eq. (12.17). This unmixing term involves only the molar volume of the solute and the properties of the solvent mixture, thus allowing evaluation of

TABLE I

Solute Partition Coefficients at 30°C on Squalane + Dinonyl Phthalate Mixtures[a]

Solute	Partition coefficient (K_R)		
	$\phi^\circ_{DNP} = 0.000$	0.452	1.000
n-Pentane	98.54	88.76	66.05
n-Hexane	304.4	274.1	197.1
n-Heptane	926.9	826.9	578.9
n-Octane	2791	2493	1685
Cyclohexane	579.2	522.6	387.9
Methylcyclohexane	1166	1046	745.0
Benzene	434.3	573.3	667.1
Toluene	1428	1844	2070

[a] Experimental values were taken from a paper by Harbison et al., (1979). Reprinted in part with permission. Copyright American Chemical Society.

$\Delta \bar{G}^{fh}_{BC}/(X^\circ_B \bar{V}_B + X^\circ_C \bar{V}_C)RT$ for a given solvent mixture from the partition coefficient of an alkane in this mixture and the pure solvents. The term can then be applied in Eq. (12.27) for studies of more complex solutes. To illustrate this application of Eq. (12.27), we consider the partition coefficients of alkanes and cycloalkanes in mixtures of squalane and dinonyl phthalate (DNP) at 30°C given in Table I. Values of $\Delta \bar{G}^{fh}_{BC}$ calculated for n-pentane, n-hexane, n-heptane, n-octane, cyclohexane, and methylcyclohexane range from 800 to 880 J/mol at $\phi^\circ_C = 0.452$ (1% uncertainty in $K_{R(BC)}$ corresponds to about 100 J/mol in the value of $\Delta \bar{G}^{fh}_{BC}$). Use of the average value (840 J/mol) in Eq. (12.27) with data for benzene and toluene as solutes in this solvent mixture leads to equilibrium constants (K'_{AC}) of 2.46 and 1.59, respectively, for complexes of benzene and toluene with DNP.

D. Gas–Liquid Chromatographic Partition Coefficients of Inert Solutes on Self-Associating Binary Solvent Systems

The form of the NIBS equation that has been most successful for describing the excess chemical potential of solutes is based on a simple mixing model of a multicomponent system

$$\Delta G^{mix} = RT \sum_{i=1}^{N} n_i \ln \phi_i + \left[\sum_{i=1}^{N} n_i \bar{V}_i \right] \left[\sum_{i=1}^{N} \sum_{j>i} \phi_i \phi_j A_{ij} \right] \quad (12.28)$$

D. Partition Coefficients on Self-Associating Solvents

Renon and Prausnitz (1967) and Nagata (1973a,b) used this simple mixing model as a starting point in their thermodynamic treatments of alcohol–hydrocarbon mixtures. The success these two approaches have shown in their ability to predict the thermodynamic excess properties of ternary alcohol plus two inert hydrocarbon systems from the measured binary data suggests that Eq. (12.28) might also provide a reasonable description of the thermochemical properties of a solute in binary solvent mixtures having a self-associating component.

The application of Eq. (12.28) to the quaternary system (A, B, C_1, and C_2) takes the form

$$\Delta G^{mix} = RT[\hat{n}_A \ln \hat{\phi}_A + \hat{n}_B \ln \hat{\phi}_B + \hat{n}_{C_1} \ln \hat{\phi}_{C_1} + \hat{n}_{C_2} \ln \hat{\phi}_{C_2}]$$
$$+ (\hat{n}_A \bar{V}_A + \hat{n}_B \bar{V}_B + \hat{n}_{C_1} \bar{V}_{C_1} + \hat{n}_{C_2} \bar{V}_{C_2})$$
$$\times [\hat{\phi}_A \hat{\phi}_B A_{AB} + \hat{\phi}_A \hat{\phi}_{C_1} A_{AC_1} + \hat{\phi}_A \hat{\phi}_{C_2} A_{AC_2} + \hat{\phi}_B \hat{\phi}_{C_1} A_{BC_1} + \hat{\phi}_B \hat{\phi}_{C_2} A_{BC_2}]$$
(12.29)

the only assumption being that the monomer–dimer interaction parameter $A_{C_1 C_2}$ equals zero. The chemical potentials of the individual components relative to the pure liquids ($\mu_i^\bullet$) are obtained through the appropriate differentiation

$$\hat{\mu}_A - \mu_A^\bullet = RT[\ln \hat{\phi}_A + 1 - \bar{V}_A/\hat{V}_{sol}]$$
$$+ \bar{V}_A[\hat{\phi}_B(1-\hat{\phi}_A)A_{AB} + \hat{\phi}_{C_1}(1-\hat{\phi}_A)A_{AC_1}$$
$$+ \hat{\phi}_{C_2}(1-\hat{\phi}_A)A_{AC_2} - \hat{\phi}_B \hat{\phi}_{C_1} A_{BC_1} - \hat{\phi}_B \hat{\phi}_{C_2} A_{BC_2}] \quad (12.30)$$

$$\hat{\mu}_B - \mu_B^\bullet = RT[\ln \hat{\phi}_B + 1 - \bar{V}_B/\hat{V}_{sol}]$$
$$+ \bar{V}_B[\hat{\phi}_A(1-\hat{\phi}_B)A_{AB} + \hat{\phi}_{C_1}(1-\hat{\phi}_B)A_{BC_1}$$
$$+ \hat{\phi}_{C_2}(1-\hat{\phi}_B)A_{BC_2} - \hat{\phi}_A \hat{\phi}_{C_1} A_{AC_1} - \hat{\phi}_A \hat{\phi}_{C_2} A_{AC_2}] \quad (12.31)$$

and

$$\hat{\mu}_{C_1} - \mu_{C_1}^\bullet = RT[\ln \hat{\phi}_{C_1} + 1 - \bar{V}_{C_1}/\hat{V}_{sol}]$$
$$+ \bar{V}_{C_1}[\hat{\phi}_A(1-\hat{\phi}_{C_1})A_{AC_1} + \hat{\phi}_B(1-\hat{\phi}_{C_1})A_{BC_1}$$
$$- \hat{\phi}_A \hat{\phi}_{C_2} A_{AC_2} - \hat{\phi}_B \hat{\phi}_{C_2} A_{BC_2} - \hat{\phi}_A \hat{\phi}_B A_{AB}] \quad (12.32)$$

where $\hat{V}_{sol}$ is the molar volume of the *true* solution

$$1/\hat{V}_{sol} = \hat{\phi}_A/\bar{V}_A + \hat{\phi}_B/\bar{V}_B + \hat{\phi}_{C_1}/\bar{V}_{C_1} + \hat{\phi}_{C_2}/\bar{V}_{C_2}$$
$$= \phi_A/\bar{V}_A + \phi_B/\bar{V}_B + (\hat{\phi}_{C_1} + \phi_C)/2\bar{V}_{C_1}$$

As shown in Chapter 7, the chemical potential of stoichiometric component C is equal to the chemical potential of the monomeric species in solution

$$\mu_C = \hat{\mu}_{C_1}$$

To obtain the customary excess properties of the solution, pure substance C must be taken as the reference state

$$\mu_C - \mu_C^* = (\hat{\mu}_{C_1} - \mu_{C_1}^\bullet) - (\hat{\mu}_{C_1}^* - \mu_{C_1}^\bullet)$$
$$= RT[\ln(\hat{\phi}_{C_1}/\hat{\phi}_{C_1}^*) - \bar{V}_{C_1}/\hat{V}_{sol} + \bar{V}_{C_1}/\hat{V}_{sol}^*]$$
$$+ \bar{V}_{C_1}[\hat{\phi}_A(1 - \hat{\phi}_{C_1})A_{AC_1} + \hat{\phi}_B(1 - \hat{\phi}_{C_1})A_{BC_1}$$
$$- \hat{\phi}_A\hat{\phi}_B A_{BC} - \hat{\phi}_A\hat{\phi}_{C_2} A_{AC_2} - \hat{\phi}_B\hat{\phi}_{C_2} A_{BC_2}] \quad (12.33)$$

Combining Eqs. (12.30–12.33), one can write the excess Gibbs free energy

$$\Delta G_{ABC}^{ex} = RT[n_A \ln(\phi_A/X_A) + n_B \ln(\phi_B/X_B) + n_C \ln(\hat{\phi}_{C_1}/\hat{\phi}_{C_1}^* X_C) + n_B$$
$$+ n_A - (n_A \bar{V}_A + n_B \bar{V}_B + n_C \bar{V}_C)/\hat{V}_{sol} + \bar{V}_C/\hat{V}_{sol}^*]$$
$$+ (n_A \bar{V}_A + n_B \bar{V}_B + n_C \bar{V}_C)[\phi_A \phi_B A_{AB} + \phi_A \phi_C A_{AC_1} + \phi_B \phi_C A_{BC_1}]$$
$$(12.34)$$

The A_{AC_2} and A_{BC_3} interaction parameters were eliminated from the basic model as a natural consequence of $\mu_C = \hat{\mu}_{C_1}$. In earlier discussions of the Kretschmer–Wiebe, Mecke–Kempter, and AEC association models, the binary interaction parameters of the alcohol polymers A_{A_iK} ($i \neq 1$) could have been eliminated in a similar manner, without ever mentioning the solubility parameter theory. Whereas the additional assumptions involving solubility parameters were not necessary, they did simplify the mathematical treatment.

For a multicomponent system obeying Eq. (12.34), the activity coefficient of component A near infinite dilution is

$$RT \ln \gamma_A^\infty = RT\{\ln[\bar{V}_A/(X_B^o \bar{V}_B + X_C^o \bar{V}_C)] + 1 - \bar{V}_A/\hat{V}_{sol}\}$$
$$+ \bar{V}_A[\phi_B^o A_{AB} + \phi_C^o A_{AC_1} - \phi_B^o \phi_C^o A_{BC_1}] \quad (12.35)$$

with the solvent compositions calculated as if the solute were not present. The A_{BC} and A_{AC_1} interaction parameters can be removed via the infinite dilution activity coefficients in the pure solvents

$$\ln \gamma_A^\infty = \phi_B^o \ln(\gamma_A)_{X_B=1}^\infty + \phi_C^o \ln(\gamma_A)_{X_C=1}^\infty + \ln[(\bar{V}_B)^{\phi_B^o}(\bar{V}_C)^{\phi_C^o}/(X_B^o \bar{V}_B + X_C^o \bar{V}_C)]$$
$$- \bar{V}_A(\hat{\phi}_{C_1} - \phi_C^o \hat{\phi}_{C_1}^*)/2\bar{V}_C - \bar{V}_A \phi_B^o \phi_C^o A_{BC_1}/RT \quad (12.36)$$

where $\hat{\phi}_{C_1}$ and $\hat{\phi}_{C_1}^*$ refer to the monomer concentration in the binary solvent and pure liquid C, respectively

$$\hat{\phi}_{C_1} = (\bar{V}_{C_1}/4K_D)[-1 + (1 + 8K_D \phi_C^o/\bar{V}_{C_1})^{1/2}]$$
$$\hat{\phi}_{C_1}^* = (\bar{V}_{C_1}/4K_D)[-1 + (1 + 8K_D/\bar{V}_{C_1})^{1/2}]$$

and K_D is the molar dimerization constant.

D. Partition Coefficients on Self-Associating Solvents

Expressed in terms of the partition coefficients in the pure solvents, Eq. (12.36) becomes

$$\ln K_{R(BC)} = \phi_B^o \ln K_{R(B)} + \phi_C^o \ln K_{R(C)} + \frac{\bar{V}_A(\hat{\phi}_{C_1} - \phi_C^o \hat{\phi}_{C_1}^*)}{2\bar{V}_C} + \frac{V_A \phi_B^o \phi_C^o A_{BC_1}}{RT} \quad (12.37)$$

Notice that in the absence of dimerization, $K_D = 0$, the previous expression reduces to

$$\ln K_{R(BC)} = \phi_B^o \ln K_{R(B)} + \phi_C^o \ln K_{R(C)} + \bar{V}_A \phi_B^o \phi_C^o A_{BC}/RT \quad (12.38)$$

an equation derived from the NIBS model for nonspecific interactions. For comparative purposes, the excess Gibbs free energy of the binary solvent was left in terms of the binary interaction parameter A_{BC}.

Calculations using various values for the dimerization constant reveal that the *correction* term is positive and becomes more significant with increasing

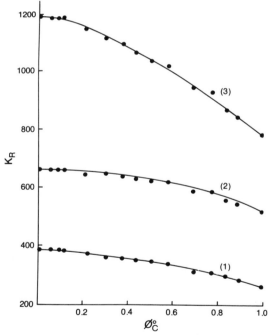

Fig. 12.1 Plot of $K_{R(BC)}$ against ϕ_C^o for n-hexane (1), cyclohexane (2), and n-heptane (3) solutes with solutions of N,N-dibutyl-2-ethylhexylamide and n-hexadecane. The solid lines are calculated from Eq. (12.37) with $K_D = 0.2$ dm^3/mol and $A_{BC_1} = 1.2$ cal/cm^3. Graph is reprinted with permission from Acree, (1982). Copyright 1982 American Chemical Society.

solute size. For values of $\bar{V}_A/\bar{V}_C = 0.5$ and $K_D/\bar{V}_C = 1.4$, partition coefficients calculated from Eq. (12.37) differ from those of Eq. (12.38) by only 2%. Judicious selection of a new binary interaction parameter A'_{BC} enables Eq. (12.38) to mathematically compensate for such a small difference, and thus, further illustrates the difficulties in separating chemical and physical effects, especially for weak association complexes.

Experimental gas–liquid partition coefficients of various solutes of n-hexadecane + N,N dibutyl-2-ethylhexylamide, and n-octadecane + N,N dibutyl-2-ethylhexylamide mixtures (Chien et al., 1981) provide an excellent opportunity to test Eq. (12.37) as studies involving dielectric properties (Kopecni, et al., 1981) suggested that N,N dibutyl-2-ethylhexylamide undergoes a weak reversible dimerization in alkane solutions. As shown in Figs. 12.1 and 12.2, Eq. (12.37) ($K_D = 0.2 \, dm^3/mol$) describes the experimental data at 30°C everywhere to within 2%. Experimental uncertainties associated with chromatographic measurements are approximately 1–1.5%.

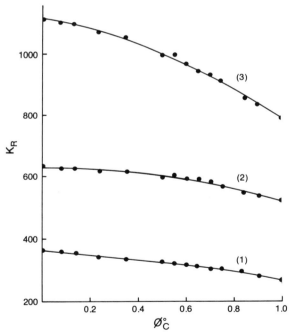

Fig. 12.2 Plot of $K_{R(BC)}$ against ϕ_C^o for n-hexane (1), cyclohexane (2), and n-heptane (3) solutes with solutions of N,N-dibutyl-2-ethylhexylamide and n-octadecane. The solid lines are calculated from Eq. (12.37) with $K_D = 0.2 \, dm^3/mol$ and $A_{BC_1} = 0.8 \, cal/cm^3$. Graph is reprinted with permission from Acree, (1982). Copyright 1982 American Chemical Society.

E. Summary

By way of conclusion, readers should be reminded that there is an important difference between using empirical models and thermodynamic models to describe partition coefficients on binary solvent mixtures. Empirical models limit gas–liquid chromatography to routine analysis and chemical separations. Thermodynamic models, in comparison, also enable the estimation of thermodynamic quantities and encourage the development of future (and hopefully better) solution models.

Problems

12.1 Using the Mecke–Kempter association model, show that the partition coefficient of an alcohol solute on binary solvent mixtures of inert hydrocarbons is given by

$$\ln K_{R(BC)} = \phi_B^o \ln K_{R(B)} + \phi_C^o \ln K_{R(C)} + \bar{V}_A \Delta \bar{G}_{BC}^{fh}/(X_B^o \bar{V}_B + X_C^o \bar{V}_C)RT$$

Is it possible to determine a solute's mode of self-association through chromatographic measurements? Explain the answer.

12.2 Although the microscopic partition model describes the chromatographic behavior of a large number of solutes in binary solvent mixtures, the model has several limitations. As seen in Section I, the microscopic partition model

$$K_{R(BC)} = \phi_B^o K_{R(B)} + \phi_C^o K_{R(C)}$$

is incapable of describing systems having either a maximum or minimum mole fraction solubility. By differentiating Eq. (12.5), determine if the microscopic partition theory will be able to describe systems in which the enthalpy of solution curve exhibits a minimum or maximum value.

Hint: The following thermodynamic relationship will be needed:

$$(\Delta \bar{H}_A^{ex})^\infty = (RT^2/\gamma_A^\infty)(\partial \gamma_A^\infty/\partial T)$$

12.3 Although thermodynamicists prefer chromatographic data to be reported in terms of thermodynamic quantities, chromatographers, particularly those interested in chemical separations, often report their data in terms of Kovat's retention indices. Kovat's retention index for a component x is related to the partition coefficient by

$$I_x = 100 \frac{\log K_{R_x} - \log K_{R_z}}{\log K_{R_{z+1}} - \log K_{R_z}} + 100z$$

where z is an n-paraffin with z carbon atoms. Under what conditions

TABLE II
Chromatographic Partition Coefficients of Selected Solutes on Dodecyl Laurate + Squalane Mixtures at 60°C[a]

Solute	K_R (as a function of dodecyl laurate mole fraction)					
	0.0000	0.3021	0.4979	0.6999	0.8456	1.0000
Cyclohexane	194.7	191.8	189.8	189.0	187.8	184.3
Methylcyclohexane	353.5	351.2	348.0	346.5	344.7	338.5
Ethylcyclohexane	981.5	970.7	965.0	960.6	954.1	937.4
n-Heptane	273.7	271.0	269.2	267.7	266.5	261.4
Octafluorotoluene	96.93	119.9	134.2	149.0	160.5	166.8

[a] The experimental data were reproduced with permission from Mathiasson and Jonsson, (1974).

can one estimate the Kovat's retention index as

$$I_{x(BC)} = \phi_B^o I_{x(B)} + \phi_C^o I_{x(C)}$$

a volume fraction average of the indices in the pure solvents ($I_{x(B)}$ and $I_{x(C)}$)?

12.4 Chromatographic partition coefficients for several solutes on dodecyl laurate + squalane binary mixtures are listed in Table II. Using this data, determine if octafluorotoluene complexes with dodecyl laurate. If complexation is indicated, calculate K'_{AC} via Eq. (12.27). The partition coefficients of the hydrocarbon solutes will be needed in the calculation of $\Delta \bar{G}_{BC}^{fh}$. The solute and solvent molar volumes (at 60°C) are:

Squalane	540 cm^3/mol
Dodecyl laurate	445 cm^3/mol
Cyclohexane	114 cm^3/mol
Methylcyclohexane	134 cm^3/mol
Ethylcyclohexane	148 cm^3/mol
n-Heptane	154 cm^3/mol
Octafluorotoluene	149 cm^3/mol

12.5 For historical reasons, the notation of Acree and Bertrand (1979) has been used in deriving expressions for K_R in binary solvent mixtures having solute–solvent complexation. By now readers should recognize that a similar expression

$$\ln K_{R(BC)} = \phi_B^o \ln K_{R(B)} + \phi_C^o \ln K_{R(C)} + \ln[1 + \bar{V}_A K_{AC}^\phi \phi_C^o/(\bar{V}_A + \bar{V}_C)]$$
$$- \phi_C^o \ln[1 + \bar{V}_A K_{AC}^\phi/(\bar{V}_A + \bar{V}_C)]$$
$$+ (\bar{V}_A \Delta \bar{G}_{BC}^{fh})/(X_B^o \bar{V}_B + X_C^o \bar{V}_C)RT$$

could be derived by combining Eqs. (10.59), (12.4), (12.6), and $\gamma_A^{sat} = \gamma_A^\infty$. How does the previous expression differ from Eq. (12.27)?

Appendix A

Solubility Parameters and Molar Volumes at 25°C[†]

Substance	δ_i (cal/cm^3)$^{1/2}$[‡]	$\bar{V}_i^\bullet$ (ml/mole)
Acetone	9.62	73.99
Acetonitrile	12.11	52.88
Benzene	9.16	89.41
Bromobenzene	9.87	105.49
1-Butanol	11.60	91.99
2-Butanol	11.08	92.39
n-Butylamine	8.66	99.58
n-Butylbenzene	8.58	156.74
Carbon disulfide	9.92	60.61
Carbon tetrachloride	8.55	97.08
Chlorobenzene	9.67	102.26
Chloroform	9.16	80.63
Cyclohexane	8.19	108.76
Cyclohexanone	10.42	104.17
Cyclopentane	8.10	94.69
Cyclopentanone	10.53	89.06
Cyclopentene	8.27	88.83
Decane	7.74	195.88
1-Decanol	10.03	191.56
2-Decanol	8.81	192.71
1-Decene	7.85	190.31
o-Dibromobenzene	10.27	119.32
Di-n-Butylamine	8.15	170.62
o-Dichlorobenzene	10.04	113.04
m-Dichlorobenzene	9.80	114.57
p-Dichlorobenzene	9.70	118.37
1,1 Dichloroethane	8.92	84.70
Dichloromethane	9.88	64.50

[†] Numerical values of the solubility parameters were taken from a tabulation by K. L. Hoy (1970).
[‡] To convert from cal/cm^3 to J/cm^3 multiply by 4.184.

Substance	δ_i (cal/cm^3)$^{1/2}$‡	$\bar{V}_i^\bullet$ (ml/mole)
Diethylamine	8.04	104.27
1,2 Diethylbenzene	8.71	153.18
2,2 Dimethylbutane	6.71	133.68
2,3 Dimethylbutane	6.97	131.12
2,2 Dimethyl-1-butanol	10.17	123.98
2,3 Dimethyl-1-butanol	10.00	123.62
3,3 Dimethyl-1-butanol	9.83	126.20
2,3 Dimethyl-2-butanol	9.58	124.73
3,3 Dimethyl-2-butanol	9.51	125.35
3,3 Dimethyl-2-butanone	8.22	125.04
1,4 Dioxane	10.13	85.68
Di-n-propylamine	7.97	138.13
Dodecane	7.92	228.50
1-Dodecanol	9.78	224.60
2-Dodecanol	8.52	225.84
Ethanol	12.78	58.69
Ethyl Acetate	8.91	98.51
Ethylbenzene	8.84	123.08
2-Ethyl-1-butanol	10.34	123.25
Ethyl butyl ether	7.68	137.21
2-Ethyl-1-hexanol	10.15	157.18
Fluorobenzene	9.11	94.52
Heptane	7.50	147.48
1-Heptanol	10.50	141.88
2-Heptanol	9.83	142.86
3-Heptanol	9.88	142.19
2-Heptanone	8.98	140.81
3-Heptanone	8.52	140.37
4-Heptanone	8.70	140.70
Hexane	7.27	131.51
1-Hexanol	10.77	125.26
2-Hexanol	10.04	126.09
3-Hexanol	9.48	125.50
2-Hexanone	8.63	124.17
3-Hexanone	8.59	123.43
1-Hexene	7.31	125.95
n-Hexylamine	8.45	132.80
Isopropanol	11.44	76.97
Isopropylamine	8.22	—
Methanol	14.50	40.73
Methyl Acetate	9.46	79.88
Methylcyclohexane	—	128.32
2-Methyl-1-pentanol	10.18	124.56
3-Methyl-1-pentanol	10.51	124.69
4-Methyl-1-pentanol	10.92	126.20
Nitromethane	12.90	53.96
Nonane	7.64	179.87

Solubility Parameters and Molar Volumes at 25°C

Substance	δ_i (cal/cm^3)$^{1/2}$‡	$\bar{V}_i^\bullet$ (ml/mole)
1-Nonanol	10.13	—
2-Nonanol	9.04	—
Octane	7.54	163.46
1-Octanol	10.30	158.47
2-Octanol	9.32	—
2-Octanone	8.81	—
Pentadecane	—	277.20
Pentane	7.02	116.12
1-Pentanol	11.12	108.68
2-Pentanol	10.77	—
3-Pentanol	10.16	—
2-Pentanone	8.99	107.45
3-Pentanone	9.06	106.41
Piperidine	9.45	99.43
1-Propanol	12.18	75.19
n-Propylamine	8.87	83.04
Pyridine	10.62	80.86
Tetradecane	—	261.28
Thiophene	9.84	79.44
Toluene	8.93	106.84
Tri-n-butylamine	7.76	239.10
Triethylamine	7.42	139.91
2,2,4 Trimethylpentane	6.86	166.09
Undecane	7.81	212.38
1-Undecanol	9.85	208.05
m-Xylene	8.88	123.48
o-Xylene	9.06	121.21
p-Xylene	8.82	124.00

Appendix B
UNIQUAC Structural Parameters[†]

Substance	r_i	q_i
Acetone	2.5735	2.336
Acetonitrile	1.8701	1.724
Anthracene	6.7738	4.480
Benzene	3.1878	2.400
Benzil	7.5860	5.520
Biphenyl	5.7112	3.680
Bromobenzene	3.9709	2.952
1-Butanol	3.9243	3.668
2-Butanol	3.9235	3.664
n-Butylamine	3.6191	3.164
Carbon disulfide	2.057	1.65
Carbon tetrachloride	3.3900	2.910
Chloroform	2.8700	2.410
Cyclohexane	4.0464	3.240
Cyclooctane	5.3952	4.320
Decane	7.1974	6.016
1-Decanol	7.2963	6.368
Dodecane	8.5462	7.096
Ethanol	2.5755	2.588
Ethylbenzene	4.5972	3.508
Heptane	5.1742	4.396
Hexadecane	11.244	9.256
Hexane	4.5000	3.856
Isopropanol	3.2491	3.124
Methanol	1.4311	1.432
Naphthalene	4.9808	3.440
Nonane	6.5230	5.476
Octane	5.8486	4.936
Pentane	3.8254	3.316
1-Pentanol	4.5987	4.208
Toluene	3.9228	2.968
2,2,4 Trimethylpentane	5.8463	5.008
o-Xylene	4.6578	3.536
m-Xylene	4.6578	3.536
p-Xylene	4.6578	3.536

[†] Structural parameters of compounds not listed above can be calculated using the UNIFAC group parameters given in Table III of Chapter 6.

Appendix C

Constants and Conversion Factors

$$
\begin{aligned}
1 \text{ calorie (cal)} &= 4.184 \text{ joules (J)} \\
R \text{ (gas constant)} &= 8.3143 \text{ J/mol-kelvin} \\
&= 1.9872 \text{ cal/mol-kelvin} \\
RT \text{ (at 298.15 K)} &= 2478.9 \text{ J/mol} \\
&= 592.47 \text{ cal/mol} \\
1 \text{ atm (pressure)} &= 101.325 \text{ kPa} \\
&= 760 \text{ torr (mm Hg)}
\end{aligned}
$$

Appendix D
Answers to Selected Problems

CHAPTER 1

1.7 Linearized form:

$$\Delta \bar{V}^{ex}/X_1 X_2 = A + B(2X_1 - 1)^2,$$

$$A = 4.389 \text{ cm}^3/\text{mol} \quad \text{and} \quad B = 0.727 \text{ cm}^3/\text{mol}$$

CHAPTER 2

2.2 $\bar{V}_1 = \bar{V}_1^\bullet + X_2(1 - X_1)\beta_{12} + X_3(1 - X_1)\beta_{13} - X_2 X_3 \beta_{23}$
$\bar{V}_2 = \bar{V}_2^\bullet + X_1(1 - X_2)\beta_{12} + X_3(1 - X_2)\beta_{23} - X_1 X_3 \beta_{13}$
$\bar{V}_3 = \bar{V}_3^\bullet + X_1(1 - X_3)\beta_{13} + X_2(1 - X_3)\beta_{23} - X_1 X_2 \beta_{12}$

2.5 (b) $(\partial \rho / \partial X_1)_{T,P} = -0.18527 + 0.44026 X_1 - 1.81254 X_1^2 + 2.15648 X_1^3 - 0.9006 X_1^4$,
$\bar{V}_1 = 38.14 \text{ ml/mol}$

2.7 Volumetric Properties of Binary Mixtures of Water (1) + N-Methylformamide (2) at 25°C

X_1	$\Delta \bar{V}_1^{app}$ (cm^3/mol)	$\Delta \bar{V}_2^{app}$ (cm^3/mol)
0.0511	16.94	59.11
0.1284	16.88	59.00
0.2108	16.88	58.86
0.3130	16.88	58.64
0.4073	16.93	58.40
0.4994	16.99	58.07
0.6398	17.17	57.60
0.6979	17.01	56.97
0.8248	17.56	56.86
0.9016	17.77	56.63
0.9506	17.92	56.61
0.9797	18.00	56.66

Answers to Selected Problems

2.8 Volumetric Properties of Sulfolane
(1) + Water (2) Mixtures at 303.15 K

X_2	$\Delta \bar{V}^{ex}$ (cm^3/mol)	$\bar{V}_2^{app}$ (cm^3/mol)
0.09998	0.048	18.56
0.19839	0.068	18.42
0.30332	0.048	18.24
0.40086	0.012	18.11
0.50410	−0.028	18.02
0.60708	−0.062	17.98
0.70744	−0.091	17.95
0.80763	−0.112	17.94
0.90380	−0.099	17.97

2.9 (b) $\rho = 0.8494$ g/cm^3.

2.10 (a) $\bar{V}_B^{app} = 152.04$ cm^3/mol

2.11 (a) 357.14 J, (b) 500 J, (c) 255.1 J

CHAPTER 3

3.3 (a) 298.9, (b) 0.7046, (c) 1.019

3.5 38.85 J

3.7 Activity Coefficients for Benzene (1) + Cyclohexane (2)
Mixtures at 298.15 K

X_1	$\gamma_1 = [y_1 P/X_1(95.05)]$	$\gamma_2 = [y_2 P/X_2(97.45)]$
0.1035	1.426	1.007
0.1750	1.363	1.018
0.2760	1.274	1.039
0.3770	1.211	1.066
0.4330	1.175	1.087
0.5090	1.134	1.124
0.5830	1.098	1.167
0.6940	1.054	1.253
0.7945	1.025	1.352
0.9005	1.007	1.489
0.9500	1.002	1.571

3.8 The expression is not consistent with the ideal solution model. The correct expression is $\rho_{mix} = \sum_{i=1}^{N} \phi_i \rho_i^{\bullet}$.

3.9 Vapor Phase Compositions for the Acetone (1) + Acetonitrile (2) System at 50°C

X_1	$y_1 = X_1(615.16)/P$	$y_2 = X_2(253.89)/P$
0.0824	0.1785	0.8205
0.1600	0.3165	0.6857
0.2531	0.4498	0.5478
0.3451	0.5589	0.4378
0.4314	0.6448	0.3507
0.4754	0.6868	0.3128
0.5077	0.7117	0.2848
0.5517	0.7491	0.2512
0.6350	0.8101	0.1922
0.7386	0.8753	0.1279
0.8138	0.9159	0.0865
0.8996	0.9571	0.0441
0.9581	0.9834	0.0177

If the solution is *truly* ideal, the vapor phase compositions will add to one, i.e., $y_1 + y_2 = 1$.

3.11 Activity Coefficients for 2-Butanone (1) + n-Heptane (2) Mixtures at 45°C

X_1	$\gamma_1 = [y_1 P/X_1(218.26)]$	$\gamma_2 = [y_1 P/X_2(115.09)]$
0.0738	3.059	1.014
0.1416	2.561	1.042
0.2764	1.932	1.126
0.3617	1.666	1.208
0.4962	1.389	1.389
0.5791	1.265	1.549
0.7008	1.139	1.881
0.8027	1.067	2.295
0.9048	1.019	2.969
0.9528	1.006	3.419

Answers to Selected Problems

CHAPTER 4

4.1 Excess Molar Volumes of Methyl Ethyl
Ketone (1) + 1-Propanol (2) + n-Heptane (3) Mixtures

X_1	X_2	Predicted Values of $\Delta \bar{V}^{ex}_{123}$ (cm^3/mol)		
		Eq. (4.4)	Eq. (4.9)	Eq. (4.14)
0.7538	0.1278	0.194	0.195	0.194
0.7211	0.1556	0.196	0.196	0.196
0.6078	0.2912	0.138	0.134	0.138
0.4794	0.4105	0.133	0.130	0.133
0.4579	0.4126	0.159	0.151	0.159
0.3235	0.5486	0.138	0.135	0.138
0.2183	0.6865	0.086	0.079	0.086
0.1108	0.7873	0.088	0.083	0.088

4.2 Predicted Excess Enthalpies for
Chloroform (1) + Ethanol (2) + n-Heptane
(3) Mixtures at 30°C

X_1	X_2	$\Delta \bar{H}^{ex}_{123}$ (J/mol)
0.25	0.25	730.0
0.33	0.33	593.5
0.25	0.50	234.7
0.50	0.25	742.4
0.33	0.50	154.6

All values calculated with Eq. (4.9).

4.4 Excess Gibbs Free Energies of
n-Hexane (1) + Cyclohexane (2) +
Benzene (3) Mixtures

X_1	X_2	$\Delta \bar{G}^{ex}_{123}$ (J/mol)
0.33	0.33	283.1
0.50	0.50	20.4
0.25	0.25	320.6
0.25	0.50	201.5
0.50	0.25	272.8
0.33	0.50	183.9

4.6 $\Delta \bar{H}^{ex}_{123} = (X_1 + X_2)^3 \Delta \bar{H}^{ex}_{12} + (X_1 + X_3)^3 \Delta \bar{H}^{ex}_{13} + (X_2 + X_3)^3 \Delta \bar{H}^{ex}_{23}$

CHAPTER 5

5.2 Calculated Values of the Excess Gibbs Free Energies for Cyclohexane (1) + OMCTS (2) Mixtures at 308.15 K

X_2	$\Delta \bar{G}_{12}^{\text{rh}}$ (J/Mol)
0.10103	151.8
0.20720	242.0
0.28156	274.8
0.36325	289.8
0.44613	287.4
0.53953	268.5
0.62624	239.0
0.72897	191.5
0.78479	153.2
0.90591	73.2

5.5 Vapor–Liquid Equilibrium Calculations for 1-Hexanol (1) + n-Hexane (2) Mixtures

X_1	γ_1	γ_2	P (kPa)
0.1074	4.543	1.079	15.60
0.1990	2.699	1.178	15.29
0.2967	1.939	1.311	14.94
0.4010	1.538	1.483	14.40
0.4942	1.329	1.668	13.68
0.5976	1.184	1.916	12.51
0.7080	1.088	2.244	10.64
0.8242	1.030	2.686	7.69
0.9176	1.006	3.137	4.24

Answers to Selected Problems

5.6 Vapor–Liquid Equilibrium Calculations for
n-Heptane (1) + 1-Propanol (2) Mixtures

X_1	γ_1	γ_2	P (kPa)
0.0699	5.782	1.012	5.147
0.1523	4.239	1.051	6.477
0.2570	3.096	1.139	7.265
0.3602	2.411	1.273	7.619
0.4789	1.904	1.511	7.806
0.6489	1.449	2.157	7.894
0.7164	1.321	2.631	7.900
0.7725	1.231	3.234	7.898
0.8374	1.142	4.420	7.882
0.8746	1.098	5.600	7.860
0.9493	1.025	11.800	7.641

5.7 Enthalpy Calculations for
n-Heptane (1) + 1-Propanol
(2) Mixtures

X_2	$\Delta \bar{H}_{12}^{ex}$ (J/mol)
0.10	340
0.20	402
0.30	414
0.40	406
0.50	383
0.60	348
0.70	298
0.80	231
0.90	136

5.9 Liquid–Vapor Equilibrium Calculations for
1-Butanol (1) + n-Hexane (2) Mixtures

X_1	γ_1	γ_2	P (kPa)
0.1291	4.163	1.073	75.638
0.2152	3.120	1.164	75.124
0.3160	2.286	1.303	73.839
0.3905	1.864	1.431	72.430
0.5114	1.391	1.698	69.050
0.5928	1.253	1.932	66.029
0.6911	1.137	2.297	60.491
0.7810	1.128	2.743	52.960
0.9125	1.011	3.705	32.201

5.11 Liquid–Vapor Equilibrium Calculations for Methyl Acetate (1) + 1-Hexene (2) Mixtures

X_1	γ_1	γ_2	P (torr)
0.1087	2.026	1.008	566.9
0.2285	1.747	1.039	626.5
0.2948	1.617	1.069	649.4
0.4147	1.419	1.148	676.2
0.5417	1.257	1.284	690.7
0.6159	1.182	1.397	693.5
0.7611	1.072	1.734	686.5
0.8163	1.043	1.920	677.8
0.8904	1.016	2.244	657.6

CHAPTER 6

6.3 $\Delta H^{ex} = \sum_{i=1}^{N} X_i \sum_{k} v_{ki}(H_k - H_k^i)$

where

$$\frac{H_k}{R} = Q_k \left[\sum_m \frac{\theta_m^g \psi_{mk} a_{mk}}{\sum_m \theta_m^g \psi_{mk}} + \sum_m \frac{\theta_m^g \psi_{mk} a_{mk}}{\sum_n \theta_n^g \psi_{nm}} - \sum_m \frac{\theta_m^g \psi_{mk}(\sum_n \theta_n^g \psi_{nm} a_{nm})}{(\sum_n \theta_n^g \psi_{nm})^2} \right]$$

6.4 It may be shown that if a mixture can be described by less than 2 different main groups and if the the components have no main groups in common, then the UNIFAC model reduces to the UNIQUAC model. Examples are benzene + n-heptane, chloroform + n-hexane, etc. mixtures.

6.5 Activity Coefficient Calculations for 1-Dodecanol (1) + n-Hexane (2) Mixtures

X_1	γ_1	γ_2
0.09998	1.797	1.036
0.19408	1.319	1.088
0.36786	1.082	1.171
0.46527	1.038	1.208
0.56916	1.016	1.237
0.67877	1.005	1.261
0.76204	1.002	1.271
0.87306	1.000	1.286

Answers to Selected Problems

CHAPTER 7

7.2 Excess Gibbs Free Energy Calculations for Chloroform (1) + Triethylamine (2)

X_1	$\Delta \bar{G}_{12}^{ex}$ (kJ/mol)
0.1	−0.41
0.2	−0.76
0.3	−1.05
0.4	−1.24
0.5	−1.31
0.6	−1.24
0.7	−1.05
0.8	−0.76
0.9	−0.41

If one were to calculate the association constant based on the experimental excess Gibbs free energies, the numerical value of K_{AB} would be 2.6.

7.5 $K_{AB} = 1.47$ and $h_{AB} = -14.5$ kJ/mol.

7.9 $h_{AB} = -15.4$ kJ/mol using $(\Delta \bar{H}_A^{ex})_{X_B=1}^{\infty}$
and
$h_{AB} = -14.0$ kJ/mol using $(\Delta \bar{H}_B^{ex})_{X_A=1}^{\infty}$.

CHAPTER 8

8.1 Monomeric Volume Fraction Concentrations Based on Different Association Models

	ϕ_{A_1}		
ϕ_A	Kretschmer-Wiebe	Mecke-Kempter	AEC
0.10	0.007298	0.009091	0.01746
0.20	0.008000	0.009524	0.02205
0.30	0.008333	0.009677	0.02489
0.40	0.008539	0.009756	0.02697
0.50	0.008682	0.009804	0.02861
0.60	0.008790	0.009836	0.02997
0.70	0.008874	0.009859	0.03113
0.80	0.008943	0.009877	0.03214
0.90	0.009000	0.009890	0.03305
1.00	0.009049	0.009901	0.03386

8.3 573.5, 524.0, and 354.2 J/mol

8.4 Calculated Values of $\Delta \bar{H}_{12}^{ex}$ for 1-Butanol (1) + n-Hexane (2) Mixtures

X_1	$\Delta \bar{H}_{12}^{ex}$ (J/mol)
0.1634	514
0.1772	525
0.3151	564
0.3323	563
0.5817	448
0.5960	437
0.5971	437
0.8173	227

8.5 Calculated Values of $\Delta \bar{H}_{12}^{ex}$ for 1-Butanol (1) + n-Hexane (2) Mixtures

X_1	$\Delta \bar{H}_{12}^{ex}$ (J/mol)
0.1634	511
0.1772	519
0.3151	546
0.3323	548
0.5817	455
0.5960	445
0.5971	445
0.8173	247

CHAPTER 9

9.1 $X_2^{az} = 0.333$

9.6 Linford and Hildebrand (1969) obtained a value of $A_{12} = 3.80$ cal/cm^3. Using this value, one calculates $\gamma_1 = 1.199$ and $\gamma_2 = 1.158$.

9.7 $X_2^{az} = 0.3514$ using $\Delta \bar{S}_1^{vap} = \Delta \bar{S}_2^{vap}$
     $X_2^{az} = 0.3565$ using $\Delta \bar{S}_1^{vap} \neq \Delta \bar{S}_2^{vap}$

CHAPTER 10

10.1 (a) $X_3^{sat} = 0.3064$,   (b) $X_3^{sat} = 0.3064$,   (c) $X_3^{sat} = 0.0355$
All values were calculated using Eq. (10.16).

Answers to Selected Problems

10.3 NIBS Predictions for the Solubility of Iodine in Cyclohexane (1) + Carbon Disulfide (2)

X_1^o	(XX)	(XV)
0.2	0.04076	0.03267
0.4	0.02920	0.02170
0.6	0.02050	0.01562
0.8	0.01399	0.01180

10.4 NIBS Predictions for the Solubility of Tetraphenyltin in Carbon Tetrachloride (1) + Cyclohexane (2)

X_1^o	(XX)	(XV)	(VV)
0.25	0.000319	0.000321	0.000321
0.50	0.000483	0.000484	0.000484
0.75	0.000721	0.000723	0.000723

The three forms of the NIBS model give identical predictions because the solvent molecules are comparable in size. Furthermore, the "unmixing" term is almost negligible because of the small $\Delta \bar{G}_{12}^{ex}$ values.

10.5 Calculated Values of the Solubility Parameter of Benzil from Solubility Measurements

Solvent	δ_3
Cyclohexane	11.38
n-Hexane	10.74
n-Heptane	10.91
Carbon tetrachloride	10.67
n-Octane	10.90
Isooctane	10.32
Benzene	10.37
Toluene	10.41
Average	10.71

10.8 Calculated values of K_{AC}^{ϕ} range from 9.5–10.5.

CHAPTER 11

11.2 (a) 5.28×10^{-7} mol (b) 1.16×10^{-5} mol

11.4 The following predictive expressions $X_{Etbz}^{Aq} \approx (2.916 \times 10^{-5})$ X_{Etbz}^{Org} and $X_{Prbz}^{aq} \approx (7.056 \times 10^{-6})$ X_{Prbz}^{Org} can be developed assuming $X_{Aq}^{Org} = 0$, $\gamma_{Etbz}^{Org} = 1$, and $\gamma_{Prbz}^{Org} = 1$. The abbreviations aq, Et bz, Org, and Pr bz indicate aqueous, ethylbenzene, organic and propylbenzene, respectively.

11.5 Partition coefficients of benzene between water and various alkanes at 25°C

Alkane	K_P^c
Methylcyclohexane	326
Cyclohexane	335
n-Hexane	267
n-Pentadecane	214
n-Tetradecane	191

How "good" the predicted values are depend to a large extent on whether the Redlich–Kister equations properly describe the benzene + alkane systems near infinite dilution.

CHAPTER 12

12.2 Differentiating Eq. (12.5), one finds

$$(\Delta \bar{H}_A^{ex})^\infty = g_B^o (\Delta \bar{H}_A^{ex})_{X_B=1}^\infty + g_C^o (\Delta \bar{H}_A^{ex})_{X_C=1}^\infty$$

where

$$g_B^o = 1 - g_C^o$$
$$= X_B^o (\gamma_A)_{X_B=1}^\infty / [X_B^o (\gamma_A)_{X_B=1}^\infty + X_C^o (\gamma_A)_{X_C=1}^\infty].$$

This expression is incapable of describing a system in which the enthalpy of solution shows a maximum or minimum value.

12.3 This derivation is found in Acree and Rytting (1980).

12.4 Equation (12.14) predicts the experimental partition coefficients to within 1%, suggesting the contributions of $\Delta \bar{G}_{BC}^{fh}$ are negligible. Using $\Delta \bar{G}_{BC}^{fh} = 0$, one calculates $K_{AC}' \approx 1.25$ at $X_{Dolr} = 0.4979$, where Dolr indicates dodecyl laurate.

References

Abrams, D. S., and Prausnitz, J. M. (1975). *AIChE J.* **21**, 116.
Acree, W. E., Jr. unpublished results.
Acree, W. E., Jr. (1981). Ph.D. Dissertation. University of Missouri-Rolla.
Acree, W. E., Jr. (1982). *J. Phys. Chem.* **86**, 1461.
Acree, W. E., Jr. and Bertrand, G. L. (1977). *J. Phys. Chem.* **81**, 1170.
Acree, W. E., Jr. and Bertrand, G. L. (1979). *J. Phys. Chem.* **83**, 2355.
Acree, W. E., Jr. and Bertrand, G. L. (1981). *J. Pharm. Sci.* **70**, 1033.
Acree, W. E., Jr. and Bertrand, G. L. (1983). *J. Solution Chem.* **12**, 101.
Acree, W. E., Jr. and Rytting, J. H. (1980). *Anal. Chem.*, **52**, 1764.
Acree, W. E., Jr. and Rytting, J. H. (1982a). *Int. J. Pharm.* **10**, 231.
Acree, W. E., Jr. and Rytting, J. H. (1982b). *J. Pharm. Sci.* **71**, 201.
Acree, W. E., Jr. and Rytting, J. H. (1982c). *J. Solution Chem.* **11**, 137.
Acree, W. E., Jr. and Rytting, J. H. (1983). *J. Pharm. Sci.* **72**, 292.
Acree, W. E., Jr., Rytting, J. H. and Carstensen, J. T. (1981). *Int. J. Pharm.*, **8**, 69.
Adjei, A., Newberger, J., and Martin, A. (1980). *J. Pharm. Sci.* **69**, 659.
Alexander, D. M. (1959). *J. Phys. Chem.* **63**, 1021.
Allen, G., Watkinson, J. G., and Webb, K. H. (1967). *Spectrochim. Acta, Part A* **23**, 2497.
Amidon, G. L., and Anik, S. T. (1981). *J. Chem. Eng. Data* **26**, 28.
Anderson, B. D. (1977). Ph.D. Dissertation. University of Kansas, Lawrence, Kansas.
Anderson, B. D., Rytting, J., Lindenbaum, S., and Higuchi, T. (1975). *J. Phys. Chem.* **79**, 2340.
Anderson, B. D., Rytting, J. H., and Higuchi, T. (1978). *Int. J. Pharm.* **1**, 15.
Anderson, B. D., Rytting, J. H., and Higuchi, T. (1979). *J. Am. Chem. Soc.* **101**, 5194.
Anderson, B. D., Rytting, J. H., and Higuchi, T. (1980). *J. Pharm. Sci.* **69**, 676.
Arich, G., Kikic, I., and Alessi, P. (1975). *Chem. Eng. Sci.* **30**, 187.
Arnett, E. M., Murty, T. S. S. R., Schleyer, P. v. R., and Joris, L. (1967). *J. Am. Chem. Soc.* **89**, 5955.
Arnett, E. M., Joris, L., Mitchell, E., Murty, T. S. S. R., Gorrie, T. M., and Schleyer, P. v. R. (1970). *J. Am. Chem. Soc.* **92**, 2365.
Arnett, E. M., Mitchell, E. J., and Murty, T. S. S. R. (1974). *J. Am. Chem. Soc.* **96**, 3875.
Arnold, D. W., Greenkorn, R. A., and Chao, K.-C. (1982). *J. Chem. Eng. Data* **27**, 123.
Ashraf, F. A., and Vera, J. H. (1981). *Can. J. Chem. Eng.* **59**, 89.
Ashworth, A. J., and Hooker, D. M. (1977) *J. Chromatogr.* **131**, 399.
Aveyard, R., and Mitchell, R. (1967). *Trans. Faraday Soc.* **65**, 2645.
Benewicz, J., Reed, C., and Levitch, M. (1957). *J. Am. Chem. Soc.* **79**, 2693.
Barker, J. A., (1953). *Aust. J. Chem.* **6**, 203.
Battino, R. (1966). *J. Phys. Chem.* **70**, 3408.
Battino, R. (1968). *J. Phys. Chem.* **72**, 4503.
Beath, L. A., and Williamson, A. G. (1969). *J. Chem. Thermodyn.* **1**, 51.
Becker, F., Fries, E. W., Kiffer, M., and Pflug, H. D. (1970). *Z. Naturforsch.* **25A**, 677.
Benson, G. C., Anand, S. C., and Kiyohara O. (1974). *J. Chem. Eng. Data* **19**, 258.
Bernatova, S., and Boublik, T. (1977). *Collect. Czech. Chem. Commun.* **42**, 2615.

Berro, C., Rogalski, M., and Peneloux, A. (1982). *J. Chem. Eng. Data* **27**, 352.
Berryman, J. M., and Heric, E. L. (1967). *J. Chem. Eng. Data* **12**, 249.
Barthelot, D. (1898). *Compt. Rend.* **126**, 1703, 1857.
Berthelot, D. J., and Jungfleisch, E. (1872). *Ann. Chim. Phys.* **4**, 26.
Bertrand, G. L., and Burchfield, T. E. (1974). In "Analytical Calorimetry," Vol. 3, (R. S. Porter and J. F. Johnson eds.) pp. 283-292. Plenum, New York.
Bissell, T. G., and Williamson, A. G. (1975). *J. Chem. Thermodyn.* **7**, 131.
Bohon, R. L., and Claussan, W. F. (1951). *J. Am. Chem. Soc.* **73**, 1571.
Bolles, T. B., and Drago, R. S. (1965). *J. Am. Chem. Soc.* **87**, 5015.
Bondi, A. (1968). "Physical Properties of Molecular Liquids, Crystals and Glasses." Wiley, New York.
Bronsted, J. N., and Koefoed, J. K. (1946). *Danske. Vid. Selsk.* Mat.-Fys. Medd. **22**, 1.
Bronsted, J. N., and Penderson, R. J. (1924). *Z. Phys. Chem.* **108**, 185.
Brown, I., Fock, W., and Smith, F. (1964) *Aust. J. Chem.* **17**, 1107.
Bruin, S. (1970). *Ind. Eng. Chem., Fundam* **9**, 305.
Bruno, T. J. (1981). Ph.D. Dissertation. Georgetown University.
Brynestad, J. (1981). *CALPHAD* **5**, 103.
Buchowski, H., Domanska, U., and Ksiazczak, A. (1979). *Pol. J. Chem.*, **53**, 1127.
Burchfield, T. E. (1977). Ph.D. Dissertation. University of Missouri at Rolla.
Burchfield, T. E., and Bertrand, G. L. (1975). *J. Solution Chem.* **4**, 215.
Butler, J. A. V. (1937). *Trans. Faraday Soc.* **33**, 229.
Casanova, C., Wilhelm, E., Grolier, J.-P. E., and Kehiaian, H. V. (1981). *J. Chem. Thermodyn.* **13**, 241.
Castagnolo, M., Inglese, A., Petrella, G., and Sacco, A. (1981). *Thermochimica Acta* **44**, 67.
Cave, G., Kothari, R., Puisieux, G., Martin, A. N., and Carstensen, J. T. (1980). *Int. J. Pharm.* **5**, 267.
Chien, C.-F., Kopecni, M. M., Laub, R. J., and Smith, C. A. (1981). *J. Phys. Chem.* **85**, 1864.
Choi, P. B., and McLaughlin, E. (1983). *AIChE J.* **29**, 150.
Colinet, C. (1967). D.E.S., Fac. des Sci. Univ. Grenoble, France.
Conder, J. R., and Young, C. L. (1979). "Physicochemical Measurements by Gas Chromatography." Wiley (Interscience), New York.
Davies, M., and Griffiths, D., (1954). *Z. Phys. Chem.* (Frankfurt/Main) **2**, 353.
Davies, M., and Griffiths, D. (1955). *J. Chem. Soc.* 132.
Davies, M., Jones, P., Patnaik, D., and Moelwyn-Hughes, E. (1951). *J. Chem. Soc.* 1249.
Deal, C. H., Derr, E. L., and Papadopoulos, M. N. (1962). *Ind. Eng. Chem. Fundam.* **1**, 17.
Delmas, G., Patterson, D., and Bhattacharya, S. N. (1964). *J. Phys. Chem.* **68**, 1468.
Derr, E. L., and Deal, C. H. (1969). "Distillation 1969," Sec. 3, p. 37, Brighton, England. Int. Conf. Distillation, Sept. 1969.
DiElsi, D. P., Patel, R. B., Abbott, M. M., and Van Ness, H. C. (1978). *J. Chem. Eng. Data* **23**, 242.
Dixon, W. D. (1970). *J. Phys. Chem.* **74**, 1396.
Dolezalek, F. (1908). *Z. Phys. Chem.* **64**, 727.
Dolezalek, F. (1918). *Z. Phys. Chem.* **93**, 585.
Dos Santos, J., Pineau, P., and Josien, M. (1965). *J. Chim. Phys. Phys-Chim. Biol.* **62**, 628.
Drucker, C. (1923). *Recl. Trav. Chim. Pays-Bas.* **42**, 552.
Duer, W. C., and Bertrand, G. L. (1970). *J. Am. Chem. Soc.* **92**, 2587.
Eduljee, G. H., and Tiwari, K. K. (1979). *Chem. Eng. Sci.* **34**, 929.
Emara, M., and Atkinson, G. (1974). *Adv. Mol. Relaxation Processes* **6**, 233.
Ewing, M. B., and Marsh, K. N. (1974). *J. Chem. Thermodyn.* **6**, 395.
Eyring, H., Henderson, D., and Jost W. (eds.) (1961). "Physical Chemistry. An Advanced Treatise," Vol. VII. Academic Press, New York.

References

Flaschner, O., and Macewen, B. (1908). *J. Chem. Soc.* **93**, 1000.
Fletcher, A. (1969). *J. Phys. Chem.* **73**, 2217.
Fletcher, A. (1972). *J. Phys. Chem.* **76**, 2562.
Fletcher, A., and Heller, C. (1967). *J. Phys. Chem.* **71**, 3742; **72**, 1841.
Flory, P. J. (1941). *J. Chem. Phys.* **9**, 660; **10**, 51 (1942).
Flory, P. J. (1953). "Principles of Polymer Chemistry." Cornell University Press, Ithaca, N. Y.
Franks, F., Gent, M., and Johnson, H. H. (1963). *J. Chem. Soc.* 2716.
Fredenslund, Aa., Jones, R. L., Prausnitz, J. M. (1975). *AIChE J.* **21**, 1086.
Fredenslund, Aa, Gmehling, J., Michelsen, M. L., Rasmussen, P., and Prausnitz, J. M., (1977a). *Ind. Eng. Chem. Process Des. Dev.* **16**, 450.
Fredenslund, Aa, Gmehling, J., and Rasmussen, P. (1977b). "Vapor–Liquid Equilibria Using UNIFAC." Elsevier, Amsterdam.
French, H. T., and Stokes, R. H. (1981). *J. Phys. Chem.* **85**, 3347.
Fried, V., Franceschetti, D. R., and Gallanter A. S., (1969). *J. Phys. Chem.* **73**, 1476.
Frost, A. A., and Pearson, R. G. (1961). "Kinetics and Mechanism," 2nd Ed. Wiley New York.
Fujii, Y., Sobue, K., and Tanaka M. (1978). *J. Chem. Soc. Faraday Trans.* 1 **74**, 1467.
Fujii, Y., Kawachi, Y., and Tanaka, M. (1981). *J. Chem. Soc. Faraday Trans.* 1 **77**, 63.
Fung, H. L. (1970). Ph.D. Dissertation. University of Kansas, Lawrence, Kansas.
Fung, H. L., and Higuchi, T. (1971). *J. Pharm. Sci.* **60**, 1782.
Garland, F., and Christian, S. D. (1975). *J. Phys. Chem.* **79**, 1247.
Garland, F., Rassing, J., and Atkinson, G. (1971). *J. Phys. Chem.* **75**, 3182.
Gaw, W. J., and Swinton, F. L., (1968). *Trans. Faraday Soc.* **64**, 2023.
Glasoe, P. K., Hallock, S., Hove, M., and Duke, J. M., (1971). *Spectrochim Acta*, **27A**, 2309.
Gmehling, J. (1983). *J. Chem. Eng. Data* **28**, 27.
Gmehling J. G., Anderson, T. F., and Prausnitz, J. M. (1978). *Ind. Eng. Chem. Fundam.* **17**, 269.
Gmehling, J., Rasmussen, P., and Fredenslund, Aa (1982). *Ind. Eng. Chem. Process Des. Dev.* **21**, 118.
Goates, J .R., Sullivan, R. J., and Ott, J. B. (1959). *J. Phys. Chem.* **63**, 589.
Goldman, S., and Krishnan, T. R. (1976). *J. Solution Chem.* **10**, 693.
Golub, V. B., Zaretsky, M. I., and Kononov N. F. (1977). *Zh. Prikl. Khim. (Leningrad)*. **30**, 453.
Gomez-Ibanez, J. D., and Shieh, J. J. C. (1965). *J. Phys. Chem.* **69**, 1660.
Gonzalez Posa C., Nunez, L., and Villar, E. (1972). *J. Chem. Thermodyn.* **4**, 275.
Gordon, L. J., and Scott R. L. (1952). *J. Am. Chem. Soc.* **74**, 4138.
Govindaswamy, S., Andiappan, AN, and Lakshmanan, SM (1976). *J. Chem. Eng. Data* **21**, 366.
Govindaswamy, S., Andiappan, AN, and Lakshmanan, SM (1977). *J. Chem. Eng. Data* **22**, 264.
Green, W. J., and Frank, H. S. (1979). *J. Solution Chem.* **8**, 187.
Guggenheim, E. A. (1952). "Mixtures," Clarendon Press, Oxford.
Haase, R. (1950). *Z. Phys. Chem.* **195**, 362.
Hala, H., Pick. J., Fried, V., and Vilim, O. (1968). "Vapour–Liquid Equilibrium Data at Normal Pressures." Pergamon, Oxford.
Hammaker, R., Clegg, R., Patterson, L., Ridder, P., and Rock, S. (1968). *J. Phys. Chem.* **72**, 1837.
Hammett, L. P. (1935). *Chem. Rev.* **17**, 125.
Handa, P. H., and Jones, D. E. (1975). *Can J. Chem.* **53**, 3299.
Hanna, J. G. (1982). *Ind. Eng. Chem. Fundam.* **21**, 489.
Harbison, M. W. P., Laub R. J., Martire, D. E., Purnell, J. H., and Williams, P. S. (1979). *J. Phys. Chem.* **83**, 1262.
Harris, K. R., and Dunlop, P. J. (1970). *J. Chem. Thermodyn.* **2**, 805.
Hepler, L. G., and Fenby, D. V. (1973). *J. Chem. Thermodyn.* **5**, 471.
Heric, E. L., and Posey, C. D. (1964a). *J. Chem. Eng. Data* **9**, 35.
Heric, E. L., and Posey, C. D. (1964b). *J. Chem. Eng. Data* **9**, 161.
Heric, E. L., and Posey, C. D. (1965). *J. Chem. Eng. Data* **10**, 25.

Hildebrand, J. H., and Scott, R. L. (1950a). "The Solubility of Nonelectrolytes," 3rd Ed. Van Nostrand-Reinhold.
Hildebrand, J. H., and Scott, R. L. (1950b). "Solubility of Nonelectrolytes," 3rd Ed. Van Nostrand-Reinhold, Princeton, New Jersey.
Hildebrand, J. H., Ellefson, E. T., and Beebe, C. W. (1919). *J. Am. Chem. Soc.* **39**, 2301.
Hildebrand, J. H., Prausnitz, J. M., and Scott, R. L. (1970). "Regular and Related Solutions." Van Nostrand-Reinhold, Princeton, New Jersey 150.
Hollo, J., and Lengyel, T. (1965). *Ind. Eng. Chem.* **51**, 957.
Hopkins, P. J., LeFevre, R. J. W., Raŋdom, L., and Ritchie, G. L. D. (1971). *J. Chem. Soc.* B 574.
Horvath, P. J. (1961). *Chem Eng.* **68**, 6, 159, 164.
Hoy K. L. (1970). *J. Paint Technol.* **42**, 76.
Huggins, M. L. (1941) *J. Phys. Chem.* **9**, 440.
Huggins, M. L. (1942). *Ann. N. Y. Acad. Sci.* **43**, 1.
Hulme, R., and Mullem, D. J. E. (1976). *J. Chem. Soc.* A 802.
Hulme, R., and Szymanski, J. T. (1969). *Acta Crystallogr.* Sect. A **25**, 753.
Jacob, K. T., and Fitzner, K. (1977). *Thermochim. Acta* **18**, 197.
Jain, D. V. S., and Lark, B. S. (1973). *J. Chem. Thermodyn.* **5**, 455.
Jain, D. V. S., and Yadav, O. P. (1971). *Indian J. Chem.* **9**, 342.
Jain, D. V. S., and Yadav, O. P. (1974). *Indian J. Chem.* **12**, 718.
Jain, D. V. S., Gupta, V. K., and Lark, B. S. (1970). *Indian J. Chem.* **8**, 815.
Jain, D. V. S., Gupta, V. K., and Lark, B. S. (1973). *J. Chem. Thermodyn.* **5**, 451.
Jones, H. K., and Lu, B. C.-Y. (1966). *J. Chem. Eng. Data* **11**, 488.
Katayama, T., Sung, E. K., and Lightfoot, E. N. (1965). *AIChE J.* **11**, 294.
Kempter, H. and Mecke, R. (1940). *Z. Phys. Chem.* B **46**, 229.
Kireev, V. A., and Sitnikov, I. P. (1941). *Zh. Fiz. Khim.* **15**, 492.
Knobeloch, J. B., and Schwartz, C. E. (1962). *J. Chem. Eng. Data* **7**, 386.
Kohler, F. (1960). *Monatsh. Chem.* **91**, 738.
Kohler, F., and Rice, O. K. (1957). *J. Chem. Phys.* **26**, 1614.
Kopecni, M. M., Laub, R. J., and Petkovic, Dj. M. (1981). *J. Phys. Chem.* **85**, 1595.
Kortum, G., and Vogel, V. M. (1955). *Z. Elektrochem.* **59**, 16.
Koshimura, H. (1977). *J. Inorg. Nucl. Chem.* **39**, 148.
Kretschmer, C. B., and Wiebe, R. (1954). *J. Chem. Phys.* **22**, 1697.
Krishnaiah, A., and Naidu, P. R. (1980). *J. Chem. Eng. Data* **25**, 135.
Krishnan, T., Duer, W. C., Goldman, S., and Fortier, J.-L. (1979). *Can. J. Chem.* **57**, 530.
Kudryavtseva, L. S., and Susarev, M. P. (1963). *Zh. Prikl. Khim.* **36**, 1231.
Kurtyka, A. (1982). *Ind. Eng. Chem. Fundam.* **21**, 489.
Kurtyka, Z. M., and Kurtyka, A. (1980). *Ind. Eng. Chem. Fundam.* **19**, 225.
Lai, T. T., Doan-Nguyen, T. H., Vera, J. H., and Ratcliff, G. A. (1978). *Can. J. Chem. Eng.* **56**, 358.
Langer, S. H., and Purnell, J. H. (1963). *J. Phys. Chem.* **67**, 263.
Langer, S. H., and Purnell, J. H. (1966). *J. Phys. Chem.* **70**, 904.
Lakhanpal, M. L., Chaturvedi, L. K., and Sharma, S. C. (1975). *Indian J. Chem.* **13**, 129.
Lakhanpal, M. L., Chaturvedi, L. K., Puri, T., and Sharma, S. C. (1976). *Indian J. Chem.* **14A**, 645.
Lana, J., and Zana, R. (1970). *Trans. Faraday Soc.* **66**, 957.
Lassetre, E. N. (1937). *Chem. Rev.* **20**, 259.
Laub, R. J., and Pecsok, R. L. (1978). "Physicochemical Applications of Gas Chromatography." Wiley, New York.
Laub, R. J., and Purnell, J. H. (1975). *J. Chromatogr.* **112**, 71.
Laub, R. J., and Purnell, J. H. (1976a). *Anal. Chem.* **48**, 799, 1720.
Laub, R. J., and Purnell, J. H. (1976b). *J. Am. Chem. Soc.* **98**, 30.
Laub, R. J., Purnell, J. H., and Williams, P. S. (1977). *J. Chromatogr.* **134**, 249.

References

Laub, R. J., Martire, D. E., and Purnell, J. H. (1978) *J. Chem. Soc. Faraday Trans.* 2, **74**, 213.
Lecat, M. (1926). *Compt. Rend.* **183**, 880.
Lecat, M. (1947). *Ann. Soc. Sci.* Bruxelles, **61**, 63.
Li, I. P.-C., Lu, B. C.-Y., and Chen, E. C. (1973). *J. Chem. Eng. Data* **18**, 305.
Li, I. P.-C., Polak, J., and Lu, C.-Y. (1974). *J. Chem. Thermodyn.* **6**, 417.
Libermann, E., and Wilhelm, E. (1975). *Monatsh. Chem.* **106**, 389.
Lien, T. R., and Missen, R. W. (1974). *J. Chem. Eng. Data* **19**, 84.
Linford, R. G., and Hildebrand, J. H. (1969). *Ind. Eng. Chem. Fundam.* **8**, 846.
Lipka, A., and Mootz, D. (1978). *Z. Anorg. Chem.* **440**, 217.
Litvinov, N. D. (1952). *Zh. Fiz. Khim.* **26**, 1144.
Lunderg, G. W. (1964). *J. Chem. Eng. Data* **9**, 193.
McAuliffe, C. (1966). *J. Phys. Chem.* **70**, 1267.
McGlashan, M. L., and Rastogi, R. P. (1958). *Trans. Faraday Soc.* **54**, 496.
McGlashan, M. L., and Williamson, A. G. (1961). *Trans. Faraday Soc.* **57**, 588.
McLaughlin, E., and Zainal, H. A. (1959). *J. Chem. Soc.* 863.
McLaughlin, E., and Zainal, H. A. (1960). *J. Chem. Soc.* 2485.
Malesinski, W. (1961). *Bull. Acad. Pol. Sci. Ser. Chim.* **9**, 137, 257, 261.
Malesinski, W. (1965). "Azeotropy and Other Theoretical Problems of Vapor–Liquid Interactions." Wiley (Interscience), New York.
Maripuri, V. C., and Ratcliff, G. A. (1971a). *Can. J. Chem. Eng.* **49**, 375.
Maripuri, V. C., and Ratcliff, G. A. (1971b). *Can. J. Chem. Eng.* **49**, 506.
Marsh, K. N. (1968). *Trans. Faraday Soc.* **64**, 883.
Martin, A., and Carstensen, J. (1981). *J. Pharm. Sci.* **70**, 170.
Martin, A., Newberger, J., and Adjei, A. (1980). *J. Pharm. Sci.* **69**, 487.
Martin, A., Wu, P. L., Adjei, A., Beerbower, A., and Prausnitz, J. M. (1981). *J. Pharm. Sci.* **70**, 1260.
Mathieson, A. R., and Thynne, J. C. J. (1956a). *J. Chem. Soc.* 3708.
Mathieson, A. R., and Thynne, J. C. J. (1956b). *J. Chem. Soc.* 3713.
Mathiasson, L., and Jonsson, R. (1974). *J. Chromatogr.* **101**, 339.
Matsui, T., Hepler, L. G., and Fenby, D. V. (1973). *J. Phys. Chem.* **77**, 2397.
Maurer, G., and Prausnitz, J. M. (1978). *Fluid Phase Equilibr.* **2**, 91.
Menshutkin, B. N. (1911). *Zh. Russ. Fiz. Khim. Obsheh.*, **43**, 395.
Meschel, S. V., and Kleppa, O. J. (1968). *J. Chem. Phys.* **48**, 5146.
Meyer, E. F., and Meyer J. A. (1981). *J. Phys. Chem.* **85**, 97.
Mikhail, S. Z., and Kimel, W. R. (1961). *J. Chem. Eng. Data* **6**, 533.
Mitra, R. C., Guhaniyogi, S. C., and Bhattacharyya, S. N. (1973). *J. Chem. Eng. Data* **18**, 147.
Monfort, J. P. (1983). *J. Chem. Eng. Data* **28**, 24.
Moore, J. W., and Pearson, R. G. (1981). "Kinetics and Mechanisms," 3rd Ed. Wiley, New York.
Morisue, T. Noda, K., and Ishida, K. (1972). *J. Chem. Eng. Jpn.* **5**, 217.
Morris, J. W., Molvey, P. J., Abbott, M. M., and Van Ness, H. C. (1975). *J. Chem. Eng. Data* **20**, 403.
Musa, R., and Eisner, M. (1959) *J. Chem. Phys.* **30**, 227.
Nagata, I. (1969). *J. Chem. Eng. Data* **14**, 418.
Nagata, I. (1973a). *Z. Phys. Chem. (Leipzig)* **252**, 305.
Nagata, I. (1973b). *Z. Phys. Chem. (Leipzig)* **254**, 273.
Nagata, I. (1973c). *Z. Phys. Chem. Neue Folge* **85**, 241.
Nagata, I. (1973d). *Z. Phys. Chem. Neue Folge* **107**, 39.
Nagata, I. (1975). *J. Chem. Eng. Data* **20**, 110.
Nagata, I. (1977). *Fluid Phase Equilibr.* **1**, 93.
Nagata, I. (1978). *Z. Phys. Chem. (Leipzig)* **259**, 1151.
Nagata, I., and Kazuma, K., (1977). *J. Chem. Eng. Data* **22**, 79.

Nagata, I., and Katoh, K. (1980). *Fluid Phase Equilibr.* **5**, 225.
Nagata, I., and Ogasawara, Y. (1982). *Thermochim. Acta*, **52**, 155.
Nagata, I., and Yamada, T. (1972). *Ind. Eng. Chem. Process Des. Dev.* **11**, 574.
Nagata, I., and Yamada, T. (1973a). *J. Chem. Eng. Data* **18**, 87.
Nagata, I., and Yamada, T. (1973b). *J. Chem. Eng. Jpn* **6**, 215.
Nagata, I., Yamada T., and Nagashima, T. (1973). *J. Chem. Eng. Jpn.* **6**, 298.
Nagata, I., Asano, H., and Fujiwara, K. (1977). *Fluid Phase Equilibr.* **1**, 211.
Nagata, I., Fujiwara K., and Ogasawara, Y. (1978). *J. Chem. Thermodyn.* **10**, 1201.
Nagata, I., Kawamura, Y., Ogasawara, Y., and Tokurika, S., (1980). *J. Chem. Thermodyn.* **12**, 223.
Naidu, G. R., and Naidu, P. R., (1981). *J. Chem. Eng. Data* **26**, 197.
Nakanishi, K., and Asakura, S. (1977). *J. Phys. Chem.* **81**, 1745.
Nernst, W. (1891). *Z. Phys. Chem.* **8**, 110.
Nguyen, T. H., and Ratcliff, G. A. (1971a). *Can. J. Chem. Eng.* **49**, 120.
Nguyen, T. H., and Ratcliff, G. A. (1971b). *Can. J. Chem. Eng.* **49**, 889.
Nguyen, T. H., and Ratcliff, G. A. (1974). *Can. J. Chem. Eng.* **52**, 641.
Nguyen, T. H., and Ratcliff, G. A. (1975). *J. Chem. Eng. Data* **20**, 256.
Nitta, T., and Katayama, T. (1973). *J. Chem. Eng. Jpn.* **6**, 1 224.
Nitta, T., and Katayama, T. (1974). *J. Chem. Eng. Jpn.* **7**, 310.
Nitta, T., and Katayama, T. (1975). *J. Chem. Eng. Jpn.* **8**, 175.
Nitta, T., Tatsuishi, A. and Katayama, T. (1973). *J. Chem. Eng. Jpn.* **6**, 475.
Nitta, T., Nakamura, Y., Ariyasu, H., and Katayama, T. (1980). *J. Chem. Eng. Jpn.* **13**, 97.
Nunez, L., Isorna, E., and Paz Andrade, M. I. (1976). *Acta Cient. Compostelana* **13**, 3.
Ohta, T., Asano, H., and Nagata, I. (1980). *Fluid Phase Equilibr.* **4**, 105.
Ohta, T., Koyabu, J., and Nagata, I. (1981). *Fluid Phase Equilibr.* **7**, 65.
Orye, R. V. and Prausnitz, J. M. (1965). *Ind. Eng. Chem.* **57**, 18.
Ott, J. B., Marsh, K. N. and Richards, A. E. (1981). *J. Chem. Thermodyn.* **13**, 447.
Perry, R. W., and Tiley, P. F. (1978). *J. Chem. Soc., Faraday Trans.* 1, **74**, 1655.
Pierotti, G. J., Deal, C. H., Derr, E. L., and Porter, P. E. (1956). *J. Am. Chem. Soc.* **78**, 2989.
Pimental, G. C., and McClellan, A. L., (1960). "The Hydrogen Bond." Freeman, San Francisco, California.
Polak, J., Murakami, S., Lam, V. T., and Benson G. C., (1970). *J. Chem. Eng. Data* **15**, 323.
Prigogine, I., and Defay, R., (1954). "Chemical Thermodynamics." (translated by D. H. Everett). Longmans, Green, New York.
Purkayastha, A., and Walkley, J. (1972). *Can. J. Chem.* **50**, 834.
Purnell, J. H., and Vargas de Andrade, J. M. (1975). *J. Am. Chem. Soc.* **97**, 3585, 3590.
Queignec, R., and Cabanetos-Queignec, M. (1981). *J. Chromatogr.* **209**, 345.
Ramalho, R. S., and Ruel, M. (1968). *Can. J. Chem. Eng.* **46**, 456.
Rao, K. V. K., and Rao, C. V. (1957). *Chem Eng. Sci.* **7**, 97.
Rassing, J., and Jensen, B. (1970). *Acta Chim. Scand.* **24**, 855.
Rastogi., K. P., Nath, J., and Misra, J. (1967). *J. Phys. Chem.* **71**, 1277.
Rastogi, R. P., Nath, J., and Das, S. S. (1977a). *J. Chem. Eng. Data* **22**, 249.
Rastogi, R. P., Nath, J., Singh, B., and Das, S. S. (1977b). *Indian J. Chem.* **15A**, 1012.
Ratcliff, G. A., and Chao, K. C., (1969). *Can. J. Chem. Eng.* **47**, 148.
Ratkovics, F., and Rehim, S. (1970). *Acta Chim. Acad. Sci. Hung.* **65**, 135.
Redlich, O., and Kister, A. T. (1948). *Ind. Eng. Chem.* **40**, 345.
Reis, J. C. R. (1982). *J. Chem. Soc., Faraday Trans.* 2, **78**, 1595.
Renon, H., and Prausnitz, J. M. (1967). *Chem. Eng. Sci.* **22**, 299. Errata **22**, 1891 (1967).
Renon, H., and Prausnitz, J. M. (1968). *AIChE J.* **14**, 135.
Robeson, J., Foster, B. W., Rosenthal, S. N., Adams, E. T., Jr., and Fender, E. J. (1981). *J. Phys. Chem.* **85**, 1254.

References

Ronc, M. and Ratcliff, G. A. (1975). *Can. J. Chem. Eng.* **53**, 329.
Rothmund, V. (1898). *Z. Phys. Chem. (Leipzig)* **26**, 433.
Rowlinson, J. S. (1959). " Liquids and Liquid Mixtures." Academic Press, New York.
Rowlinson, J. S. (1969) "Liquids and Liquid Mixtures," 2nd Ed. Butterworth, London.
Rytting, J. H., Anderson, B. D., and Higuchi, T. (1978). *J. Phys. Chem.* **82**, 2240.
Salamon, T., Liszi, J. and Ratkovicx, F. (1975a). *Acta Chim. Acad. Sci. Hung.* **87**, 137.
Salamon, T., Liszi, J. and Ratkovics, F. (1975b). *Magy. Kem. Foly.* **81**, 151.
Saluja, P. P. S., Young, T. M., Rodewald, R. F., Fuchs, R. F., Kohli, F. H., and Fuchs, R. (1977). *J. Am. Chem. Soc.* **99**, 2949.
Salvani, C., Winterhalter, D., and Van Ness, H. (1965). *J. Chem. Eng. Data* **10**, 168.
Sameshima, J. (1918). *J. Am. Chem. Soc.* **40**, 1482.
Savini, C. G., Winterhalter, D. R., and Van Ness, H. C. (1965). *J. Chem. Eng. Data* **10**, 168.
Sayegh, S. G., and Ratcliff, G. A. (1976). *J. Chem. Eng. Data* **21**, 71.
Sayegh, S. G., Vera, J. H., and Ratcliff, G. A. (1979). *Can J. Chem. Eng.* **57**, 513.
Scatchard, G. (1931a). *Chem. Rev.* **8**, 321.
Scatchard, G. (1931b). *J. Am. Chem. Soc.* **53**, 3186.
Scatchard, G. (1932). *Dan. Kemi.* **13**, 77.
Scatchard, G. (1934). *J. Am. Chem. Soc.* **56**, 10, 995.
Scatchard, G. (1937). *Trans. Faraday Soc.* **33**, 160.
Scatchard, G., and Raymond, C. L. (1938). *J. Am. Chem. Soc.* **60**, 1278.
Scatchard, G., Wood, S. E., and Mochel, J. M. (1939). *J. Phys. Chem.* **43**, 119.
Scatchard, G., Wood, S. E., and Mochel, J. M. (1940). *J. Am. Chem. Soc.* **62**, 712.
Scatchard, G., Ticknor, L. B., Goates, J. R. and McCartney, E. R. (1952). *J. Am. Chem. Soc.* **74**, 3721.
Schrier, E., Pottle, M. and Scheraga, H. (1964). *J. Am. Chem. Soc.* **86**, 3444.
Scott, R. L. (1949). *J. Chem. Phys.* **17**, 268.
Servanton-Gadouleau, M., Biais, J., and Lemanceau, B. (1975). *J. Chim. Phys. Physicochim Biol.* **72**, 831.
Sharma, B. R., and Singh, P. P. (1975). *J. Chem. Eng. Data* **20**, 360.
Shatas, J.P., Jr., Abbott, M. M., and Van Ness, H. C. (1975). *J. Chem. Eng. Data* **20**, 406.
Showmik, B. B. (1971). *Spectrochim Acta* **27A**, 321.
Siman, J. E., and Vera, J. H. (1979). *Can J. Chem Eng.* **57**, 355.
Singh, P. P., Sharma, B. R., and Sidhu, K. S. (1979). *Can. J. Chem.* **57**, 387.
Skjøld-Jorgensen, S., Kolbe, B., Gmehling, J., and Rasmussen, P. (1979). *Ind. Eng. Chem. Process Des. Dev.* **18**, 714.
Smith, F., and Brown. I. (1973). *Aust. J. Chem.* **36**, 691.
Smith, V. C., and Robinson, R. L., Jr. (1970). *J. Chem. Eng. Data* **15**, 391.
Smith, E. B., Walkley, J., and Hildebrand, J. H. (1950). *J. Phys. Chem.* **63**, 703.
Smutek, M., Fris, M., and Fohl, J. (1967). *Collect. Czech. Commun.* **32**, 931.
Sodovyev, V., Montrose, C., Watkins, M., and Litovitz, T. (1968). *J. Chem. Phys.* **48**, 2155.
Sondern, Chr., and Perkampus, H.-H. (1982a). *Ber. Bunsenges. Phys. Chem.* **86**, 562.
Sondern, Chr., and Perkampus, H.-H. (1982b). *Ber. Bunsenges, Phys. Chem.* **86**, 941.
Staverman, A. J. (1950). *Rec. Trav. Chim. Pays-Bas* **69**, 163.
Stokes, R. H., and Adamson, M. (1977). *J. Chem. Soc., Faraday Trans 1* **73**, 1232.
Swinton, R. J. (1980). *J. Chem. Thermodyn.* **12**, 489.
Sytilin, M. S. (1974). *Russ. J. Phys. Chem. (Eng. Trans.)* **48**, 1353.
Taft, R. W. (1952). *J. Am. Chem. Soc.* **74**, 2729.
Takeo, M., Nishii, K., Nitta, T., and Katayama, T. (1979). *Fluid Phase Equilibr.* **3**, 123.
Tasic, A., Djordjevic, B., and Grozdanic D. (1978). *Chem. Eng. Sci.* **33**, 189.
Tewari, Y. B., Martire, D. E., Wasik, S. P., and Miller, M. M. (1982) *J. Solution Chem.* **11**, 435.

Timmermans, J. (1959). "Physico-chemical Constants of Binary Systems." Wiley (Interscience), New York.
Tobolsky, A. V. and Thatch, R. E. (1962). *J. Colloid Sci.* **17**, 410.
Tochigi, K., Lu, B. C.-Y., Ochi, K., and Kojima, K. (1981). *AIChE J.* **27**, 1022.
Tomlins, R. P., and Marsh, K. N. (1976). *J. Chem. Thermodyn.* **8**, 1185.
Toop, G. W. (1965). *Trans. TMS-AIME* **233**, 850.
Treszczanowicz, A. (1973). *Bull. Acad. Polon. Sci., Ser. Sci. Chim.* **21**, 189.
Treszczanowicz, A. J., and Benson G. C. (1978). *J. Chem. Thermodyn.* **10**, 967.
Treszczanowicz, A. J., and Benson G. C. (1980). *J. Chem. Thermodyn.* **12**, 173.
Treszczanowicz, A. J. and Treszczanowicz, T. (1981). *Bull. Pol. Sci., Ser. Sci. Chim.* **29**, 269.
Treszczanowicz, A. J., Treszczanowicz, T., and Rogalski, M. (1981). *Bull. Pol. Sci., Ser. Sci. Chim.* **29**, 277.
Tsao, C. C., and Smith, J. M. (1953). "Applied Thermodynamics." Chem. Eng. Prog. Symp. No. 7. p. 107.
Tucker, E. and Becker, E. (1973). *J. Phys. Chem.* **77**, 1783.
Tucker, E. E., Farnham, S. B., and Christian, S. D., (1969). *J. Phys. Chem.* **73**, 3820.
Valvani, S. C., Yalkowsky, S. H., and Amidon, G. L. (1976). *J. Phys. Chem.*, **80**, 829.
van der Waals, J. C. (1890). *Z. Physik. Chem.* **5**, 133.
Van Duyne, R., Taylor, S., Christian, S., and Affsprung, H. (1967). *J. Phys. Chem.* **71**, 3427.
van Laar, J. J. (1906). "Sechs Vortage uber das Thermodynamicshe Potential." Braunschweig.
van Laar, J. J. (1910). *Z. Physik. Chem.* **72**, 723.
van Laar, J. J., and Lorenz, R. (1925). *Z. Anorg. Chem.* **146**, 42.
Van Ness, H. C. (1959). *Chem. Eng. Sci.* **11**, 118.
Van Ness, H. C., and Mrazek, R. V. (1959). *AIChE J.* **5**, 209.
Van Ness, H., van Winkle, J., Richtol, H. and Hollinger, H. B. (1967). *J. Phys. Chem.* **71**, 1483.
Vernon, F. (1971). *J. Chromatogr.* **63**, 249.
Vesely, F., Dohnal, V., and Prchal, M. (1982). *Collect. Czech. Chem. Commun.* **47**, 1045.
de Visser, C., Pel, P., and G. Somsen, (1977) *J. Solution Chem.* **6**, 571.
Vitoria, M., and Walkley, J. (1969). *Trans. Faraday Soc.* **65**, 57.
Wade, J., and Merriman, R. W. (1911) *J. Chem. Soc.* **99**, 997.
Warner, J. C., Scheib, R. C., and Svirbely, W. J., (1934). *J. Chem. Phys.* **2**, 590.
Weimer, R. F., and Prausnitz, J. M. (1965). *J. Chem. Phys.* **42**, 3643.
Weissman, S., and Wood, S. E. (1960). *J. Chem. Phys.* **32**, 1153.
Wieczorek, S. A. (1978). *J. Chem. Thermodyn.* **10**, 187.
Wieczorek, S. A., and Stecki, J. (1978). *J. Chem. Thermodyn.* **10**, 177.
Wilson, G. M. (1964). *J. Am. Chem. Soc.* **86**, 127.
Wilson, G. M., and Deal, C. H. (1962). *Ind. Eng. Chem. Fundam.* **1**, 20.
Wohl, K. (1946). *Trans. Am. Inst. Chem. Eng.* **42**, 215. Errata Chem. Engr. Progress, **49**, 218 (1953).
Wolff, H. P. P., and Shadiakly, A. (1981). *Fluid Phase Equilibr.* **7**, 309.
Yalkowsky, S. H., and Valvani, S. C. (1976). *J. Med. Chem.* **19**, 727.
Yalkowsky, S. H., and Valvani, S. C. (1979). *J. Chem. Eng. Data* **24**, 127.
Yalkowsky, S. H., and Valvani, S. C. (1980). *J. Pharm. Sci.* **69**, 912.
Yalkowsky, S. H., Flynn, G. L., and Slunick, T. G. (1972). *J. Pharm. Sci.* **61**, 852.
Yalkowsky, S. H., Orr, R. J., and Valvani, S. C. (1979). *Ind. Eng. Chem. Fundam.* **18**, 351.
Young, F. E., and Hildebrand, J. H. (1942). *J. Am. Chem. Soc.* **64**, 839.
Young, K. L., Mentzer, R. A., Greenhorn, R. A., and Chao, K. C. (1977). *J. Chem. Thermodyn.* **9**, 979.
Zana, R., and Lang, J. (1975). *Adv. Mol. Relaxation Processes* **7**, 21.

Index

A

Acenaphthene
 solubility in
  benzene, 204
  carbon tetrachloride, 206
Acetic acid, naphthalene solubility in, 234
Acetic acid–n-decane system, azeotropic properties, 188
Acetic acid–n-heptane system, azeotropic properties, 188
Acetic acid–n-hexane system, azeotropic properties, 188
Acetic acid–n-nonane system, azeotropic properties, 188
Acetic acid–n-octane system, azeotropic properties, 188
Acetic acid–pyridine system, azeotropic properties, 188
Acetic acid–n-undecane system, azeotropic properties, 188
Acetic acid–water mixtures, excess molar volumes for, 15
Acetone
 naphthalene solubility in, 234
 solubility parameters and molar volume, 84, 271
 values for ASOG method, 120
Acetone–acetonitrile mixtures
 vapor–liquid equilibrium data for, 60
 vapor phase compositions, 278
Acetone–carbon disulfide system
 azeotropic properties, 188
 entropy of mixing, 77
Acetone–chloroform system
 azeotropic properties, 188
 entropy of mixing, 76
 vapor–liquid equilibrium data for, 52, 149

Acetone–cyclohexane mixtures, vapor–liquid equilibrium in, 51
Acetone–n-hexane system, azeotropic properties, 188
Acetonitrile
 solubility parameter and molar volume of, 84, 271
 UNIFAC parameters for functional groups in, 127
Acetonitrile–carbon tetrachloride system, entropy of mixing, 76
Acetonitrile–ethanol–cyclohexane system, Effective UNIQUAC equation applied to, 108
Activity, as basic thermodynamic property, 33–37
Activity coefficient(s), 33–34
 derivation, 177, 259
 gas–liquid chromatography determination of, 115
 partial molar excess Gibbs free energy and, 55
 solubility and, 198
 of ternary systems, 179–181
AEC association model, 266
Alcohol(s)
 ternary, gas–liquid partition coefficients in binary solvents, 261–262
 polymers, 156
n-Alkane–alcohol system, entropy of mixing, 76
n-Alkanol–isooctane systems, vapor pressures, 151
Ammonia, equilibrium concentration studies using, 247
Analytical Group Solution Method (AGSM), 124–125
 primary aim of, 131–132

295

Analytical Solution of Groups model (ASOG), 117–124
  parameters for, 119
  primary aim of, 131–132
Aniline, naphthalene solubility in, 234
Aniline–p-cymene system, azeotropic properties, 188
Aniline–n-decane system, azeotropic properties, 188
Aniline–n-dodecane system, azeotropic properties, 188
Aniline–n-nonane system, azeotropic properties, 188
Aniline–pseudocumene system, azeotropic properties, 188
Aniline–n-tetradecane system, azeotropic properties, 188
Aniline–n-tridecane system, azeotropic properties, 188
Aniline–n-undecane system, azeotropic properties, 188
Anthracene
  solubility in binary solvent mixtures, 221, 224, 226, 227, 233
    complex with chloroform, 232
    mole fraction solubility, 232
  solubility in various solvents, 198, 200, 203
  solute properties, 225
Apparent partial molar quantities, equations for, 19–21
Argon–nitrogen systems, deviations in, 134
Athermal solutions
  definition, 88
  of polymers, 74
Attenuated Equilibrium Constant (AEC) model, 157, 165–171
Average molar volume, 14
Azeotrope(s)
  in binary mixtures, 185–189
  definition of, 54, 185
  positive–negative types, 185
  ternary, prediction from binary data, 189–191
  vapor–liquid equilibrium and, 176–196
  of binary mixtures under isobaric conditions, 185–189

B

Benzene
  acenaphthene solubility in, 204
  anthracene solubility in, 198

  benzoic acid solubility in, 229
  in binary solutions, properties, 63
  bi-phenyl solubility in, 204
  fluoranthene solubility in, 204
  fluorene solubility in, 204
  gas–liquid chromatography of, 264
  hydrocarbon solubility in, 203–205, 206
  naphthalene solubility in, 204, 234
  partition coefficient studies on, 250, 252
  partition coefficients in water, solvent systems, 286
  phenanthrene solubility in, 198, 204
  solubility parameter and molar volume, 84, 271
  solvent properties, 225
  o-terphenyl solubility in, 204
  1,3,5-triphenylbenzene solubility in, 204
  UNIFAC parameters for functional groups in, 127
  values for ASOG method, 120
Benzene–acetonitrile system, phase equilibrium studies on, 56–57
Benzene–antimony trichloride systems, thermodynamic studies on, 143–144
Benzene–carbon tetrachloride
  deviation from ideality, 75
  mixing properties, 83–85
  naphthalene solubility in, 235
  nonelectrolyte solubility in, 222, 224, 226
Benzene–chloroform system, entropy of mixing, 76
Benzene–cyclohexane systems
  activity coefficients, 277
  azeotropic properties, 188
  deviation from ideality, 75
  liquid–liquid equilibrium, 243
  naphthalene solubility in, 235
  nonelectrolyte solubility in, 222, 224, 226
  vapor–liquid equilibrium data, 59
Benzene–cyclohexene system, azeotropic properties, 188
Benzene–cyclooctane system, nonelectrolyte solubility in, 224, 226
Benzene–2,2-dimethylpentane system, azeotropic properties, 188
Benzene–diphenyl system, deviation from ideality, 75
Benzene–n-heptane system
  activity coefficient by ASOG method, 120–121
  azeotropic properties, 188

Index

copper acetylacetonate solubility in, 238
iodine solubilities in, 238
mixing properties, 84, 85
nonelectrolyte solubility in, 224, 226
Benzene–n-hexadecane system
  naphthalene solubility in, 235
  nonelectrolyte solubility in, 222, 226
Benzene–n-hexane system
  deviation from ideality, 75
  naphthalene solubility in, 235
  nonelectrolyte solubility in, 222, 224, 226
Benzene–n-hexene system, azeotropic properties, 188
Benzene–isooctane system
  ASOG predictions of properties, 123
  nonelectrolyte solubility in, 224, 226
Benzene–naphthalene systems, solid–liquid phase diagram for, 205
Benzene–toluene system
  naphthalene solubility in, 235
  nonelectrolyte solubility in, 222, 223, 226
Benzil
  solubilities in binary solvent mixtures, 210, 213–215, 219–221, 223–224, 228
  solubility in various solvents, 239, 285
  solute properties, 225
Benzoic acid
  distribution in benzene and water, 248
  solubility in binary solvent mixtures, 228
p-Benzoquinone
  solubility in binary solvent mixtures, 210, 215, 224, 226
  solute properties, 225–226
Binary data
  multicomponent properties derived from, 62–73
  of ternary alcohol–hydrocarbon systems, thermodynamic excess properties from, 150–175
Binary solvent mixtures
  containing nonspecific interactions, 74–114
  solubility parameter, 207–208
  vapor–liquid equilibrium, 176–179, 181–184
Binary 1–hexanol–n-alkane systems, equilibrium constants, 171
Biphenyl
  solubility
    in benzene, 204
    in carbon tetrachloride, 206
Biphenyl–p-dichlorobenzene systems, crystallization temperatures, 237
Boiling point isobar, equations, 182, 184
Boltzmann factors, 91, 99, 102
Bromobenzene, solubility parameter and molar volume, 271
Brönsted equation, 110
Brönsted–Kofoed congruence principle, 124
Bubble-point curve, derivation, 53
Bubble-point surface equation, 179
1-Butanol
  Kretschmer–Wiebe association constant, 159
  Mecke–Kempter association constant, 165
  naphthalene solubility in, 234
  solubility parameter and molar volume, 271
1-Butanol–n-hexane systems
  excess enthalpies, 174
  vapor–liquid equilibria, 113, 281
2-Butanol, solubility parameter and molar volume, 271
n-Butanol–benzene–cyclohexane system, azeotropic composition, 192
sec-Butanol
  naphthalene solubility in, 234
t-Butanol, values for ASOG method, 120
$tert$-Butanol, naphthalene solubility in, 234
Butanols, Henry's law constants, 43
2-Butanone
  UNIFAC parameters for functional groups in, 127
  values for ASOG method, 120
2-Butanone–benzene system, vapor–liquid equilibria data, 105
2-Butanone–chloroform systems, vapor–liquid equilibria, 146–148
2-Butanone–ethyl acetate–n-hexane system, azeotropic composition, 192
2-Butanone–n-heptane systems
  activity coefficients, 278
  vapor–liquid equilibrium data, 61
2-Butanone–nitrobenzene, ASOG predictions of properties of, 122
2-Butanone–water–cyclohexane, azeotropic composition, 192
Butyl p-hydroxybenzoate, solubility in n-propanol, 24
n-Butylamine, solubility parameter and molar volume, 271

n-Butylbenzene, solubility parameter and molar volume, 271
2-*n*-Butylbenzimidazole–benzotrizole–benzene system, self-association studies on, 165
n-Butylcyclohexane, solubility parameter and molar volume, 84

C

Carbazole, solubility in isooctane–*n*-butyl ether, 211–213
Carbon disulfide
  anthracene solubility in, 198
  naphthalene solubility in, 234
  phenanthrene solubility in, 198
  solubility parameter and molar volume, 84, 271
Carbon disulfide–benzene system, vapor–liquid equilibrium in, 49, 50
Carbon disulfide–cyclopentane system, azeotropic properties, 188
Carbon disulfide–isooctane system, mixing properties, 83
Carbon disulfide–methyl formate system, azeotropic properties, 188
Carbon disulfide–nitromethane system, entropy of mixing, 77
Carbon disulfide–*n*-pentane system, azeotropic properties, 188
Carbon tetrachloride
  acenaphthene solubility in, 206
  anthracene solubility in, 198
  benzil solubility in binary solvent mixtures containing, 213, 214
  benzoic acid solubility in, 229
  in binary solutions, 213–215
    properties, 63
  biphenyl solubility in, 206
  fluoranthene solubility in, 206
  fluorene solubility in, 206
  naphthalene solubility in, 198, 234
  phenanthrene solubility in, 198, 206
  solubility parameter and molar volume, 84, 271
  solvent properties of, 225, 226
  o-terphenyl solubility in, 206
  UNIFAC parameters for functional groups in, 127
Carbon tetrachloride–acetonitrile system
  mixing properties, 95
  comparison of two models, 94, 95

Carbon tetrachloride–cyclohexane system
  naphthalene solubility in, 235
  nonelectrolyte solubility in, 222
Carbon tetrachloride–*n*-hexadecane system
  naphthalene solubility in, 235
  nonelectrolyte solubility in, 222
Carbon tetrachloride–*n*-hexane system
  naphthalene solubility in, 235
  nonelectrolyte solubility in, 222
Carbon tetrachloride–isooctane systems, benzil solubilities in, 219, 220
Carbon tetrachloride–*n*-octane system, benzil solubilities in, 221
Carbon tetrachloride–OMCTS system, nonelectrolyte solubility in, 223
Carbon tetrachloride–tetrachloroethylene mixtures, vapor pressures, 40–41
Carbon tetrachloride–toluene system
  naphthalene solubility in, 235
  nonelectrolyte solubility in, 222
Carboxylic acids, monomeric and dimeric treatment of, 227–229
Cavity formation, in dissolution, 198
Chain rule, mathematical relationships involving, 3–4
Chemical potential
  as important thermodynamic quantity, 30
  of solute in binary solvents, 218
Chlorobenzene
  naphthalene solubility in, 234
  solubility parameter and molar volume, 271
Chloroform
  naphthalene solubility in, 234
  solubility parameter and molar volume, 84, 271
  UNIFAC parameters for functional groups in, 127
Chloroform–ethanol–acetone system, azeotropic composition, 192
Chloroform–ethanol–*n*-heptane systems, excess enthalpies, 279
Chloroform–ethyl formate–2-bromopropane system, azeotropic composition, 192
Chloroform–methanol systems
  ASOG predictions of properties, 122–123
  thermodynamic excess properties, 48
Chloroform–methanol–acetone system, azeotropic composition, 192
Chloroform–methanol–2-butanone system, azeotropic composition, 192

Index

Chloroform–methanol–ethyl acetate system, liquid–vapor equilibrium data for, 130–131
Chloroform–triethylamine systems
  excess Gibbs free energy, 283
  molar excess enthalpies, 138–139, 144, 146
Clausius–Clapeyron equation, 182, 184, 187, 190
Colinet equation, 66, graphical representation, 67
Complex liquid, definition, 74
Complex mixture(s)
  definition, 74
  entropy of mixing, 76
Concentration variables, derivation, 1
Copper acetylacetonate
  partition coefficients in benzene–n-heptane system, 251, 253–254
  solubility in binary solvent system, 238
Cumene, UNIFAC parameters for functional groups in, 127
Cycle rule, mathematical relationships involving, 4
Cyclohexane
  benzoic acid solubility in, 229
  in binary solutions, properties, 63
  gas–liquid chromatography of, 264, 267–268, 270
  in liquid–liquid equilibrium studies, 250–252
  solubility parameter and molar volume, 84, 271
  solvent properties, 225
  values for ASOG method, 120
Cyclohexane–carbon disulfide system, mixing properties, 84, 86
Cyclohexane–carbon tetrachloride system
  benzoic acid solubility in, 228
  benzil solubility in, 214
  nonelectrolyte solubility in, 222–223
Cyclohexane–cyclohexanol systems, phase compositions, 47
Cyclohexane–cyclooctane system, nonelectrolyte solubility in, 223–224, 226
Cyclohexane–ethylbenzene system, nonelectrolyte solubility in, 226
Cyclohexane–n-heptane system
  benzoic acid solubility in, 228
  mixing properties, 84–85
  nonelectrolyte solubility in, 223–224, 226

Cyclohexane–n-hexadecane system
  naphthalene solubility in, 235
  nonelectrolyte solubility in, 222, 226
Cyclohexane–n-hexane system
  benzoic acid solubility in, 228
  naphthalene solubility in, 235
  nonelectrolyte solubility in, 222–224, 226
Cyclohexane–isooctane system, nonelectrolyte solubility in, 224, 226
Cyclohexane–methyl iodide system, phenanthrene solubility in, 208–209
Cyclohexane–octamethylcyclotetrasiloxane system, excess Gibbs free energies, 111
Cyclohexane–n-octane system, nonelectrolyte solubility in, 224, 226
Cyclohexane–OMCTS systems
  excess Gibbs free energies, 280
  nonelectrolyte solubility in, 223
Cyclohexane–toluene system
  naphthalene solubility in, 235
  nonelectrolyte solubility in, 222
Cyclohexanol, naphthalene solubility in, 234
Cyclohexanone, solubility parameter and molar volume, 271
Cyclooctane, solvent properties, 225
Cyclooctane–carbon tetrachloride system, benzil solubility in, 214
Cyclopentane, solubility parameter and molar volume, 84, 271
Cyclopentane–cyclopentanol mixtures, phase compositions, 46
Cyclopentane–tetrachloroethylene mixtures, phase compositions, 46
Cyclopentanone, solubility parameter and molar volume, 271
Cyclopentene, solubility parameter and molar volume, 271

D

Dalton's law, 49
  expression for, 40, 41
Decane, solubility parameter and molar volume, 84, 271
1-Decanol
  Kretschmer–Wiebe association constant for, 159
  Mecke–Kempter association constant for, 165
  solubility parameter and molar volume, 271

2-Decanol, solubility parameter and molar volume, 271
1-Decene, solubility parameter and molar volume, 271
Dew-point curve, derivation, 53
n-Dibromobenzene, solute properties, 225
o-Dibromobenzene, solubility parameter and molar volume, 271
p-Dibromobenzene, solubility in binary solvent mixtures, 210, 222, 226, 228
1,1-Dibromoethane, naphthalene solubility in, 234
Di-n-butylamine, solubility parameter and molar volume, 271
n,n-Dibutyl-2-ethylhexamide–n-hexadecane, use in gas–liquid chromatography, 267
Dichlorobenzenes, solubility parameters and molar volumes, 271
1,1-Dichloroethane, solubility parameter and molar volume, 271
1,2-Dichloroethane–n-nonane mixtures, excess molar volumes, 11
Dichloromethane, solubility parameter and molar volume, 271
Dieterici equation of state, 9
Diethyl ether
 anthracene solubility in, 198
 phenanthrene solubility in, 198
 UNIFAC parameters for functional groups in, 127
 values for ASOG method, 120
Diethyl ether–chloroform system, entropy of mixing, 76
 thermodynamic properties, 148
Diethylamine
 solubility parameter and molar volume, 272
 UNIFAC parameters for functional groups in, 127
1,2-Diethylbenzene, solubility parameter and molar volume, 272
3,3-Dimethyl-2-butanone, solubility parameter and molar volume, 272
Dimethylbutanols, solubility parameters and molar volumes, 272
Diisopropylamine, UNIFAC parameters for functional groups in, 127
Diisopropylether
 UNIVAC parameters for functional groups in, 127
 values for ASOG method, 120

Dimethylamine, UNIFAC parameters for functional groups in, 127
2,2-Dimethylbutane, solubility parameter and molar volume, 84, 272
2,3-Dimethylbutane
 solubility parameter and molar volume, 84, 272
 values for ASOG method, 120
2,3-Dimethyl-2-butene, UNIFAC parameters for functional groups in, 127
Dimethyl ether, UNIFAC parameters for functional groups in, 127
1,4-Dioxane, solubility parameter and molar volume, 272
p-Dioxane–chloroform system
 two complexes in, 141
 properties, 142, 144
Di-n-propylamine, solubility parameter and molar volume, 272
Dodecane, solubility parameter and molar volume, 84, 272
n-Dodecane–n-hexane system, deviation from ideality, 75
1-Dodecanol–n-hexene systems, activity coefficients and excess Gibbs free energies, 133
1-Dodecanol–n-hexane systems, activity coefficients, 282
Dodecanols, solubility parameters and molar volumes, 272
Dodecyl laurate squalene mixture, chromatographic partition of solutions on, 270

E

Effective UNIQUAC equation, 107, 108
Entropy of mixing equation, 38
Equilibrium constant, molarity based, expression, 210
Ethanol
 anthracene solubility in, 198
 Kretschmer–Wiebe association constant, 159
 Mecke–Kempter association constant, 165
 naphthalene solubility in, 234
 phenanthrene solubility in, 198
 solubility parameter and molar volume, 272
 values for ASOG method, 120
Ethanol–benzene system, vapor–liquid equilibria data, 104–105

Ethanol–benzene–2-butanone system
  mixing properties, 106
  UNIQUAC representation, 107
Ethanol–benzene–cyclohexane system,
  azeotropic composition, 192
Ethanol–2-butanone system, vapor–liquid
  equilibria data, 104–105
Ethanol–chloroform system, azeotropic
  properties, 188
Ethanol–cyclohexane system, UNIQUAC
  equation applied to, 108
Ethanol–n-heptane systems, heat of mixing
  data on, 150
Ethanol–n-hexadecane systems, heat of mixing
  data on, 151
Ethanol–n-hexane system, azeotropic
  properties, 188
Ethanol–isooctane system, ASOG predictions
  for properties, 122
Ethanol–toluene systems, heat of mixing data,
  150
Ethanol–toluene–cyclohexane system,
  equilibrium constants, 172
Ethanol–p-xylene–cyclohexane system,
  equilibrium constants, 172
Ethyl acetate, solubility parameter and molar
  volume, 272
Ethylbenzene
  solubility parameter and molar volume, 84,
    272
  solvent properties, 225
  UNIFAC parameters for functional groups
    in, 127
  values for ASOG method, 120
Ethylbenzene–benzene system, naphthalene
  solubility in, 235
Ethylbenzene–carbon tetrachloride system,
  nonelectrolyte solubility in, 222
Ethylbenzene–cyclohexane system,
  nonelectrolyte solubility in, 222
2-Ethyl-1-butanol, solubility parameter and
  molar volume, 272
Ethyl butyl ether, solubility parameter and
  molar volume, 272
Ethyl cyclohexane
  gas–liquid chromatography of, 270
  values for ASOG method, 120
Ethylene dibromide, naphthalene solubility in,
  234
2-Ethyl-1-hexanol, solubility parameter and
  molar volume, 272

Euler's theorem, derivation, 5–7
Exact differentials, mathematical relationships
  involving, 7
Excess chemical potentials of third component
  in binary solvent mixtures, 197
Excess partial molar enthalpies of third
  component in binary solvent mixtures, 197

F

Flory–Huggins model for entropy of mixing,
  75, 88–91, 116–117, 125, 152, 156, 162,
  166, 168, 172, 217, 225, 231, 260, 262
  deviations from, 75
  expression for, 118
Fluoranthene
  solubility
    in benzene, 204
    in carbon tetrachloride, 206
Fluorene
  solubility
    in benzene, 204
    in carbon tetrachloride, 206
Fluorobenzene, solubility parameter and molar
  volume, 272
Fugacity
  as basic thermodynamic property, 30–33
  from Henry's law, 249
  solubility and, 198–199
  standard state of, definition, 199
Fugacity coefficient, 34

G

Gas, standard state, 34
Gas–liquid chromatography
  activity coefficients by, 115
  of inert solutes on self-associating binary
    solvent systems, 264–267
  in studies of thermodynamic properties of
    binary systems, 197
Gas–liquid partition coefficients, derivation
  of, 259
Gibbs–Duhem equation, 24, 32, 37, 43, 135,
  143
  derivation, 19
  for ideal vapor, 55
Gibbs free energy, 16–17, 19, 62–63, 89,
  97–98, 103, 107, 109, 176, 216, 243,
  247

activity coefficients and, 177, 180
  of apparent binary mixture, 137
  of binary solvent mixture, 210
  dissolution and, 197, 201
  excess molar, calculation, 178, 184
  fugacity and, 32
  for miscible binary mixtures, 244
  of ternary alcohol–hydrocarbon mixture, 261
  of a ternary liquid mixture, 87, 88
    alcohol and two hydrocarbons, 151–152, 154, 162–163
    at the UCST, 245
Gibbs free energy of mixing, 90, 230, 243, 245
  equation for, 38, 45
  of binary solutions, 76, 231, 260, 267
  of ternary solution, 65, 155, 166–167
  UCST from, 246
Gibbs–Helmholtz relationship, 95
Group contribution methods for prediction of thermodynamic excess properties, 115–149
Guggenheim–Staverman equation, 107

H

Hammett equation, 110
Helmholtz free energy, 16
Henry's law, 250
  application to butanols, 43
  fugacity from, 249
  equation for, 42, 44
Heptane, solubility parameter and molar volume of, 84, 272
n-Heptane
  gas–liquid chromatography of, 264, 267–268, 270
  phosphorus solubility in, 207
  solvent properties, 225
n-Heptane–carbon tetrachloride system
  benzoic acid solubility in, 228
  equilibrium constant, 214
  nonelectrolyte solubility in, 223–224
n-Heptane–n-dodecane system, nonelectrolyte solubility in, 224
n-Heptane–n-hexadecane system, nonelectrolyte solubility in, 223
n-Heptane–1-propanol systems
  calorimetric data, 113
  vapor–liquid equilibrium calculations, 281
  vapor pressures, 112

n-Heptane–1-propanol–chlorobutane systems, vapor–liquid equilibria of, 114
Heptanols, solubility parameters and molar volumes, 272
Heptanones, solubility parameters and molar volumes, 272
n-Hexadecane, solvent properties, 225
n-Hexadecane–n-heptane system, deviation from ideality, 75
Hexafluorobenzene–benzene system, liquid–vapor equilibria, 194
Hexane
  solubility parameter and molar volume, 84, 272
n-Hexane
  anthracene solubility in, 198, 200
  gas–liquid chromatography of, 264, 267–268
  naphthalene solubility in, 234
  phenanthrene solubility in, 198, 200
  solvent properties, 225
  UNIFAC parameters for functional groups in, 127
  values for ASOG method, 120
n-Hexane–benzene–1-butanol, vapor–liquid equilibrium data, 97
n-Hexane–benzene–$tert$-butyl alcohol system, vapor–liquid equilibrium data for, 100
n-Hexane–carbon tetrachloride system
  benzoic acid solubility in, 228
  benzil solubility in, 214
  nonelectrolyte solubility in, 223
n-Hexane–chloroform system
  anthracene-chloroform association constant, 232
  mole fraction solubility in, 232
n-Hexane–cyclohexane system
  deviation from ideality of, 75
  equilibrium constants, 169
  nonelectrolyte solubility in, 223
n-Hexane–cyclohexane–benzene systems, excess Gibbs free energies, 279
n-Hexane–n-hexadecane system
  deviation from ideality, 75
  naphthalene solubility in, 235
  nonelectrolyte solubility in, 222, 226
n-Hexane–n-heptane systems
  equilibrium constants, 169
  nonelectrolyte solubility in, 223
n-Hexane–n-hexadecane system, nonelectrolyte solubility in, 226

Index

n-Hexane–methanol systems, activity
  coefficients, 117
1-Hexanol
  Kretschmer–Wiebe association constant,
    159
  Mecke–Kempter association constant, 165
1-Hexanol–cyclohexanol systems,
  experimental enthalpies, 158
1-Hexanol–*n*-heptane systems
  equilibrium constants, 169
  experimental enthalpies, 158, 164
1-Hexanol–*n*-hexane system
  activity coefficients of, 128–130
  vapor–liquid equilibria, 112, 280
Hexanols, solubility parameters and molar
  volumes, 272
Hexanones, solubility parameters and molar
  volumes, 272
1-Hexene
  solubility parameter and molar volume,
    272
  UNIFAC parameters for functional groups
    in, 127
  values for ASOG method, 120
2-Hexene, UNIFAC parameters for functional
  groups in, 127
n-Hexylamine, solubility parameter and molar
  volume, 272
High dilution technique, complexation reaction
  study by, 197
Higuchi stoichiometric complexation model,
  210–215
Homogeneous functions, nature of, 4–5
Hydrate formation, self-association constants
  and, 249
Hydrogen-bonded complexes, enthalpies of
  formation, 197
Hydrogen bonding
  in alcohols, 155
  effects on properties of, 150
  classification of organic liquids according to
    potentiality of, 193
Hydroxyl groups, in aliphatic alcohols,
  standard states, 117

I

Ideal Associated Solution model, 134–135
  AB-type complex, 137–140
  AB_2-type complexes, 140
Ideal gas law, 5

Ideal solution(s)
  dilute, thermodynamic behavior, 42
  solid–liquid equilibrium in, 199
  thermodynamic properties, 30–61
  vapor–liquid equilibrium, 39–41
Implicit differentiation, 2–3
Inert cosolvent, effect on entropy of mixing, 77
Interaction parameters of various organic
  compounds, 129
Iodine
  equilibrium concentration studies using, 247
  solubility in binary solvent mixtures, 210,
    221, 226, 233, 238
  NIBS predictions, 285
  solute properties of, 225
Isooctane
  solvent properties, 225
  UNIFAC parameters for functional groups
    in, 127
Isooctane–*n*-butyl ether system, carbazole
  solubility in, 211–213
Isooctane–carbon tetrachloride system, benzil
  solubility in, 214
Isooctane–cyclohexane system, nonelectrolyte
  solubility in, 223
Isooctane–*n*-hexadecane system, nonelectrolyte
  solubility in, 223
Isooctane–nitroethane system, liquid–liquid
  equilibrium, 241
Isopropanol
  naphthalene solubility in, 234
  solubility parameter and molar volume, 272
Isopropanol–2-butanone–cyclohexane system,
  azeotropic composition, 192
Isopropylamine
  solubility parameter and molar volume, 272
  UNIFAC parameters for functional groups
    in, 127

J

Jacob–Fitzner equation, 66–67
  graphical representation, 67
  illustration, 69

K

Kohler equation, 65, 72–73
  graphical representation, 67
  illustration, 68

Kohler-type expression, 160
Kovat's retention index, 270
Kretschmer–Wiebe association model, 151–161, 163, 166, 172, 261–262, 266, 283
　constants for alcohols, 159
　predictive ability of, 172–173
　two-constant type, 171–173

L

Lakhanpal equation, 66
Least squares method, applications, 8–9
Lever principle, statement of, 54
Lewis acid–base adducts, complexation reactions involving, 197
l'Hôpital's rule, 20
Linear free energy relationships, 110
Liquid–liquid equilibrium, 240–256
　between two immiscible solvents, 247–252
　of binary organic phase and water, 252–254
　in binary systems, 246–247, 250
Liquid solution, composition, 1
London dispersion forces, 134
Lower critical solution temperature (LCST), 241

M

Malesinski equation, predictive abilities of, 188–190
Mathieson–Thynne equation, 65
Mecke–Kempter association model, 157, 161–166, 266, 283
　association constant, 164
Methanol
　hydroxyl group in standard state, 117
　Kretschmer–Wiebe association constant, 159
　Mecke–Kempter association constant, 165
　naphthalene solubility in, 234
　UNIFAC parameters for functional groups in, 127
Methanol–acetone–cyclohexane system, azeotropic composition, 192
Methanol–benzene system, mixing properties, comparison of two models, 93
Methanol–carbon disulfide system, liquid–liquid equilibrium, 240
Methanol–carbon tetrachloride system, entropy of mixing, 76

Methanol–cyclohexane systems, proton magnetic resonance data on, 151
Methanol–ethanol–water system, liquid–vapor equilibrium data, 131
Methanol–methyl acetate–n-hexane system, azeotropic composition, 192
Methyl acetate, solubility parameter and molar volume, 272
Methyl acetate–1-hexene systems, vapor–liquid equilibria, 114, 282
Methylamine, UNIFAC parameters for functional groups in, 127
2-Methyl-1-butene
　UNIFAC parameters for functional groups in, 127
　values for ASOG method, 120
2-Methyl-2-butene, UNIFAC parameters for functional groups in, 127
Methylcyclohexane
　gas–liquid chromatography of, 264, 270
　solubility parameter and molar volume, 272
Methyl ethyl ketone–1-butanol–n-heptane mixtures
　estimation of properties by use of binary data, 62–73
　excess enthalpies, 72
　excess molar volumes, 71
Methyl ethyl ketone–1-propanol–n-heptane systems, excess molar volumes, 279
n-Methylformamide–water systems, volumetric properties, 276
Methyl lithocholate–carbon tetrachloride system, self-association studies on, 165–166
Methylpentanols, solubility parameters and molar volumes, 272
2-Methyl piperidine–water system, liquid–liquid equilibrium, 242
2-Methylpropane, UNIFAC parameters for functional groups in, 127
Mixing rules, van der Waals' equation of state, 63
Mixing sign, enthalpy of, 75
Molar volumes, table, 271–273
Mole fraction, derivation, 1
Molecular size, effect on properties of solution, 75
Multicomponent properties, derived from binary data, 62–73

Index

N

Naphthalene
  solubility
    in benzene, 204
    in binary solvent mixtures, 210, 219, 222, 226, 235
    in carbon tetrachloride, 206
    in pure solvents, 234
  solute properties, 225
Nearly Ideal Binary Solvent Theory (NIBS), 216–229, 252, 254, 260, 264, 267
  application to solute-solvent complexation systems, 229–233
  gas–liquid chromatography and, 258–264
Negative azeotrope, definition, 54
Nitrate salts, ternary-fused, enthalpies of mixing, 67
Nitromethane, solubility parameter and molar volume, 272
Nitromethane–benzene system, entropy of mixing, 77
Nitromethane–p-dioxane system, entropy of mixing, 76
2-Nitro-5-methylphenol, in binary solvent mixtures, 215
Nonane, solubility parameter and molar volume, 272
n-Nonane–1,2-dichloroethane mixtures, excess molar volumes, 11
Nonanols, solubility parameters and molar volumes, 272
Nonelectrolyte solutions, classification, 74
Nonelectrolytes
  gas–liquid chromatography of, 257–270
  solubility behavior of, 197–239
Nonideal solutions
  thermodynamic properties, 30–61
  vapor–liquid equilibrium, 48–57
Nonrandom two-liquids (NRTL) model, 97–101, 111, 134, 178, 259
Nucleoside bases, model to explain stacking, 166

O

Octafluorohexane, gas–liquid chromatography of, 270
n-Octane
  gas–liquid chromatography of, 264
  solubility parameter and molar volume, 84, 273
  solvent properties, 225
n-Octane–carbon tetrachloride system benzil solubility in, 214
  nonelectrolyte solubility in, 224
n-Octane–tetraethylamine system, deviation from ideality, 75
1-Octanol
  Kretschmer–Wiebe association constant, 159
  Mecke–Kempter association constant, 165
1-Octanol–n-decane systems, infrared data, 151
Octanols, solubility parameters and molar volumes, 273
2-Octanone, solubility parameter and molar volume, 273
OMCTS, solvent properties, 225

P

Partial derivatives, mathematical relationships for, 1–2
Partial molar quantities, 12–15
  derivation, 21–25
Partition coefficient(s)
  derivation, 249
  solute association and dissociation, 248
Partition law, 248
Pentadecane, solubility parameter and molar volume, 273
Pentane
  gas–liquid chromatography of, 264
  solubility parameter and molar volume, 273
n-Pentanol–2-methylpentane systems
  AGSM predictions, 125
  Kretschmer–Wiebe association constant, 159
  Mecke–Kempter association constant, 165
Pentanols, solubility parameters and molar volumes, 273
3-Pentanone, UNIFAC parameters for functional groups in, 127
Pentanones, solubility parameters and molar volumes, 273
Perfluoromethylcyclohexane–methylcyclohexane system, deviation from ideality, 75
Phase transition, effect on solubility of a solid, 202

Phenanthrene
  in binary solvent mixtures, 215
  solubility
    in benzene, 198, 204
    in carbon tetrachloride, 206
    in cyclohexane–methyl iodide, 208–209
    in various solvents, 198, 200, 203
Phenol–cresol system, entropy of mixing, 76
Phosphorus, solubility in n-heptane, 207
Piperidine, solubility parameter and molar volume, 273
Polar liquids, mixing effects, 75
Polyethylene, CH_2 unit in, normalization factors, 126
Polymers, alcohols, 150, 152
Polyoxyethylene glycol, mixing of, in binary solvent, 66
Positive azeotrope, definition, 54
Prigogine equation, 187
  predictive ability, 189
Propanol, hydroxyl group in, standard state, 117
Propanol–benzene–cyclohexane system, azeotropic composition, 192
1-Propanol
  Kretschmer–Wiebe association constant, 159
  Mecke–Kempter association constant, 165
  naphthalene solubility in, 234
  solubility parameter and molar volume, 273
1-Propanol–2-propanol system entropy of mixing, 76
1-Propanol–p-xylene–cyclohexane system equilibrium constants, 172
2-Propanol
  UNIFAC parameters for functional groups in, 127
  values for ASOG method, 120
2-Propanol–benzene–methylcyclohexane system, equilibrium constants, 172
2-Propanol–toluene–cyclohexane system, equilibrium constants, 172
Propionic acid–butanoic acid system, entropy of mixing, 76
Propionic acid–n-decane system, azeotropic properties, 188
Propionic acid–n-heptane system, azeotropic properties, 188
Propionic acid–n-nonane system, azeotropic properties, 188
Propionic acid–n-octane system, azeotropic properties, 188

Propionitrile, UNIFAC parameters for functional groups in, 127
Propylamine
  UNIFAC parameters for functional groups in, 127
n-Propylamine
  solubility parameter and molar volume, 273
n-Propylbenzene–ethylbenzene, aqueous solubilities, 255
Proton transfer reactions, complexation reactions involving, 197
Pyridine, solubility parameter and molar volume, 273
Pyridine–chloroform mixtures, phase compositions, 47
Pyridine–n-heptane system, azeotropic properties, 188
Pyridine–n-nonane system, azeotropic properties, 188
Pyridine–n-octane system, azeotropic properties, 188
Pyridine–2,2,4-trimethylpentane system, azeotropic composition, 189
Pyridine–water system, entropy of mixing, 76

Q

q-fractions, definition, 64

R

Raoult's law, 63, 137, 176, 226, 249–250
  expression for, 40–41, 43–44
  negative deviations from, 75, 86, 89, 141, 219
  positive deviations from, 80, 83, 86
Rastogi equation, 67
Redlich–Kister equation, 64, 69, 70, 72, 94, 177, 179
  illustration of, 68
  one- and two-parameter types, 178
  Scatchard modification of, 64
  Wilson equation compared to, 93–94
Redlich–Kwong equation of state, 9
Regular Solution equations, 82–84, 86, 88, 134, 180, 225, 246
Relative apparent partial molar property, definition, 21

Index

S

Saddle azeotropes, occurrence, 185
Scatchard equation, 64
Scatchard–Hildebrand model, 81–88, 90, 154, 159, 205–210, 219, 227, 261–262
Similis similibus solvantur, 197
Simple associated solutions, 134–149
　thermodynamic properties, 135–136
Simple liquid, definition, 74
Simple mixture (simple solution), definition, 74
Solid–liquid equilibrium, in ideal solution, 199
Solubility
　factors contributing to, 198
　ideal, equation for estimation, 203
　intermolecular forces and, 197
Solubility behavior of nonelectrolytes, 197–239
Solubility parameters, table, 271–273
Solute–solvent complexation systems, NIBS model applied to, 229–233
Solution nonideality, interpretation of, 77
Solvent structuring, in dissolution, 198
Squalene–dinonyl phthalate mixtures, partition coefficients of hydrocarbons on, 264
Stannic iodide
　solubility in binary solvent mixtures, 210, 222
　solute properties, 225
Stoichiometric complexation model of Higuchi, 210–215
Styrene
　UNIFAC parameters for functional groups in, 127
　values for ASOG method, 120
Sulfolane–water systems, volumetric properties, 277
Sulfur–benzene system, liquid–liquid equilibrium, 241, 243
Sulfuric acid, equilibrium concentration studies using, 247
Symmetrical Reference System, for liquid and solid solutions, 35

T

Taft equation, 110
Ternary alcohol–hydrocarbon systems
　thermodynamic excess properties, 150–175
　experimental data, 161
Ternary mixtures
　containing nonspecific interactions, 74–114
　of nearly symmetrical binary solutions, properties, 63
　vapor–liquid equilibrium
　　isobaric, 184
　　isothermal, 179–181
o-Terphenyl
　solubility
　　in benzene, 204
　　in carbon tetrachloride, 206
Tetrachloroethylene–carbon tetrachloride mixtures, vapor pressures of, 40–41
Tetradecane, solubility parameter and molar volume, 273
Thermodynamic classification of nonelectrolyte solutions, 74
Thermodynamic excess functions, 45–48
Thermodynamic excess properties of liquid mixtures, 115–149
Thiophene, solubility parameter and molar volume, 273
Tie lines, definition, 53
Toluene
　gas–liquid chromatography of, 264
　naphthalene solubility in, 234
　solubility parameter and molar volume, 84, 273
　solvent properties, 225
　UNIFAC parameters for functional groups in, 127
　values for ASOG method, 120
Toluene–cyclohexane system, deviation from ideality, 75
Toluene–*n*-hexadecane system, naphthalene solubility in, 235
Toluene–*n*-hexane system, naphthalene solubility in, 235
Toluene–methylcyclohexane system, deviation from ideality, 75
Toop equation, 65
　graphical representation, 67
Total derivatives, mathematical relationships for, 2
Tri-*n*-butylamine, solubility parameter and molar volume, 273
1,1,1-Trichloroethane, UNIFAC parameters for functional groups in, 127
Trichloroethylene–benzene–cyclohexane system, azeotropic composition, 192

Triethylamine, UNIFAC parameters for functional groups in, 127
Triethylamine–water system, liquid–liquid equilibrium, 241–242
Trimethylamine, UNIFAC parameters for functional groups in, 127
Trimethylbenzenes, solubility parameters and molar volumes, 84, 273
2,2,4-Trimethylpentane, values for ASOG method, 120
1,3,5-Triphenylbenzene, solubility in benzene, 204
Tsao–Smith equation, 65
  graphical representation, 67
  illustration, 69
  Knobeloch–Schwartz modification, 65
Two-constant Kretschmer–Wiebe association model, 171–173
  predictive ability, 172–173

U

Undecane, solubility parameter and molar volume, 273
1-Undecanol, solubility parameter and molar volume, 273
UNIFAC group contribution method, 128
  solubility predictions using, 233–236
UNIQUAC Functional Group Activity Coefficients Model (UNIFAC), 125–131
  primary aim, 131–132
UNIQUAC models, 101–108, 233, 259
  applied to various systems, 104–105
  Effective UNIQUAC Equation, 107
UNIQUAC parameters, 102
Unsymmetrical Reference System for liquid and solid solutions, 35
Upper critical solution temperature (UCST), definition, 241, 245

V

van der Waals equation of state, 1–2, 62–63, 78, 81
van der Waals theory, 79
van Laar equations, 62, 77–81
  Regular Solution equations and, 82–83
  simplifying assumptions in, 78, 80
van't Hoff equation, 155
Vapor–liquid equilibrium (VLE)
  azeotropic systems and, 176–196
  in binary systems
    isobaric, 181–184
    isothermal, 176–179
  in ideal solution, 39
  in nonideal solution, 48
  in ternary systems, 179–181, 184
Vapor-phase composition, equation, 178
Vapor pressure of supercooled liquid solute, graphical extrapolation of, 200
Volume fraction, derivation, 1

W

Water–deuterium oxide mixtures, 25–26
Wilson–Deal description of liquid mixtures, 116–117
Wilson equation 63, 90–97, 118, 137, 178, 233, 246–247, 259
  advantages of, 96
  Redlich–Kister equation compared to, 93–94

X

Xylene isomers, solubility parameters and molar volumes, 84, 273